Micro Mechatronics: Science and Technology

Micro Mechatronics: Science and Technology

Edited by **Eve Versuh**

NY RESEARCH PRESS
New York

Published by NY Research Press,
23 West, 55th Street, Suite 816,
New York, NY 10019, USA
www.nyresearchpress.com

Micro Mechatronics: Science and Technology
Edited by Eve Versuh

International Standard Book Number: 978-1-63238-322-8 (Hardback)

Printed in China.

Contents

Preface

This book has been an outcome of determined endeavour from a group of educationists in the field. The primary objective was to involve a broad spectrum of professionals from diverse cultural background involved in the field for developing new researches. The book not only targets students but also scholars pursuing higher research for further enhancement of the theoretical and practical applications of the subject.

This book presents a concise summary of the widespread aspects of various fundamentals and applications of micro mechatronics. The technique of nano mechatronics has widespread applications across multiple spheres like sensors, automobiles and machinery. Its present day utilization has extended to aerospace equipment, biomedical apparatus, water engineering and other such emerging fields. Essentially, the technology for miniaturization is critical for achieving high performance and low consumption of resources, giving rise to small space instrumentation. This book presents a summary on the research work carried out at Center of Excellence for Education and Research of Micro-Nano Mechatronics. The research work has created a means to create multiple applications of micro-nano mechatronics, on the basis of mechanical engineering, control systems engineering, and advanced medical engineering. Various topics in this book talk about advancements in research on nano measurement and control, designing and manufacture, and their applications in the field of biomedical equipment.

It was an honour to edit such a profound book and also a challenging task to compile and examine all the relevant data for accuracy and originality. I wish to acknowledge the efforts of the contributors for submitting such brilliant and diverse chapters in the field and for endlessly working for the completion of the book. Last, but not the least; I thank my family for being a constant source of support in all my research endeavours.

Editor

Chapter 1

Research and Technology on Micro-Nano Mechatronics

Toshio Fukuda, Masahiro Nakajima and
Masaru Kojima

1. Introduction

In our daily life, various devices are applied for automobiles, computer peripheries, printers, cameras, amusements, robotics, automation, environmental monitoring, energy resource, biological/medical treatments, and so on "Microtechnology" was commonly used to realize high-efficiency, high-integration, high-functionality, low-energy consumption, low-cost, miniature, and so on By miniaturizing the elemental devices on sensors, actuators, and computers in micro-scale, "Micromechatronics" came up as the one of the important technology Recently, "Nanotechnology" has an important role in the industrial applications as an advanced field of mechatronics named as "Nanomechatronics" The micro-nano mechatronics is basically defined to integrate major three technologies "Controller", "Sensor" and "Actuator" based on the electronics and mechanical engineering as depicted in Figure 1.

Figure 2 shows the demands of micro-nano mechatronics for various social and industrial applications For various applications for industry, some techniques are important, especially micro/nano fabrication, assembly, control, material, and evaluation techniques Micro-nano mechatronics is based on various technologies, especially life science, medicine, sensing/ actuating, material science, energy/power, and design/control From social aspects, human resource, environmental issue, saving energy, safety/security, medical/health, and aging population are currently demanded From industrial aspects, service robots, dependable products, tailor-made products, alternative energy, techno-care service, environmental friendly products, are particularly demanded The micro-nano mechatronics is a key technology to solve those problems/issues and leading conventional technologies for future.

The applications of micro-nano mechatronics are mainly categorized into the "Mechanical", "Electrical", and "Biological/Medical" applications The key point for the categorization is inorganic (wet) and organic (dry) "Mechanical" applications are relatively based on the inorganic materials or technologies, such as lithography technique On the other hand, "Biological/Medical" applications, the organic materials or technologies are used, such as self-

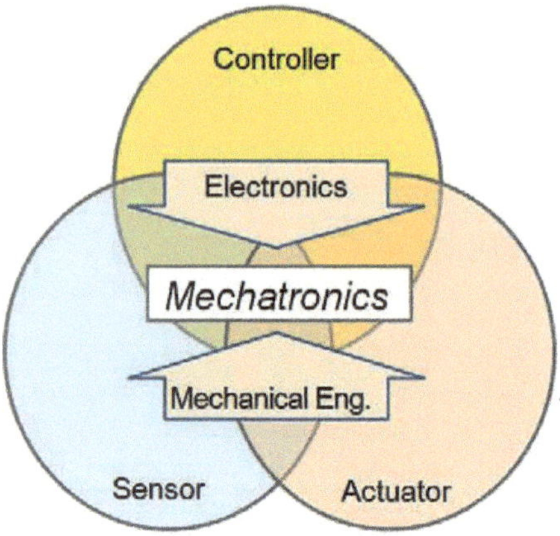

Figure 1. Micro-Nano mechatronics

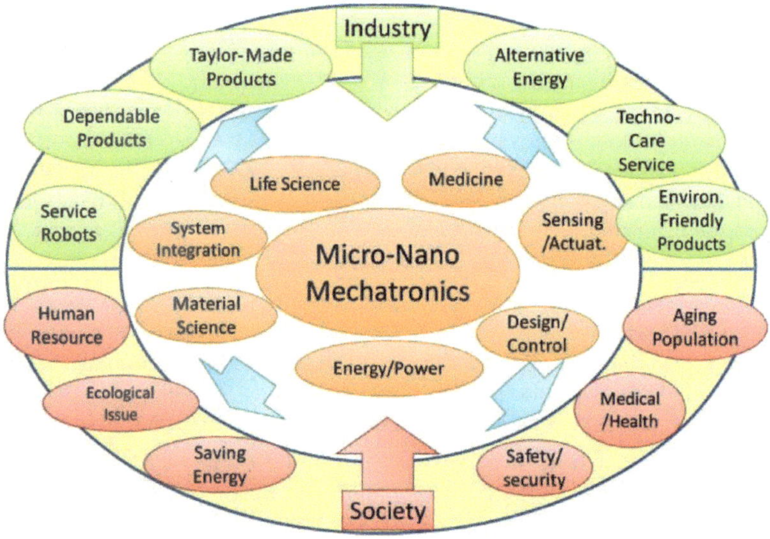

Figure 2. Micro-Nano mechatronics for social and industrial demands

assembly technique In between them, "Electrical" applications are placed for delivering or calculating information and so on Since the micro-nano mechatronics is the composite academic fields, the required technologies are mainly categorized in to basic/middle/high integration levels as depicted in Figure 3.

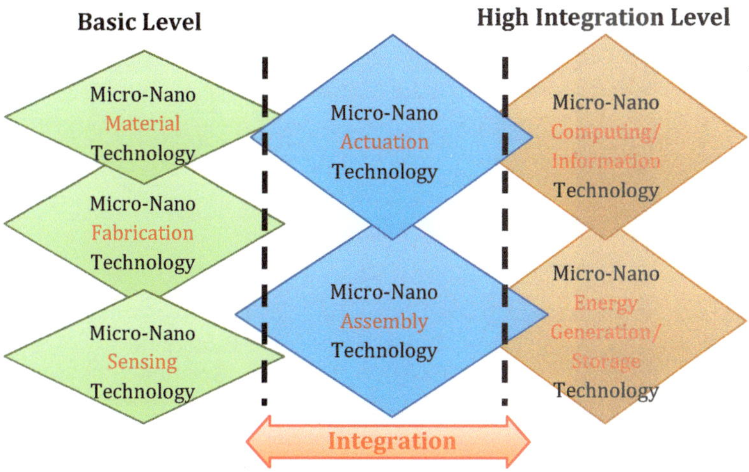

Figure 3. Required technologies for micro-nano mechatronics

Basically, nanotechnology is placed in the combinations of the top-down and bottom-up approaches The possibility to control the structure of matter atom by atom was first discussed by Richard Feynman in 1959 seriously [1] One of the approaches to fill the gap between top-down and bottom-up approaches is "Nanomanipulation", which realizes controlling the position at the micro/nanometer scale, is considered to be one of the promising ways It might be a key technology to lead the appearance of replication-based assemblers The top-down fabrication process, or micro machining, provides numbers of nanometer structures at once On the other hand, the bottom-up fabrication process, or chemical synthesis such as self-assembly [2] , also provides numerous nanometer structures In fact, both approaches reach nanometer scale with the limitations of physical/chemical aspects at present Hence, the technology to fill its gap is considered to be one of the important at this moment for micro-nano mechatronics Especially, current research directions are mainly two flows, "green innovation" and "life innovation" as depicted in Figure 4 These innovations will be achieved in various research and developments Table 1 and 2 show the challenging issues by categorized fields.

2. Micromechatronics for industrial and research applications

In micro scale, the important technologies are Micro Machine, Micro Mechatronics, Micro Fabrication and Assembly for micromechatronics Recently, borderless applications are investigated such as Micro Biology, Wet Mechatronics, Micro Total Analysis System, Micro Medical Engineering, and Regenerative Medical Engineering Some examples of micro devices mainly in research field are micro-actuator [3] , micro-ink-jet head [4] , micro-force sensor [5] , micro-tactile sensor [6] , micro-fuel battery[7] , micro fluidics device[8] , blood vessel simulator [9] , and so on.

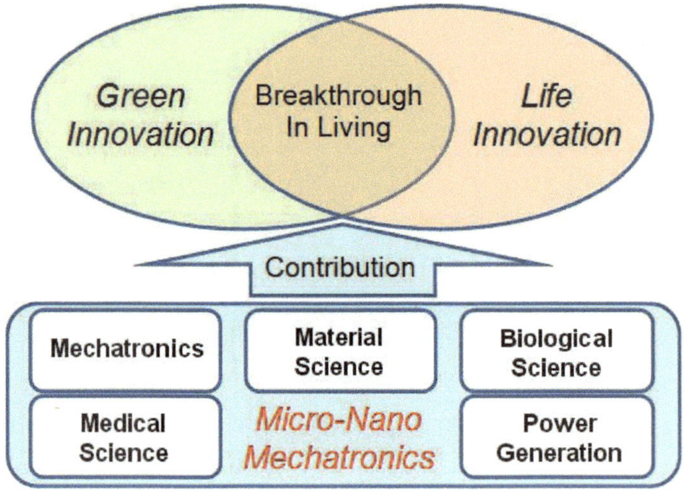

Figure 4. Green and life innovations based on micro-nano mechatronics

Green innovation	Technological Challenges
Natural Resources	Micro/nano devices for Discovery of oil resource, Management of water resources, Prevention of forest destruction, etc
Environmental Pollution	Monitor, control and management of Environment, Distributed sensing & control, Pollution control, Green vehicle, etc
Energy Development	Energy saving, harvesting and alternatives , Energy grid and management, Power control and green electronics, etc
Food and Agriculture	Safety, testing and tracing, Efficient harvesting, nutritious products, genetically-modified products, etc

Table 1. Challenges for green innovation by micro-nano mechatronics

Life innovation	TechnologicalChallenges
Medicine for life	Inspection and diagnosis , Re-generative medicine, Gene therapy and life science, monitoring diseases, Neuro Science, In-situ diagnostics, Cell diagnosis and surgery, New drug and medicine, DDS, Minimally invasive surgery, Rehabilitation, Techno-care, Wearable robots, Cyborg, QoL, etc
Biology–Analysis and Synthesis	Sensing , manipulation and automation, New species, DNA diagnosis & manipulation, Cell screening, transport, cultivation, and function and differentiation control, Artificial cell, Life in chip, Cloning of stem cells, etc

Table 2. Challenges for life innovation by micro-nano mechatronics

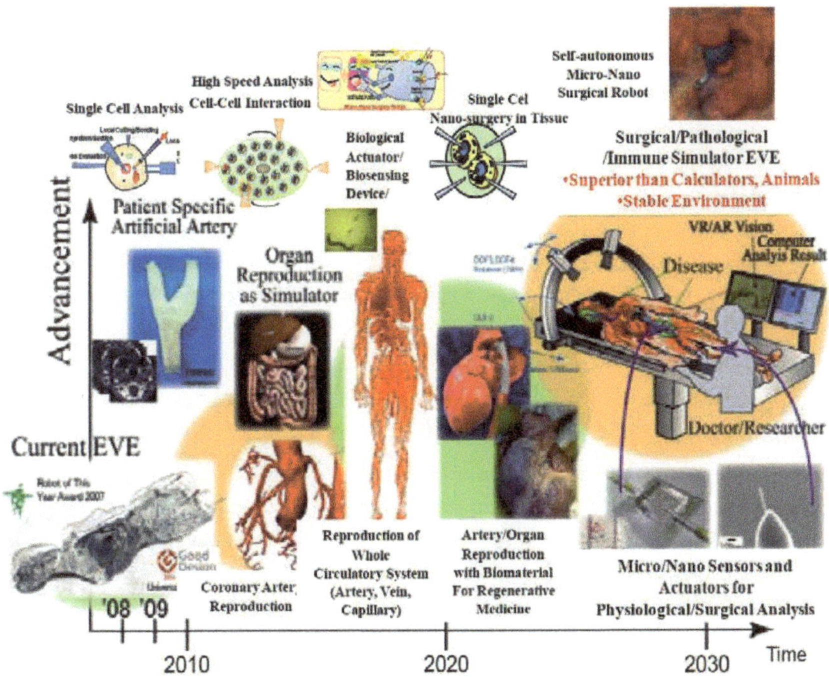

Figure 5. Blood vessel simulator and surgical operation system

In medical applications, surgical supporting or simulating technologies are important We developed the Patient blood vessel simulator "EVE (Endo Vascular Educator)" as challenging the frontier of the surgical simulation since 1989 This simulator is fabricated and assembled based on the micro-fabrication technologies of rapid prototyping technique [9] .

Recently, we have interests to realize the "In vitro" realization of "In vivo" environment For the purpose, it is needed to understand our biological system in detail The investigation approach using micro/nano devices can measure and control inside "In vivo" biological system Moreover, "In vitro" constructions and measurements of biological system can reveal it clearly in single/multiple cell It is quite important for the significant improvement of regeneration engineering/biology, inducing the evaluation of clinical condition of blood vessel We have been worked blood vessel simulator and surgical operation system to improve and integrate for the "In vivo" realization Based on the EVE system, we are trying to develop novel type of surgical simulation system with the "In vitro" realization of "In vivo" environment The human cells can be used to construct as blood vessel or other organs including diseases by 3D cell assembly technique The surgical simulator can be integrated with micro-nano sensors and devices to evaluate the surgical operations and other applications such as drug delivery system.

3. Nanomechatronics for industrial and research application

In nano scale, the important technologies are Nano-measurement, Nano-fabrication and Nano-assembly for nanomechatronics The advanced applications are investigated for quantum dots [10] , quantum processing [11] , photonic crystal [12] , drug delivery system (DDS) [13] , field emission display [14] , nano-field emission electron source [15] , nano-X-ray sources [16] , nano-actuator[17] , nano-temperature sensor[18] , nano-IR sensor[19] , super-molecules for solar energy conversion [20] , and so on.

The wide scale controlled devices from atomic scale to meter scale is expected to realize in the near future For the high integrated, miniaturized, and functionalized NEMS, one of the effective ways is to use the bottom-up fabricated nanostructures or nanomaterials directly As typical example, the nanodevices are investigated based on the carbon nanotubes (CNTs) It has interesting mechanical, electronic and chemical properties which have been under investigation in various studies [21] There is possibility to use their fine structures directly For example, "telescoping carbon nanotube", which is fabricated by peeling off the outer layers of multi-walled carbon nanotubes (MWNTs), is one of the most interest nanostructures As previous works, the pulling out of the inner core was pulled out mechanically inside a TEM [22] The MWNTs were used them as the rotation axis of silicon chip as rotational actuators [23] .We reported on the direct measurements of electrostatic actuation of a telescoping MWNT inside SEM and TEM [17] [24] [25] Those applications are newly developed using the bottom-up structures of CNTs.

4. Research on center for micro-nano mechatronics in Graduate School of Engineering, Nagoya University

We established a "Center for Micro-nano Mechatronics" at Graduate School of Engineering, Nagoya University in 2008 with the aim of applying nanotechnology to practical systems in micro-nano scale from a system approach viewpoint [26] Our Center strives to foster "young researchers who dare to challenge unexploited fields" by building a novel interdisciplinary field based on micro-nano mechatronics This field will promote "the world-highest-level of micro-nano mechatronics research with an emphasis on originality" from a viewpoint of not only the acquisition of advanced technology, but also social issues.

Our center aims not only to create novel functional materials and advanced mechatronics, but also to discover breakthroughs in next-generation medicine we promote researches in four basic fields, Nano control engineering, Nano measurement engineering, Nano design and manufacturing, and Nano materials science and conduct an applied research encompassing all these basic research fields to attend to the needs of the advanced medical engineering.

Our center implements a strategy to realize applications of micro-nano mechatronics, which are based on mechanical engineering or materials science, control systems engineering, and advanced medical engineering As shown in Figure 6, by establishing joint research and international collaborations between the above research fields, we are creating the most advanced micro-nano mechatronics and train researchers who can comprehend industrial circles and social issues using an open cluster system as well as conduct research to solve problems spanning these four basic fields In particular, we will initially focus on tasks in the bio- or medical welfare technologies using a number of unexploited fields, which may consequently produce venture enterprises Some research results are figured as shown in Figure 7 in the four basic research fields Detail information or more recent results will be given by the following chapters in this book.

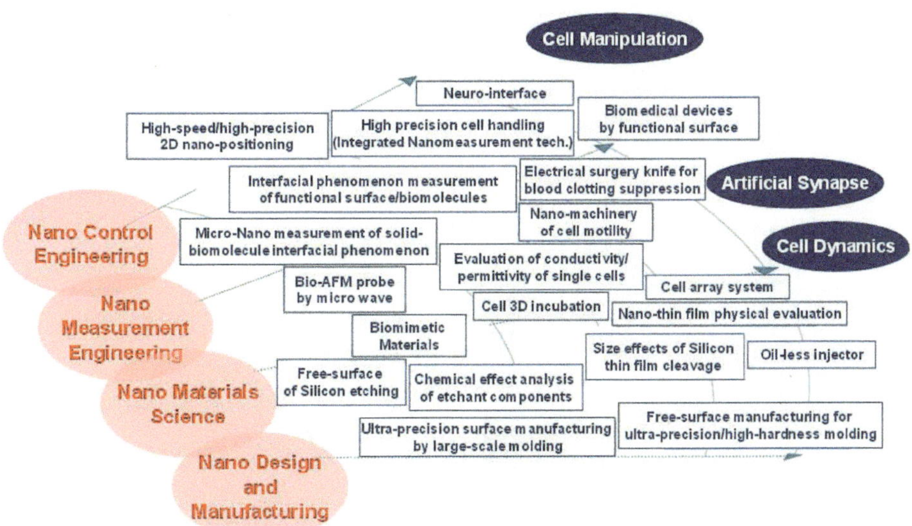

Figure 6. Milestone of Micro-Nano mechatronics research fields

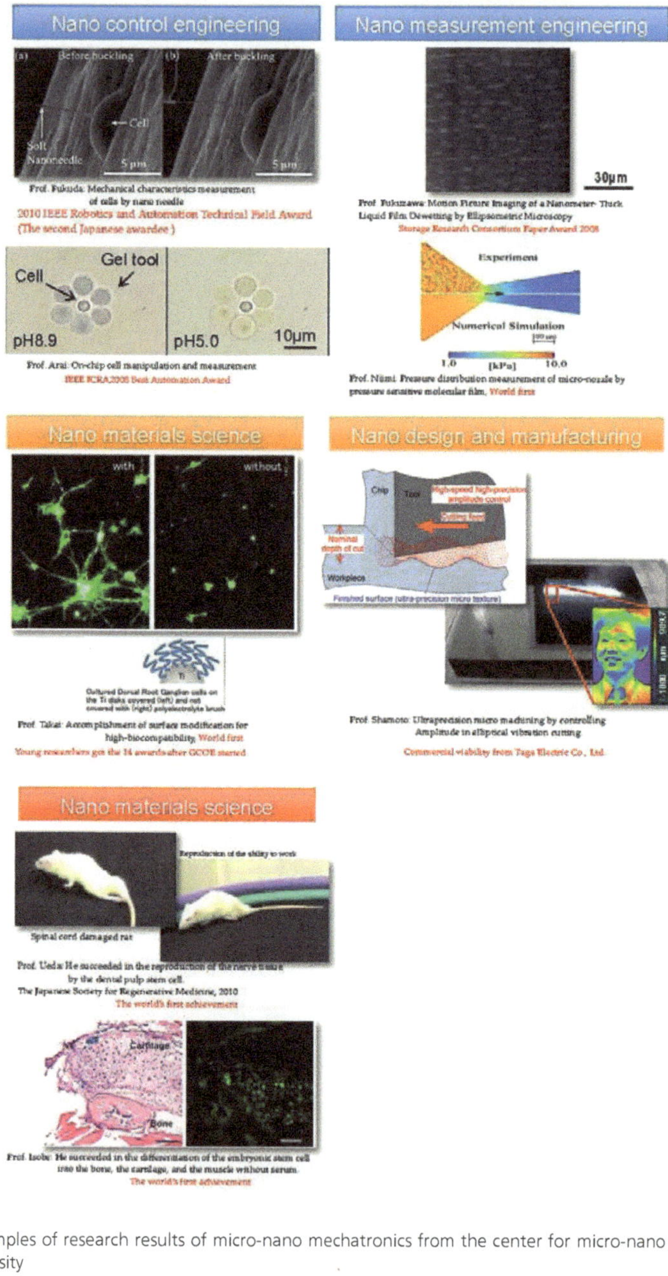

Figure 7. Examples of research results of micro-nano mechatronics from the center for micro-nano mechatronics of Nagoya University

5. Conclusion

This chapter presents the brief introduction of research and technology on micro-nano mechatronics For industrial applications, various devices are developed and available, such as applications to the automobiles, computer peripheries, amusements, printers, cameras, robotics automation, environmental monitoring, biological/medical treatments, energy resource, and so on Those devices are investigated based on the micro-and nano-mechatronics technologies to realize high-efficiency, high-integration, high-functionality, low-energy consumption, low-cost, miniature, and so on Micro-nano mechatronics technologies can be applied to break though the advanced industrial field including the nanobiology and medical applications.

Author details

Toshio Fukuda[1], Masahiro Nakajima[2] and Masaru Kojima[1]

1 Department of Micro-Nano Systems Engineering, Nagoya University, Nagoya, Japan

2 Graduate School of Engineering Center For Micro-nano Mechatronics, Nagoya University, Nagoya, Japan

References

[1] Feynman R. P. There's Plenty of Room at the Bottom. Caltech's Engineering and Science 1960;23 22-36.

[2] Whitesides G. M, Grzybowski B. Self-Assembly at All Scales. Science 2002;295 2407-2409.

[3] Kanno I, Tazawa Y, Suzuki T, Kotera H. Piezoelectric unimorph miroactuators with X-shaped structure composed of PZT thin film. Microsystem Technologies 2007;13(8) 825-829.

[4] Ishida Y, Hakiai K, Baba A, Asano T. Electrostatic Inkjet Patterning Using Si Needle Prepared by Anodization. Japanese Journal of Applied Physics 2005;44(7B) 5786-5790.

[5] Sun Y, Nelson B. J. MEMS Capacitive Force Sensors for Cellular and Flight Biomechanics. Biomedical Materials 2007;2(1) 16-22.

[6] Motoo K, Arai F, Fukuda T, Katsuragi T, Itoigawa K. High sensitive touch sensor with piezoelectric thin film for pipetting works under microscope. Sensors and Actuators A 2006;126 1-6.

[7] Yoshida K, Hagihara Y, Tanaka S, Esashi M. Normally-closed Electrostatic Micro Valve with Pressure Balance Mechanism for Portable Fuel Cell Application. In: proceedings of 19th IEEE International Conference on Micro Electro Mechanical Systems, MEMS2006, 22-26 January 2006, Istanbul, Turkey.

[8] Motoo K, Toda N, Arai F, Fukuda T, Sekiyama K, Nakajima M. Generation of Concentration Gradient from Two-Layered Flow with High-Speed Switching Microvalve. Biomedical Microdevices 2008;10 329-335.

[9] Ikeda S, Arai F, Fukuda T, Negoro M. An In Vitro Soft Membranous Model of Individual Human Cerebral Artery Reproduced with Visco-Elastic Behavior. In: proceedings of the 2004 IEEE International Conference on Robotics and Automation, ICRA2004, 18-22 April 2005, Barcelona, Spain.

[10] Invitrogen Corporation. http://www.invitrogen.com/ (accessed 26 July 2012).

[11] Niskanen A. O, Harrabi K, Yoshihara F, Nakamura Y, Lloyd S, Tsai J. S. Quantum Coherent Tunable Coupling of Superconducting Qubits. Science 2007;316 723-726.

[12] Vlasov Y. A, O'Boyle M, Hamann H. F, McNab S. J. Active control of slow light with photonic crystal waveguides. Nature 2005;438 65-69.

[13] Kakudo T. S, Futaki S. Transferrin-modified liposomes equipped with a pH sensitive fusogenic peptide: an artificial viral like delivery system. Biochemistry 2004;43(19) 5618-5628.

[14] Saito Y, Uemura S. Field Emission from Carbon Nanotues and its Application to Electron Sources. Carbon 2000;38 169-182.

[15] Yabushita R, Hata K, Sato H, Saito Y. Development of compact field emission scanning electron microscope equipped with multiwalled carbon nanotube bundle cathode. Journal of Vacuum Science and Technology B 2007;25 640-642.

[16] Zhang J, Cheng Y, Lee Y. Z, Gao B, Qiu Q, Lin W. L., Lalush D., Lu J. P., Zhou O. A nanotube-based field emission x-ray source for microcomputed tomography. Review of Scientific Instruments 2005;76 094301.

[17] Nakajima M, Arai S, Saito Y, Arai F, Fukuda T. Nanoactuation of Telescoping Multiwalled Carbon Nanotubes inside Transmission Electron Microscope. Japanese Journal of Applied Physics 2007;42 L1035-1038.

[18] Elshimy H. M, Nakajima M, Imaizumi Y, Arai F, Fukuda T. Fabrication of FIB-CVD Nano Temperature Sensors for Local Temperature Sensing in Water Environments. Journal of Robotics and Mechatronics 2007;19(5) 512-518.

[19] Lee J. U. Photovoltaic effect in ideal carbon nanotube diodes. Applied Physics. Letters 2005;87 073101.

[20] Mandalia H. C, Jain V. K, Pattanaik B. N. Application of Super-molecules in Solar Energy Conversion- A Review. Research Journal of Chemical Sciences 2012;2 89-102.

[21] Iijima S. Helical Microtubules of Graphitic Carbon. Nature 1991;354 56-58.

[22] Cumings J, Zettl A. Low-Friction Nanoscale Linear Bearing Realized from Multiwall Carbon Nanotubes. Science 2000;289 602-604.

[23] Fennimore A. M, Yuzvinssky T. D, Han W. Q, Fuhrer M. S, Cummings J, Zettl A. Rotational Actuators Based on Carbon Nanotubes. Nature 2003;424 408-410.

[24] Dong L. X, Nelson B. J, Arai F, Fukuda T. Towards Nanotube Linear Servomotors. IEEE Transaction on Automation Science and Engineering 2006; 3 228-235.

[25] Nakajima M, Arai F, Fukuda T. In situ Measurement of Young's Modulus of Carbon Nanotube inside TEM through Hybrid Nanorobotic Manipulation System. IEEE Transaction on Nanotechnology 2006;5 243-248.

[26] Center for Micro-nano Mechatronics, Graduate School of Engineering, Nagoya University. http://www.mech.nagoya-u.ac.jp/cmm/index_e.html (accessed 26 July 2012)

Chapter 2

High Knudsen Number Flow — Optical Diagnostic Techniques

Tomohide Niimi

1. Introduction

Along with the development in micro- and nano-technology with small characteristic length and high technology under high vacuum environments, it has been strongly desired to understand thermo-fluid phenomena around a micro- and nano-system such as a magnetic head slider of hard disk drive, micro thruster for micro satellite, semiconductor thin film fabrication system and so on. In cases where the characteristic dimension of the flow is of the order of the mean free path (about 60 nm at the atmospheric condition), we have to analyze the flows at the atomic or molecular level, because the continuum approach becomes invalid. The atomic or molecular gas flow had been used synonymously with the rarefied gas flow so far, but has to be applied also to the flow around the micro- and nano-devices working under atmospheric conditions as mentioned above. We call these flows with large *Kn* "High Knudsen number flows". As an indicator of rarefaction of gas-flows, the Knudsen number (*Kn*) is defined by the ratio of the mean free path (λ) divided by the characteristic dimension (L) of the flow. The flow regimes are classified on a scale of *Kn* number, as follows.

- $Kn < 10^{-3}$: *Continuum flow regime*

  Compressible Navier-Stokes equation with no-slip boundary condition

- $10^{-3} < Kn < 10^{-1}$: *Slip flow regime*

  Compressible Navier-Stokes equation with velocity slip and temperature jump at the boundary

- $10^{-1} < Kn < 10$: *Transition flow regime*

  Boltzmann equation, where intermolecular collisions should be taken into account

- $10 < Kn$: *Free molecular flow regime*

  Boltzmann equation, where intermolecular collisions are negligible compared with interactions between the gas molecules and the walls

For high Knudsen number flows, we have to take into account the followings that may be neglected for the continuum flow regime. In the case of large λ, there appear the strong nonequilibrium phenomena because of few intermolecular collisions. For extremely small L, on the other hand, the flow field is strongly influenced by interaction of molecules with a solid boundary rather than intermolecular collisions.

Experimental analyses of thermo-fluid phenomena related to the high Knudsen number flows need the optical diagnostic techniques based on atoms or molecules, such as their emission and absorption of photons. However, the experimental techniques are behind on development compared with the molecular simulation techniques such as the molecular dynamic method (MD), the direct simulation Monte-Carlo method (DSMC) and so on.

In this chapter, the optical diagnostic techniques for the high Knudsen number flows are mainly described, such as laser induced fluorescence (LIF), resonantly enhanced multiphoton ionization (REMPI) and pressure-sensitive molecular film (PSMF), and some experimental results obtained by the use of the techniques, i.e., applications of LIF to visualization of rarefied gas flows including complicated shock wave system and to measurement of rotational temperature, establishment of a REMPI system and its application to detection of rotational nonequilibrium in highly rarefied gas flows, and development of the PSMF for micro gas flow measurements.

2. Laser Induced Fluorescence (LIF)

Some atoms and molecules emit fluorescence when irradiated by a laser beam whose wavelength corresponds to an absorption line of molecules, as shown in Figure 1. In Figure 2 [1], LIF of iodine molecules (I_2) seeded in argon gas was applied to visualization of a flow field of a supersonic free jet issued from a sonic nozzle with exit diameter of 0.5 mm into a vacuum chamber, clearly showing the barrel shock waves and Mach disk. Figure 3 shows an example of the visualization of complicated flowfield structure and shock wave systems of two interacting supersonic free jets [2] by using I_2-LIF. NO-LIF has been also employed to visualize the low density high speed flows [3].

A method for planar measurement of rotational temperature using two-line I_2-LIF was proposed for the rarefied gas flows [4]. If the fluorescence intensities at a point in the flow field are designated by F_1 and F_2 when the iodine molecules in the rotational levels J''_1 and J''_2 are excited, respectively, the ratio F_1 / F_2 depends on rotational temperature and two rotational quantum numbers, J''_1 and J''_2. Therefore, once two absorption lines are selected, the rotational temperature can be deduced from the ratio of the fluorescence intensities.

Figures 4 are typical images of a supersonic free jet visualized by the use of irradiation of laser beams with different wavelengths corresponding to the respective absorption lines. These

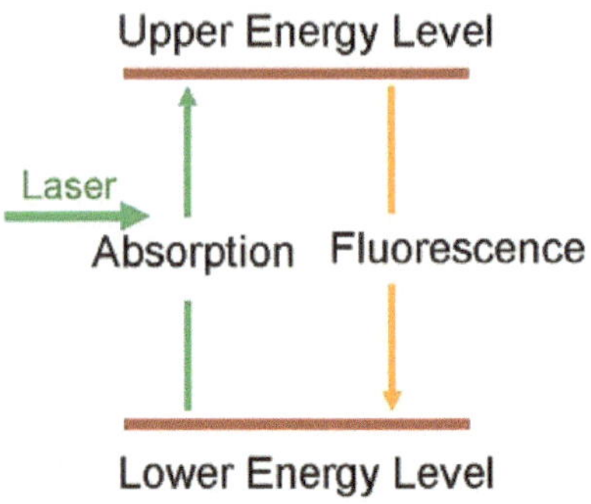

Figure 1. Principle of LIF

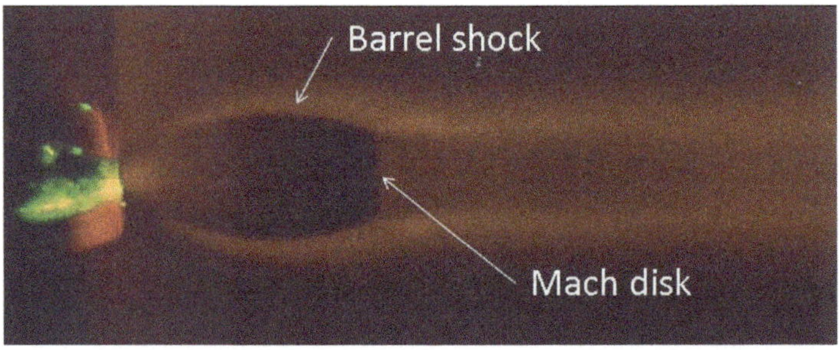

Figure 2. Supersonic free jet visualized by I_2-LIF [1]

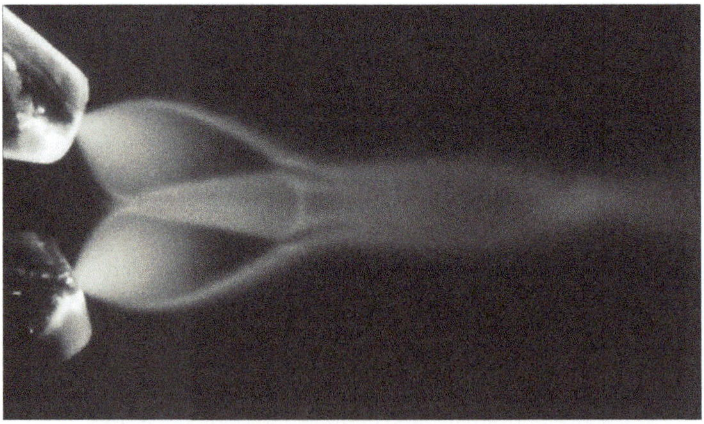

Figure 3. Interacting supersonic free jets visualized by I_2-LIF [2]

images are obtained at the same pressure condition: source pressure P_s = 16kPa and background pressure P_b = 100Pa. It can be seen that the fluorescence intensity distributions are very different, depending on the absorption lines. If using two images among them, we can deduce the temperature distribution two-dimensionally as shown in Figure 5.

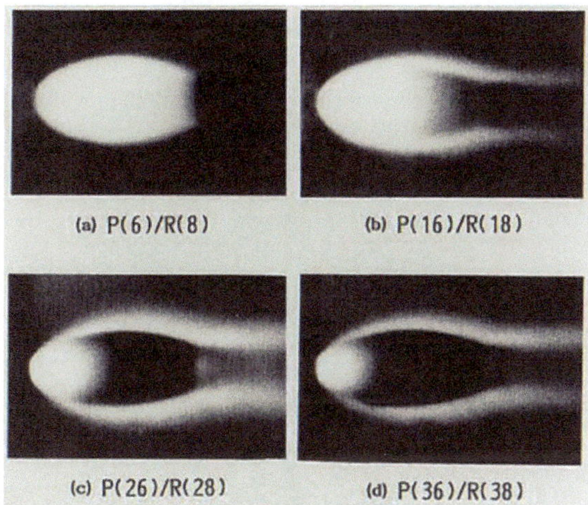

Figure 4. Supersonic free jets visualized by I_2-LIF with different wavelengths. P and R means P- and R-branches, respectively, and the number the rotational energy level [4].

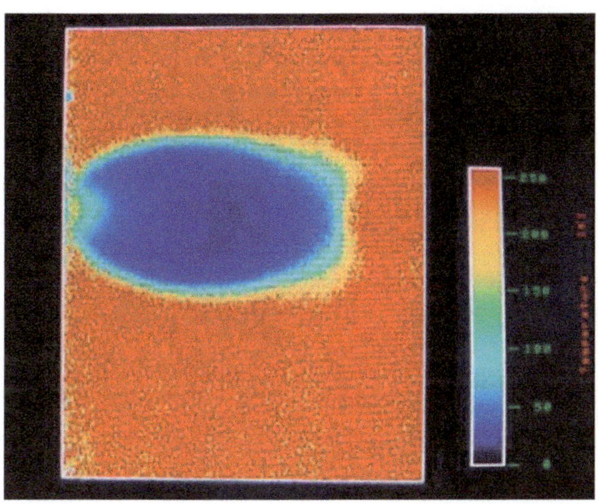

Figure 5. Two-dimensional temperature distribution of a supersonic free jet [4].

3. Resonantly Enhanced Multi-Photon Ionization (REMPI)

To measure the rotational population in rarefied supersonic nitrogen free jets and finally to confirm the non-Boltzmann distribution of the rotational levels experimentally, a REMPI (Resonantly Enhanced Multi-Photon Ionization) method has been applied to detection of nitrogen ions directly as an ion current. REMPI is known to have high detection sensitivity, which allows obtaining the signal under the very low number density condition. Here the principle of REMPI is introduced and the (2+2) N_2-REMPI is applied to a supersonic nitrogen free jet to detect the non-Boltzmann distribution of the rotational levels.

Figure 6 depicts the schematic energy level diagram for (2+2) N_2-REMPI, illustrating the relevant processes. In this process, nitrogen molecules at the ground state ($X^1\Sigma_g^+$) are excited to the resonant state ($a^1\Pi_g$) by two-photon absorption. Then the excited molecules are ionized by additional two-photon energy. Since four photons participate in this process, the ion current is proportional to the fourth power of laser flux in principle. However, when the laser flux is sufficiently high so that almost all the excited molecules are ionized, the ion current is proportional to square of laser flux, because the REMPI process reflects the two-photon transition process from the ground to the resonant state. In other words, the REMPI spectrum reflects directly the rotational population in the ground state of neutral molecules, so the rotational temperature can be deduced from the REMPI spectra, provided that the flow is in equilibrium, that is, the rotational energy distribution follows the Boltzmann distribution.

The REMPI technique is applied to detection of the rotational non-equilibrium in a nitrogen free-molecular flow. All experiments are carried out in a vacuum chamber evacuated by a turbomolecular pump and a dry pump as a backing pump, allowing an oil-free vacuum environment. Nitrogen gas is issued from a sonic nozzle with D = 0.50mm diameter, and expanded into the chamber. Stagnation temperature is kept at 293 K and source pressures P_0 is set at 30 Torr (3.9×10^3 Pa) to 1 Torr (1.3×10^2 Pa). A Nd:YAG-pumped dye laser operated

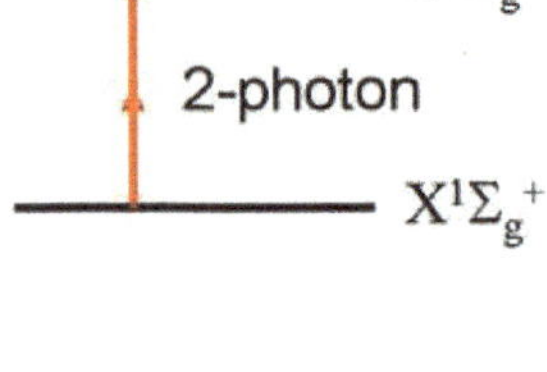

Figure 6. (2+2) N_2-REMPI process.

with Rhodamine 6G dye is used as a laser source, and the output is frequency-doubled by a BBO crystal. The wavelength of the laser beam is ranged from 283 to 284.1 nm. The beam is focused with a quartz lens on a centerline of a nitrogen free-molecular flow. The ionized nitrogen molecules are detected by a secondary electron multiplier or a tungsten Langmuir probe. A wavelength is scanned by a step of 0.001 nm, and the signal intensity is integrated for 100 laser pulses per each step.

Figure 7 represents an experimental (2+2) N_2-REMPI spectrum of the $(v', v'') = (1,0)$ vibrational band, measured for $P_0D = 15$ Torr-mm and at a focal point of $x = 2.5$ mm downstream from the nozzle exit ($x/D = 5$) along the centerline of the jet. In this figure, the horizontal axis indicates the wavelength of the laser, and the vertical one the signal intensity normalized by a peak. Numbers on rulers in the figure correspond to spectral positions for O, P, Q, R and S branches. When the population number is designated by $N(J'')$, the rotational line intensity $I_{J',J''}$ in (2+2) N_2-REMPI spectra is given by

$$I_{J',J''} = Ag(J'')S(J', J'')N(J'')/(2J''+1) \tag{1}$$

where J' and J'' are the rotational quantum number of the resonant and ground state, respectively, A is a proportional constant independent of the rotational quantum number, and $g(J'')$ is the nuclear spin statistical weight of nitrogen molecules formed by N^{14} atoms. $S(J',J'')$ is the two-photon Hönl-London factor [5] for the $a^1\Pi_g \leftarrow X^1\Sigma_g^+$ transition. $N(J'')$ is proportional to $(2J''+1)\exp(-E_{rot}/kT_{rot})$ (E_{rot}: rotational energy, T_{rot}: rotational temperature, k: Boltzmann's constant), provided that the rotational energy distribution follows the Boltzmann distribution. In this case, $I_{J',J''}$ is given by [6].

$$I_{J',J''} = Ag(J'')S(J', J'')\exp(-E_{rot}/kT_{rot}) \tag{2}$$

and the rotational temperature can be deduced from a slope of Boltzmann plot of $\ln(I_{J',J''} / gS(J',J''))$ versus E_{rot}/k. If there appears nonlinearity in the Boltzmann plot, therefore, the rotational energy distribution deviates from the Boltzmann distribution and the rotational temperature cannot be defined.

Figure 8 shows a result of Boltzmann plots at several x / D's for $P_0D = 15$ Torr-mm by using the measured REMPI spectra. In this figure, the horizontal axis indicates the E_{rot}/k and the vertical one the $\ln(I_{J',J''} / gS(J',J''))$. Numbers attached to the data points are the rotational quantum numbers of the ground state. It is found from Figure 8 that all the Boltzmann plots demonstrate the nonlinearity even at $x / D = 1.0$ for $P_0D = 15$ Torr-mm and especially the data in higher rotational levels deviate from a line, confirming the non-Boltzmann distribution of the rotational levels.

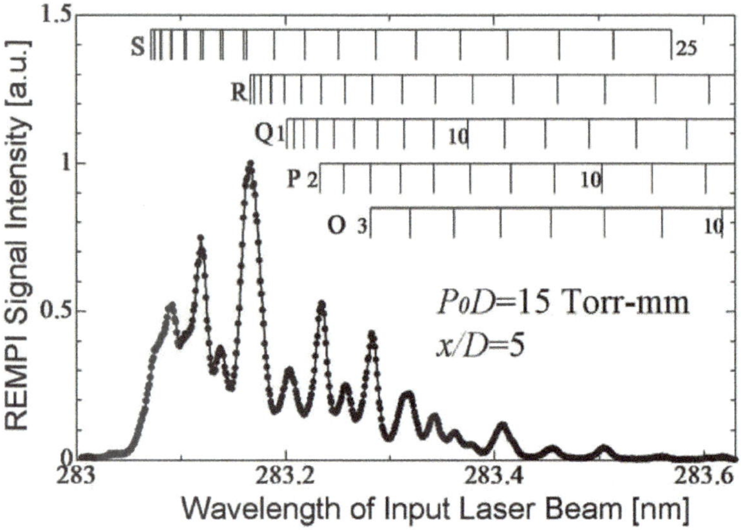

Figure 7. N_2-REMPI spectrum [6].

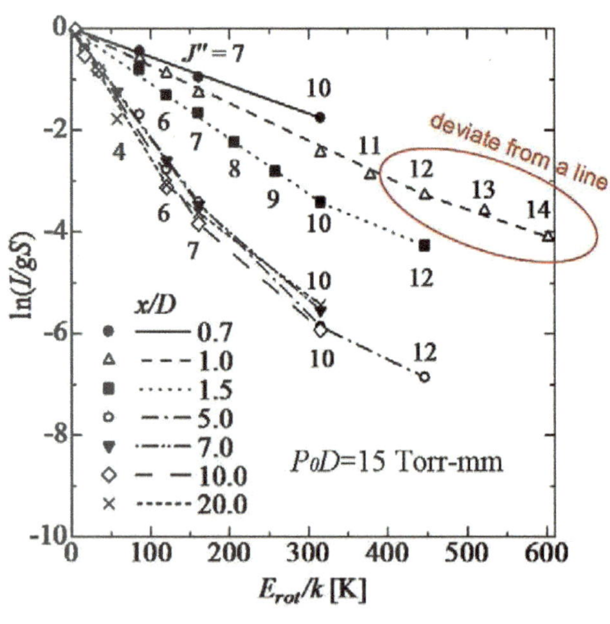

Figure 8. Boltzmann plots [6].

The rotational temperature is deduced from the Boltzmann plots of Figure 8 by using only the linear portion of the plots lying at smaller rotational quantum numbers. Figure 9 shows the rotational temperature distribution along the centerline of a supersonic free jet, where the horizontal axis indicates x / D and the vertical one the rotational temperature. Solid circles indicate the measured rotational temperatures, and the broken lines the translational temperature distribution calculated from isentropic relations using the Mach numbers [7]. The rotational temperature distribution measured by REMPI is compared with other data measured by different ways, i.e. the rotational temperatures measured by using only some rotational lines constituting a linear portion of the Boltzmann plots obtained by the electron beam method [8] and calculated by using an equation of the rotational temperature distribution for the rotational relaxation [9]. As you can see, all results agree with each other, if using only the data at smaller rotational quantum numbers for the REMPI and electron beam techniques.

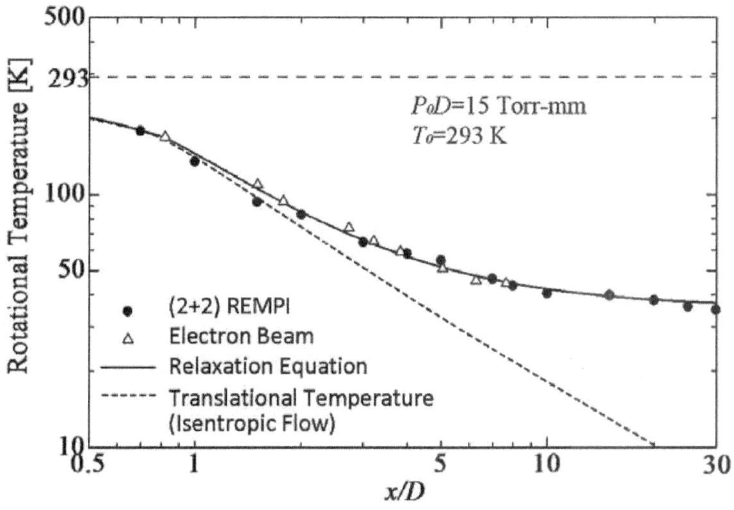

Figure 9. Rotational temperature distribution along the centerline of a supersonic free jet [6].

To clearly reveal the non-Boltzmann distribution, the rotational population obtained by REMPI for P_0D = 15 Torr-mm is presented in Figure 10, in which the horizontal axis indicates the rotational quantum number and the vertical one the population ratio of each rotational level to the total. Symbols such as solid circles correspond to the population ratio for each measurement point and each curve to a Boltzmann distribution at temperature deduced from a Boltzmann plot using only the data for smaller rotational quantum numbers as mentioned above. In the condition of P_0D = 15 Torr-mm, the rotational population at x / D = 1.0 follows almost the Boltzmann distribution, whereas at x / D = 7.0 the experimental data deviate from the Boltzmann distribution evidently. Comparing between the distributions of x / D = 7.0 and 20.0, for $J'' \geq 7$ the both show the same tendency, suggesting partial freezing of the rotational

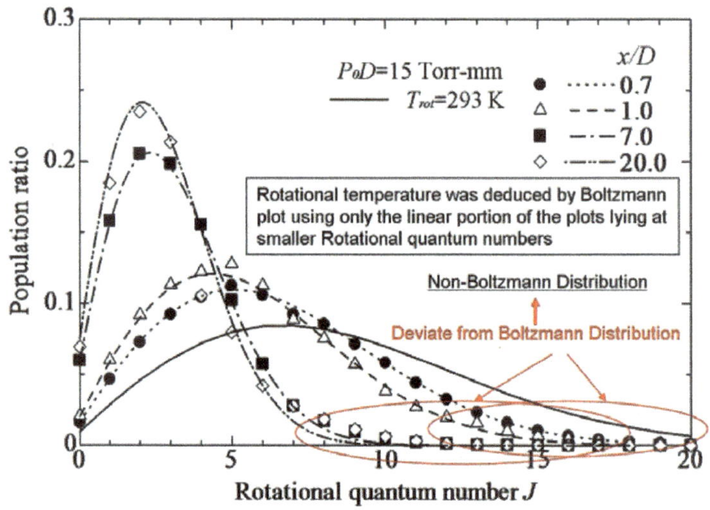

Figure 10. Rotational population obtained by REMPI [6].

population. Generally the higher rotational quantum number the molecules have, the lower the transition probability becomes, resulting in the deviation from the Boltzmann distribution particularly for the higher rotational levels. This is the reason why the freezing of the rotational population starts partially at the higher rotational levels.

4. Pressure Sensitive Paints (PSP) and Pressure Sensitive Molecular Film (PSMF)

There have been no appropriate techniques for the measurement of gas pressure on a solid surface inside micro-systems. To measure pressure distributions in a low density gas-flows with high Knudsen number, a pressure sensitive paint (PSP) technique [10, 11] have been adopted.

The pressure measurement technique using PSP is based on the oxygen quenching of luminescent molecules. PSP is composed of luminescent molecules and a binder material to fix the luminescent molecules to a solid surface. Figure 11 depicts the schematic energy level diagram for PSP and oxygen molecules. When PSP layer applied to the surface is illuminated by UV light, the luminescent molecules are excited by absorption of photon energy, and then the molecules emit luminescence. On the other hand, oxygen molecule with triplet ground state acts as a quencher of the luminescence. As a result, the phosphorescence intensity decreases as an increase in partial pressure of oxygen. Pressure on the solid surface can be deduced from the relationship between pressure and the luminescence intensity (Stern-Volmer plot [11]).

$$\frac{I_{ref}}{I}=\sum_{n=1}^{N}A_n\left(\frac{P}{P_{ref}}\right)^n \tag{3}$$

where I is the luminescence intensity and P is the oxygen pressure. I_{ref} is the luminescence intensity at a known reference pressure P_{ref}. The coefficients A_n are the constants called as the Stern-Volmer coefficients determined by calibration tests. The luminescence intensity I of the ideal PSP ($N = 1$) depends inversely on P following to Eq. 3, but actual PSPs have nonlinear dependence of I^{-1} on P. In practice, a second-order polynomial ($N = 2$) is commonly used.

Because PSP works as a so-called "molecular sensor", it seems also suitable for analyses of high Knudsen number flows, which require diagnostic tools in the molecular level. However, an application of PSP to micro-devices is very difficult, because conventional PSPs are very thick compared to the dimension of micro-devices owing to the use of polymer binders. Moreover, they do not have sufficient spatial resolution for the pressure measurement of micro-flows due to the aggregation of luminescent molecules in polymer binders [12].

Pressure-sensitive molecular films (PSMFs) have been developed by using the Langmuir-Blodgett (LB) technique [13] to fabricate thin films, and tested to confirm the feasibility of the pressure measurement around micro-devices [14]. PSMF with nanometer order thickness and high spatial resolution is suitable for analyses of micro-flows.

LB films are fabricated according to the following procedure. First, a drop of a dilute solution of amphiphilic molecules in a volatile solvent is spread on the interface between air and sub-phase. After the solvent is evaporated, a monolayer of the molecules remains on the interface. The monolayer is transferred to a substrate with compressing the monolayer so as to control the order of the molecules.

In some cases, arachidic acid (AA) has been adopted as a spacer molecule to control the intermolecular spacing of luminescent molecules, and to form a stable LB film with highly ordered structure. However, it is desirable for high pressure sensitivity to adjust the molar ratio of luminophore and AA [14].

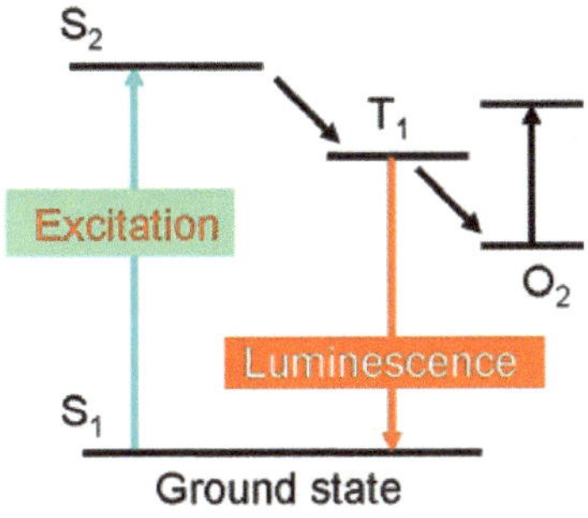

Figure 11. Schematic energy level diagram. S: Singlet state, T: Triplet state.

The PSMF sample was applied to a glass slide, and the sample was set inside a vacuum chamber evacuated by vacuum pumps to examine the pressure sensitivity. Pure oxygen gas was supplied into the chamber, and the pressure in the chamber was monitored by a capacitance manometer. The temperature of the sample was controlled by a Peltier element. A xenon-arc lamp with a band-pass filter (400 ± 20 nm) was used as an excitation light source illuminating the sample via an optical fiber. The luminescence was filtered by a long-pass filter (600 nm) to eliminate the light from the xenon-arc lamp, and was detected by an image intensified CCD camera (512 × 512 pixels, 14 bit).

The PSMF based on Pd(II) Mesoporphyrin IX (PdMP), which is used as luminescent molecules, has been fabricated and it is clarified that the pressure sensitivity of PSMF of PdMP is sufficiently high [14]. However, the PSMF composed of PdMP cannot be applicable to the pressure range higher than 130 Pa due to saturation of oxygen quenching of PdMP; this is caused by the long life time of PdMP, which is so long (1.0 msec [15]) that most of luminescent molecules are quenched. It is desired for PSMF to work around an atmospheric pressure, because the most micro-devices are used in an atmospheric pressure. In order to fabricate a useful PSMF for an atmospheric pressure range, four samples of PSMFs are prepared, composed of Pt(II) Mesoporphyrin IX (PtMP), Pt(II) Mesoporphyrin IX dimethylester (PtMPDME), Pt(II) Protoporphyrin IX (PtPP) and Cu(II) Mesoporphyrin IX dimethylester (CuMPDME). Those luminescent molecules have shorter life time compared with PdMP (e.g. the life time of PtMPDME and CuMPDME are 0.14 msec and 0.1 msec, respectively [15]). Figure 12 shows the pressure sensitivities of four PSMFs below 21 kPa (equal to the partial pressure of oxygen in atmospheric pressure) [16]. The horizontal axis of the Stern-Volmer plot is the normalized pressure P/P_{ref} and the vertical axis is the inverse luminescent intensity ratio I_{ref}/I, where I_{ref} is the reference luminescent intensity at the reference pressure $P_{ref} = 1.0 \times 10^{-2}$ Pa. It is clarified that the pressure sensitivity of PtMP is highest among the four tested PSMFs and is comparable to that of conventional PSPs. The temperature dependency of PSMF also has to be studied, because that of PSP is the main factor of the measurement error. PSMF was applied to the measurement of micro gas-flows [17].

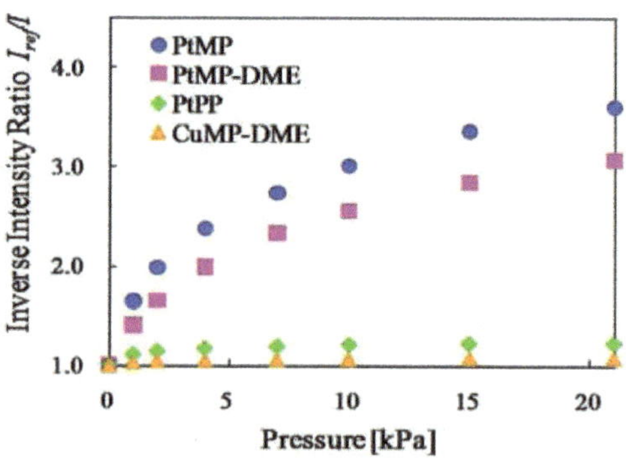

Figure 12. Stern-Volmer plots of PSMFs [16]

The flow in a converging-diverging nozzle was examined for the demonstration, because the pressure considerably changes in a nozzle and it was easy to compare the result with a numerical one.

Figure 13 shows the geometry of a converging-diverging micro-nozzle, which was two dimensional geometry, and the thickness of the micronozzles was 1.0 mm, and the width of the throat was 103 μm, the diverging length was 492 μm and the diverging angle was 23.5 degrees. The micro-nozzle was put on a holder with an inlet and outlet port, and then it was covered with the PSMF deposited glass (see Figure 13b). Air was supplied from the inlet port, and the outlet port was evacuated by a scroll pump. PtMP was employed as a luminophore of PSMF. Luminescent intensity of PSMF was detected through a fluorescent microscope. The images were taken with the exposure time of 8.0 s, and 32 pictures were averaged to reduce the noise.

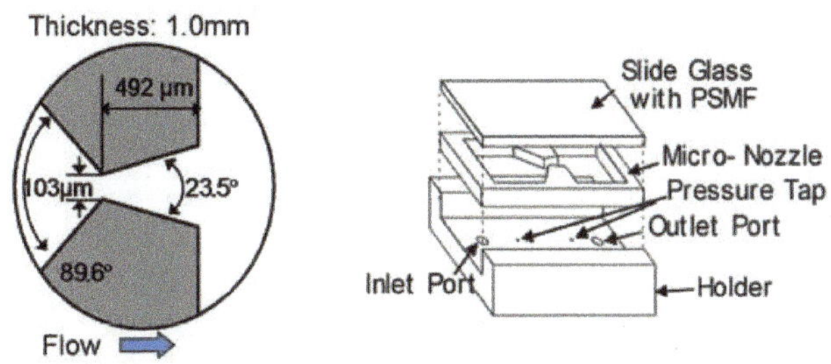

(a) Geometry of a micro-nozzle

(b) Schematic image of micro-nozzle assemblies

Figure 13. Micro-nozzle.

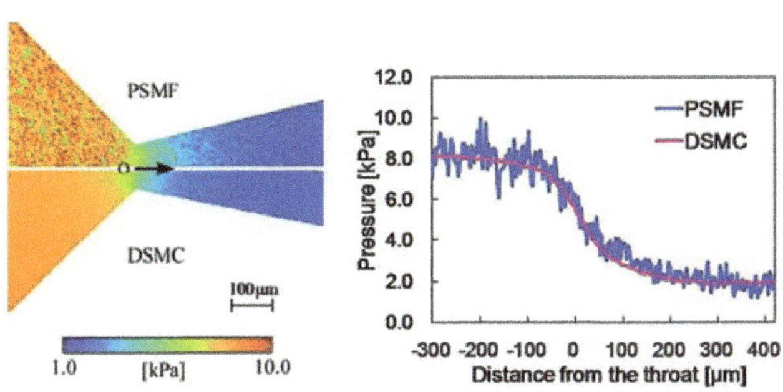

(a) Pressure distribution inside the micro-nozzle measured by PSMF (upper) and calculated by DSMC (lower)

(b) Pressure profile inside the micro-nozzle along the centerline of the nozzle

Figure 14. Pressure distribution inside the micro-nozzle [17].

The micro-nozzle flow was analyzed by the direct simulation Monte Carlo (DSMC) method [18], which is a popular method for high Knudsen number flow analysis. The numerical conditions were adjusted to the PSMF experimental conditions.

Figure 14 shows the pressure distribution of the micro-nozzle with the inlet pressure P_{in} = 10.0 kPa and the outlet pressure P_{out} = 1.0 kPa, where the wind-off images were taken at $P_{in} = P_{out}$ = 1.0 kPa. The numerical result by DSMC is also shown in the lower side of Figure 14. Moreover, the pressure profiles along the centerline are shown in Figure 14b. Though the pressure profile contains some noise as shown in Figure 14b, it is clearly observed that the pressure distribution obtained by PSMF is quantitatively in good agreement with that by DSMC. Especially, the sharp pressure drop near the throat was captured by PSMF, and this result indicates that the spatial resolution of PSMF is high enough for micro-scale measurement. The pressure value at -300 μm from the throat is about 8 kPa as shown in Figure 14b and is smaller than P_{in} = 10.0 kPa, which was measured at -10 mm. This pressure drop may be caused by wall friction and the acceleration at the converging area of the nozzle.

Author details

Tomohide Niimi

Department of Micro-Nano Systems Engineering, Nagoya University, Nagoya, Japan

References

[1] Fujimoto T, Niimi T. Three Dimensional Structures of Interacting Freejets. Rarefied Gas Dynamics 1988, Space-Related Studies, AIAA; 391-406.

[2] Niimi T, Fujimoto T, Taoi N. Flow Fields of Interacting Parallel Supersonic Free Jets. JSME International Journal Series B 1996;39(1) 95-100.

[3] Mori H, Taniguchi M, Nishihira R, Niimi T. Experimental Analyses of Linear-Type Aerospike Nozzles with and without Sidewalls. AIAA paper 2005; 1350.

[4] Niimi T, Fujimoto T, Shimizu N. Planar Measurement of Temperature in Rarefied Gas Flow by LIF Images. Proceedings of 17th International Symposium on Rarefied Gas Dynamics 1991: 1482-1489.

[5] Bray R. G, Hochstrasser R. M. (1976). Two-photon absroption by rotating diatomic molecules. Molecular Physics 1976;31(4) 1199-1211.

[6] Mori H, Niimi H, Akiyama I, Tsuzuki T. Experimental Detection of Rotational Non-Boltzmann Distribution in Supersonic Free Molecular Nitrogen Flows. Physics of Fluids 2005;17(11) 117103.

[7] Ashkenas H, Sherman F. S. The structure and utilization of supersonic free jets in low density wind tunnels. Proceedings of 4th International Symposium on Rarefied Gas Dynamics 1966;2 84-105.

[8] Marrone P. V. Temperature and density measurements in free jets and shock waves. Physics of Fluids 1967;10(3) 521-538.

[9] Gallagher R. J, Fenn J. B. Relaxation rates from time of flight analysis of molecular beams. Journal of Chemical Physics 1974;60(9) 3487-3491.

[10] Liu T, Champbell T, Burns S. P, Sullivan J. P. Temperature- and pressure-sensitive paints in aerodynamics. Applied Mechanics Reviews 1997;50 227-246.

[11] Bell J. H, Schairer E. T, Hand L. A, Mahta R. D. Surface pressure measurements using luminescent coatings. Annual Review of Fluid Mechanics 2001;33 155-206.

[12] Mori H, Niimi T, Hirako M, Uenishi H. Molecular number flux detection using oxygen sensitive luminophore. Physics of Fluids 2005;17 100610.

[13] Ulman A. An Introduction to Ultrathin Organic Films From Langmuir-Blodgett to Self-Assembly. San Diego: Academic Press; 1991.

[14] Matsuda Y, Mori H, Niimi T, Uenishi H, Hirako M. Development of pressure sensitive molecular film applicable to pressure measurement for high Knudsen number flows. Experiments in Fluids 2007;42 543-550.

[15] Becker R. S. Metalloporphyrins. Electronic spectra and nature of perturbations. I. Transition metal ion derivatives. The Journal of Physical Chemistry 1963;67(12) 2662-2669.

[16] Matsuda Y, Mori H, Sakazaki Y, Uchida T, Suzuki S, Yamaguchi H, Niimi T. Extension and characterization of pressure-sensitive molecular film. Experiments in Fluids 2009;47(6) 1025-1032.

[17] Matsuda Y, Uchida T, Suzuki S, Misaki R, Yamaguchi H, Niimi T. Pressure-sensitive molecular film for investigation of micro gas flows. Microfluidics and Nanofluidics 2011;10(1) 165-171.

[18] Bird G. A. Molecular Gas Dynamics and the Direct Simulation of Gas Flows, New York: Oxford University Press; 1994.

Chapter 3

Precision Micro Machining Methods and Mechanical Devices

Eiji Shamoto, Norikazu Suzuki, Takashi Kato and
Burak Sencer

1. Ultra-precision/micro machining of die steel by elliptical vibration cutting [1]

A novel method to attain ultra-precision sculpturing in micro/nano scale for difficult-to-cut materials is introduced. Elliptical vibration cutting technology is well-known for its excellent performance in achieving ultra-precision machining of steel materials with single crystal diamond tools. Elliptical vibration locus is generally controlled and held to a constant in practice. On the contrary, the proposed method utilizes the variations of the elliptical vibration locus in a positive manner. Depth of cut can be actively controlled in elliptical vibration cutting by controlling vibration amplitude in the thrust direction. By utilizing this as a fast tool servo function in elliptical vibration cutting, high performance micro/nano sculpturing can be attained without using conventional fast tool servo technology. Following sections describe the development of the high performance micro/nano sculpturing system and ultra-precision/micro machining applications.

1.1. Introduction

The authors have developed "elliptical vibration cutting" technology [2], and have demonstrated that ultra-precision machining of difficult-to-cut materials, such as hardened steel and hard/brittle materials, can be attained practically by applying ultrasonic elliptical vibration to single crystal diamond tools [3]. Several ultrasonic elliptical vibration tools and their control systems were also developed in the past studies [4]. Since variation of vibration amplitudes causes deterioration of machining accuracy and surface quality, most research efforts were dedicated to keeping the elliptical vibration locus ultra-precisely constant. Otherwise, ultra-precision cutting cannot be achieved in practice.

On the other hand, the authors focus on utilizing the variation in vibration amplitude in a positive manner, in contrast with conventional studies. It is considered as a unique function, i.e., the depth of cut can be actively controlled by controlling the vibration amplitude in the depth of cut direction while machining. By utilizing this function to serve as a sort of fast tool servo (FTS), the ultra-precision sculpturing of difficult-to-cut materials in micro/nano scale is achieved efficiently. Note that it is redundant and disadvantageous to combine the elliptical vibration tool with the conventional FTS, since both devices have actuators and the vibration tool is too heavy to be actuated at high frequency by the FTS.

Following sections introduce the proposed machining method and a vibration control system of a two-degree-of-freedom (2-DOF) elliptical vibration tool, which enables precise amplitude control of ultrasonic elliptical vibration at high speed. Subsequently, experimental verifications in textured grooving and in nano sculpturing are described.

1.2. Ultra-precision micro machining method for difficult-to-cut materials

Figure 1 shows the proposed machining with depth of cut control in elliptical vibration cutting. The tool is fed at a nominal cutting speed and vibrated elliptically at the same time. Because of this intermittent process at an ultrasonic frequency, tool wear and adhesion are restricted, and the ultra-precision cutting of hardened steel can be attained with single crystal diamond tools. The vibration amplitude in the depth of cut direction is controlled simultaneously in the proposed machining process as shown in the figure. The trajectory of the cutting edge, then, changes dynamically, and its envelope is transferred to the finished surface.

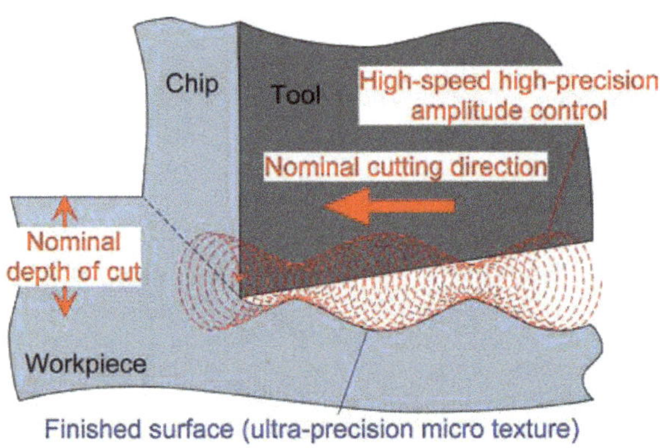

Figure 1. Machining by controlling amplitude in elliptical vibration cutting

By controlling the amplitude ultra-precisely at high speed, the ultra-precision sculpturing of the difficult-to-cut materials can be achieved efficiently without using conventional FTS technology [5]. In other words, the elliptical vibration cutting technology is equipped with a FTS function by itself. Although amplitude control command is not identical with the envelope of the cutting edge trajectory, as shown in Figure 1, their difference is not crucial to the present study. The difference is insignificant in practice when the slope is not steep. The depth of cut can be controlled within half of the maximum amplitude in the depth of cut direction, and available frequency range of the amplitude control is limited to that which is relatively lower than the elliptical vibration frequency. Therefore, performance in the role as FTS strongly depends on the specifications of the vibrator.

1.3. Elliptical vibration tool and high-speed amplitude control system

Figure 2 shows the developed system of the high-speed amplitude control of elliptical vibration at a frequency of about 36 kHz. A two-degree-of-freedom (2-DOF) elliptical vibration tool [6], which was designed to generate arbitrary elliptical vibration, is utilized in the present study. The vibrator is actuated by using some PZT actuators, which are sandwiched with metal cylindrical parts, namely a bolt clamped Langevin type transducer (BLT). As the vibrator is designed to have same resonant frequencies in second resonant mode of longitudinal vibration and fifth resonant mode of bending vibration, it can generate large longitudinal and bending vibrations simultaneously at the same ultrasonic frequency by applying exciting voltages to the actuators. Thus, an arbitrary 2-DOF elliptical vibration can be obtained at the diamond tool tip attached to the vibrator by combining both resonant vibration modes with some phase shift.

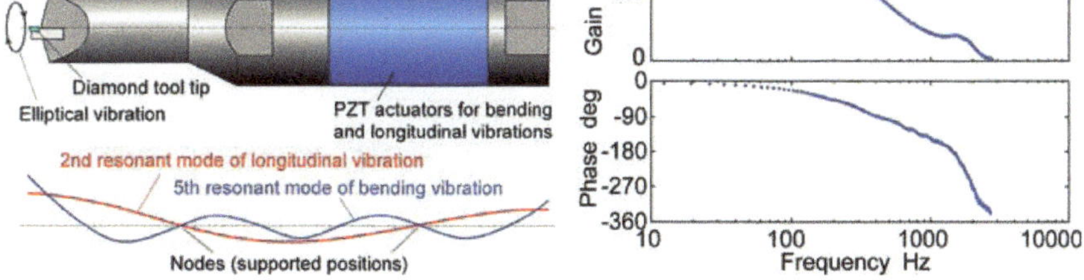

Figure 2. Elliptical vibration control system and frequency response of amplitude control

Gain of the amplifier can be controlled by external input in the developed system, and thus the exciting voltage supplied to the actuator is changed. The amplitude is, consequently, controlled by the external input. As the maximum amplitude in the depth of cut direction is 4 $\mu m_{p\text{-}p}$, the vibration amplitude can be controlled to change the depth of cut within 2 μm by this system. Measured frequency response of amplitude control is shown in Figure 2. The developed system is able to control the vibration amplitude with a frequency bandwidth of more than 300 Hz. This frequency bandwidth is relatively narrow as compared with that of conventional FTS. It might not, however, be a big problem because the elliptical vibration cutting technology is available only at relatively low cutting speed.

1.4. Ultra-precision micro/nano sculpturing of textured grooves

The developed control system was applied to grooving experiments. The ultrasonic elliptical vibration tool was mounted on an ultra-precision planing machine, NIC-300 (Nagase Integrex Co., Ltd.), and was fed straight for grooving. The vibration amplitude was controlled with sinusoidal and zigzag wave commands at the same time, and then, micro textured grooves were formed on the surface of a hardened steel workpiece. The vibration amplitude was controlled to change from 2 $\mu m_{p\text{-}p}$ to 4 $\mu m_{p\text{-}p}$. This corresponds to the depth of cut variation of 1 μm. On the other hand, the amplitude in the cutting direction is set to be constant, 4 $\mu m_{p\text{-}p}$.

Figure 3 shows photographs of grooves machined at the cutting speed of 0.2 m/min by the single crystal diamond tool with a nose radius of 1 mm and their surface profiles measured by a laser microscope, VK-9500 (Keyence Corp.). It was confirmed that the grooves with various ultra-precision micro textures can be machined successfully on the hardened steel, and mirror surface quality can be obtained on all grooves. Measured surface profiles of sinusoidal grooves agreed precisely with the command waves. On the other hand, measured corners of zigzag grooves are rounded, and their step heights were relatively smaller than the variation width of command waves of 1 μm. This is considered to be caused by cutting off high-frequency components in the amplitude control commands by the LPF, which the zigzag waves include at their sharp corners.

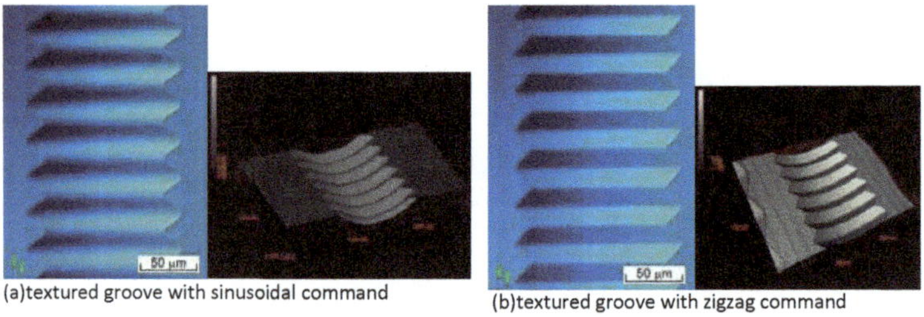

(a)textured groove with sinusoidal command

(b)textured groove with zigzag command

Figure 3. Microphotographs and surface profiles (cutting speed: 0.2 m/min, freq.: 100 Hz)

1.5. Ultra-precision micro/nano sculpturing on plane surfaces

In order to attain arbitrary sculpturing, an ultra-precision micro/nano sculpturing system was developed by using the developed vibration control system and the ultra-precision planing machine. Figure 4 shows the developed sculpturing system, where the planing machine is simply controlled to machine a plane surface at constant cutting speed. The ultrasonic elliptical vibration tool is attached on a Z-axis table of the machine tool. The vibration amplitude in the depth of cut direction along the Z-axis is controlled in synchronization with cutting feed motion in the X-axis, and then, arbitrary micro/nano sculpturing can be attained on a flat top surface of a steel workpiece by merely combining simple planing operations at constant cutting speed with high-speed depth of cut control. An industrial computer is utilized to detect the coordinate positions and generate the voltage signal for controlling vibration amplitude. The coordinate values are constantly monitored in the developed system by directly communicating with NC control system or by using an external optical sensor. The dynamic command signal for vibration amplitude control is generated to change the depth of cut in accordance with the CAD data.

The developed machining system was applied to nano sculpturing experiments of picture images. CAD data for sculpturing were produced from gray scale images, where the gray values of 8 bits (256 gradations) in pixels were converted into the amplitude commands. As the vibration amplitude in the experiments was changed within a range from 2 μm_{p-p} to 4 μm_{p-p}, the depth of cut was changed within 1 μm. The resolution of the depth of cut control is, therefore, about 4 nm. Hardened steel workpieces (64x48 mm, JIS: SUS420J2, HRC53) were machined with single crystal diamond tools with a nose radius of 1 mm. The size of original images is set to 3200x2400 pixels, and thus, 1 pixel corresponds to 20x20 μm.

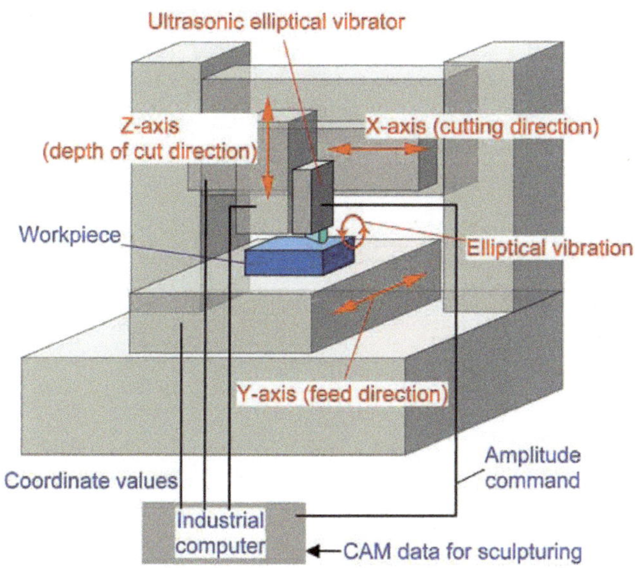

Figure 4. Developed micro/nano sculpturing system using elliptical vibration cutting

Figure 5 shows an example of the nano sculpturing of picture images. The depth of cut was controlled in nano scale in accordance with the image data. As shown in Figure 5, the gray scale picture image, the microphotograph of the machined surface and its profile, which was measured by optical surface profiler (ZYGO NewView 6200), correspond well to each other.

The result shows that picture images can be printed successfully on hardened steel as nano-scale sculptures.

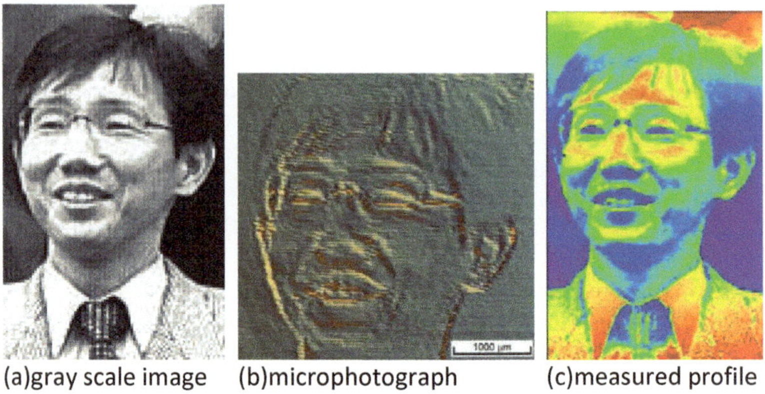

Figure 5. Gray scale image for amplitude command, machined surface and measured surface profile

Nano sculpturing experiments involving dimple patterns were also carried out with the developed sculpturing system. Sinusoidal commands to control the vibration amplitude were input to the elliptical vibration control system during machining, and the phase of the sinusoidal commands was changed by 180 degrees in every cutting feed, so that precisely aligned patterns were sculptured.

Figure 6 shows microphotographs. The hexagonal dimple patterns, whose borders are sharp, can be observed on the left. On the other hand, isolated circular dimple patterns were also sculptured successfully on the right, as the maximum depth of cut was considerably smaller than the amplitude variation. The results show that a variety of dimple patterns can be obtained ultra-precisely on the steel materials by using the developed sculpturing system.

1.6. Conclusion

Novel ultra-precision sculpturing technology for difficult-to-cut materials at micro/nano scale was proposed by utilizing elliptical vibration cutting technology. In the proposed method, the depth of cut is controlled without the conventional FTS technology by actively manipulating the vibration amplitude in the depth of cut direction. In order to verify the proposed method, the vibration amplitude control system and a high performance micro/nano sculpturing system were developed and applied to sculpturing experiments on hardened steel. Consequently, micro textured grooves, an image of a picture and various dimple patterns were manufactured on the hardened steel workpiece successfully as nano-scale sculptures. These were done by merely combining a simple feed motion at a constant speed with high-speed depth of cut control.

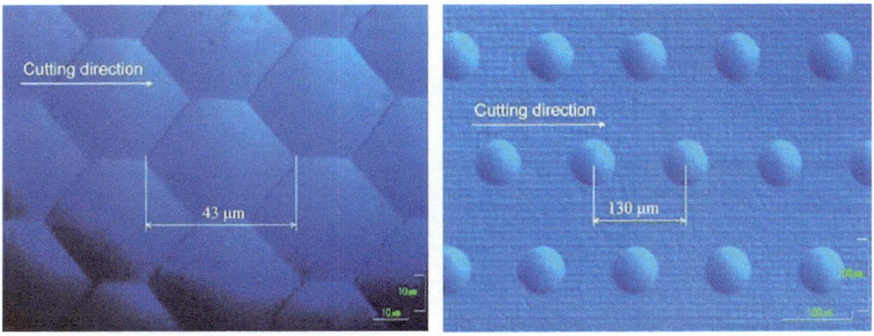

Figure 6. Microphotographs of sculptured dimples

2. Analytical prediction of chatter stability in ball end milling with tool inclination [7]

A new analytical model to predict chatter vibration in ball end milling with consideration of tool inclination is introduced in this section.

2.1. Introduction

The ball end milling is an important precision machining process, which is used in production of dies, molds, impellers, screw propellers and parts with free-form surfaces. As the slender ball end mills or the thin workpiece structures are flexible, the self-excited chatter vibration often occurs and causes severe problems such as short tool life and deterioration of surface quality. Many researchers investigated prediction of the self-excited chatter vibration in the milling process. Altintas and Budak [8] developed an analytical model to solve the chatter stability for the cylindrical end mills. Altintas, Shamoto, Lee and Budak extended the analytical model for the cylindrical end mills to predict the stability limits in ball end milling [9], but the tool inclination has not been considered due to geometric complexity of the ball end milling process.

On the other hand, 5-axis machining technology has been widely spread recently, and it enables arbitrary tool inclination for better machining efficiency, accuracy and stability.

Therefore, an analytical model of the ball end milling process with the self-excited chatter vibration is developed and verified with consideration of the tool inclination.

2.2. Outline of analytical model to predict chatter stability in ball end milling with tool inclination

The ball end milling process with the tool inclination is illustrated in Figure 7. The Cartesian coordinates xyz are fixed to the ball end mill and aligned with the workpiece coordinates uvw before the inclination. The origins of both the coordinate systems are placed at the present ball center. u, v and w are the cutting feed direction, the pick feed direction and the normal direction to the finished surface respectively. The tool is inclined around x by i_x and then around y by i_y, as shown in the figure. The sky-blue region shows engagement of the ball end mill with the workpiece. If vibration occurs in the process, it changes the uncut chip thickness and generates the dynamic cutting force in this engagement region. This dynamic cutting force excites the mechanical structure and generates vibration again.

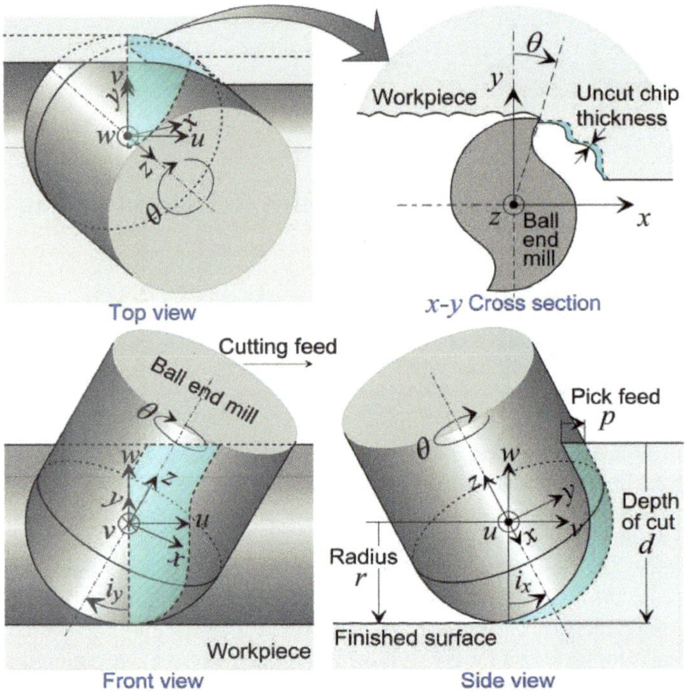

Figure 7. Ball end milling process with tool inclination.

This closed loop system can be expressed by the block diagram shown in Figure 8. If the vibration grows up through this closed loop, the system is unstable and generates the self-excited chatter vibration. The diagram contains two self-excitation mechanisms, i.e. regeneration and mode coupling. Note that the original or static milling forces and the forced vibrations, which change in synchronization with the tooth passage, are neglected in the present study, because they do not basically affect the self-excited chatter vibration.

In the ball end milling process, most of the surface removed by the present tooth has been cut by the previous tooth, as shown in the x-y cross section in Figure 7, and thus the present uncut chip thickness fluctuates by not only the present vibration but also the previous vibration. This regenerative effect is represented generally by the delay term μe^{-Ts}, where T is the tooth passing period. The overlap factor μ can be approximated by 1, since the feed rate is usually small enough compared with the ball radius r.

The multi-directional vibrations can be coupled when their modes in different directions have close resonant frequencies. The present study, therefore, deals with the xyz vibrations and their mode coupling effect as shown in the diagram.

The transfer functions are supposed to be measured, while the directional milling force coefficients, which are called process gains in the present paper, need to be calculated for the prediction of the chatter stability. The process gains are time-dependent in the milling process, but it is known that they can be approximated by their averages, i.e. DC components [8]. Therefore, the average process gains are derived numerically by assuming small displacement in each direction [7].

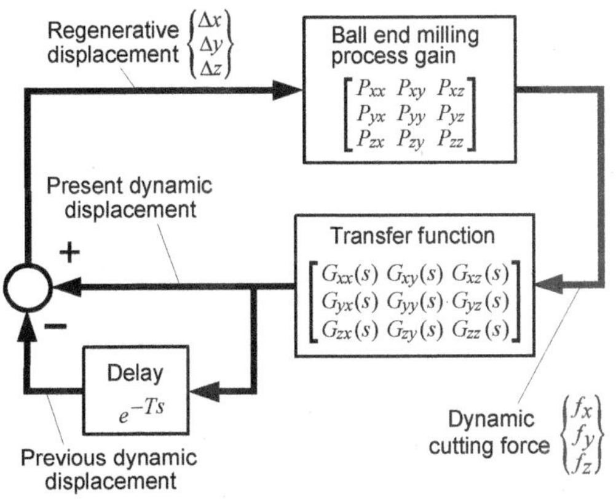

Figure 8. Block diagram of ball end milling process with regenerative and mode-coupling chatter vibrations.

2.3. Experimental setup and measurement

The machining experiments were conducted on the inclined workpiece surface with a slender ball end mill and a vertical machining center (Millac 3VA-DS, Okuma Corp.), as shown in Figure 9. The vibration was measured at the tool holder with the displacement sensor. The transfer function matrix was measured by the impulse response method, and the specific cutting forces, which are required to compute the process gains, are identified using a dynamometer and an oblique cutting model [10].

2.4. Analytical and experimental results

Figure 10 shows the predicted gain margins g_m, the predicted stability limits, i.e. contour lines of g_m=1, and experimental results of chatter vibrations. The experimental vibrations were classified by using the chatter frequency component of the measured displacement s_0, as shown

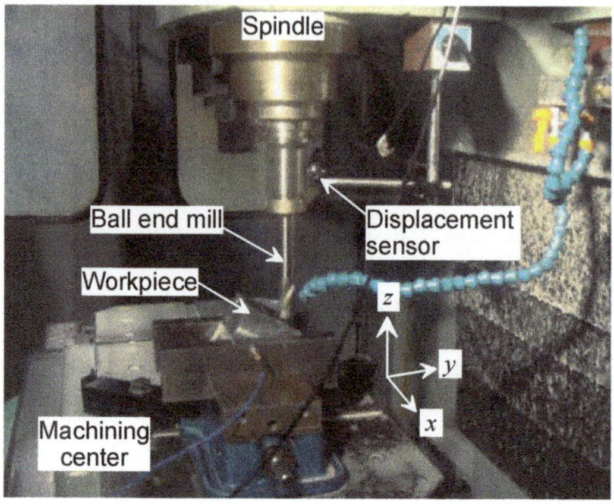

Figure 9. Ball end milling experiment with tool inclination.

under the figure. For example, severe chatter was detected at d>2 mm and i_x=10 deg, where the measured displacement has a large chatter frequency component of s_0>1.3 μm. On the other hand, only the spindle speed harmonics were observed at d=2.5 mm and i_x=45 deg. The analytical and experimental results are all in a good agreement as shown in the figure. There were no chatter vibrations at g_m>1.2, while severe chatter vibrations were detected at g_m<0.6.

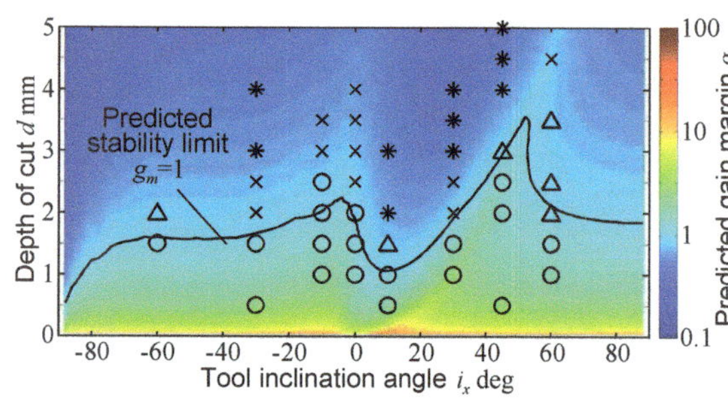

Figure 10. Predicted gain margins and chatter stability limits, and experimental results at varied tool inclination angle i_x. Workpiece: aluminum alloy (JIS:A5052); Cutter: HSS ball end mill (EBD80820, OSG Corp.), Cutter: number of flutes n_f = 2, radius r = 10 mm; Cutting conditions: i_y = 0 deg, pick feed p = 1 mm, n = 6240 min-1, feed rate of 0.01 mm/tooth; Cutting fluid: soluble. Chatter vibrations were classified as follows; o: no chatter ($s_0 \leq 0.12$ μm), Δ: slight chatter (0.12 μm < $s_0 \leq 0.24$ μm), x: chatter (0.24 μm < $s_0 \leq 1.2$ μm), *: severe chatter (1.2 μm < s_0).

2.5. Conclusion

The analytical model of the ball end milling process with the self-excited chatter vibration was developed with consideration of the tool inclination, and it was applied to predict the chatter stability at varied spindle speed and tool inclination in the present research. The predicted chatter stability agreed well with the experimental results, and it is expected that the present analysis will contribute to improve machining efficiency, surface quality and tool life in the free-form machining with the flexible structures like slender ball end mills and thin workpieces.

3. A new fluid bearing utilizing traveling waves [11]

A new non-contact fluid bearing utilizing traveling waves is proposed in this research. Conventional hydrostatic bearings utilize externally compressed fluid, which requires plumbing and compressors. In contrast, on proposed bearing system the moving part is supported with a thin fluid film compressed by the waves traveling radially on the bearing surface. The proposed bearing realizes non-contact smooth motion without such a large apparatus, and furthermore it has a capability to electrically control the bearing force or clearance.

3.1. Introduction

Bearings are one of the most fundamental and necessary elements for machines to generate motion between to surfaces smoothly with low friction. There have been constant demands for low friction or non-contact bearings for many years [12-14].

As for non-contact bearings, L. -D. Girard invented a hydrostatic bearing in 1865 [15]. An active magnetic bearing was invented around the same time, and its actual system was developed as early as 1950 [16]. The feasibility of a squeeze bearing was first demonstrated and reported in 1964 [17]. These non-contact bearings have been improved and utilized in practice according to their different characteristics. However, it seems that no fundamental principles of non-contact bearings have been proposed after those inventions.

A new principle of non-contact fluid bearings, which utilizes traveling waves, is proposed in this research, and a prototype device is developed on the basis of the proposed principle.

3.2. Working principle of the new fluid bearing system

According to this new principle the non-contact fluid bearing is realized by generating the traveling waves radially on the bearing surface as shown in Figure 11.

The working principle of the first prototype is illustrated in Figure 11a. Three sets of piezo actuators are placed radially around the bearing surface, and sinusoidal voltage is applied to the actuators with a phase shift to generate traveling waves on the flexible surface (Figure 11b). As the waves are generated radially, the fluid is transported from outside to the center, which in return generates pressure and a floating force to support an object. When the amplitude and frequency of the voltage are increased, more fluid is pumped under the bearing surface allowing higher loads to be supported.

The following key properties show that this new bearing design is promising for high precision applications:

i) It is very compact, does not require any pump or bulky tubes to supply air, ii) the bearing gap can be controlled by adjusting driving voltages, iii) it has a natural resistance to moment of force with a single pad, and iv) can be used during high speed motion.

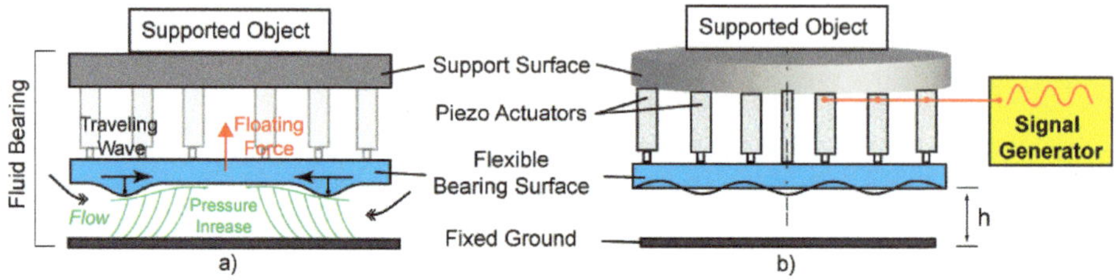

Figure 11. Principle of proposed fluid bearing – a) Working Principle b) Traveling Wave

3.3. Evaluation of the developed bearing system

In order to study the properties of the proposed bearing system a prototype bearing is developed with respect to the bearing structure illustrated in Figure 11. Figure 12 shows the prototype bearing structure floating on the guide surface. Atmospheric air is utilized as the bearing fluid in this prototype.

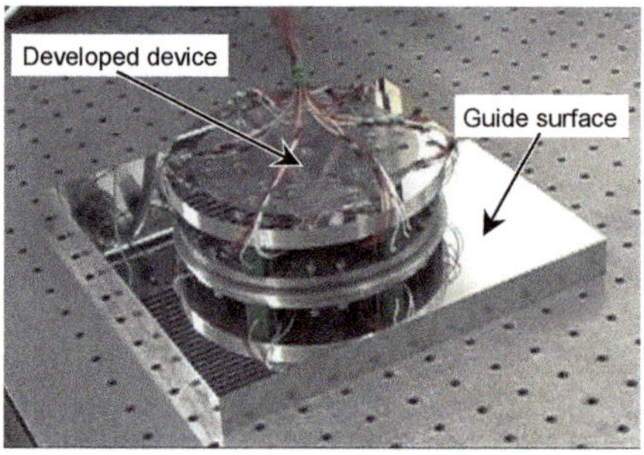

Figure 12. Prototype fluid bearing system

The developed fluid bearing is controlled with a piezo driving apparatus, and it is confirmed that the device can be floated successfully and can move smoothly with almost no friction like a hydrostatic air bearing. The performance of the developed bearing is evaluated experimentally in the following.

3.3.1. Traveling wave on the bearing surface

Displacement distribution on the bearing surface, and its transient change are measured with an optical fiber sensor in the circular direction. The measured displacements are synchronized by considering phase shifts from the applied voltages, and one cycle of their transient change is shown in Figure 13. It shows that the traveling wave is generated with amplitude of about 6 μm and it is absorbed near the center. As shown, a nearly perfect wave to transport the fluid could be generated on the bearing surface.

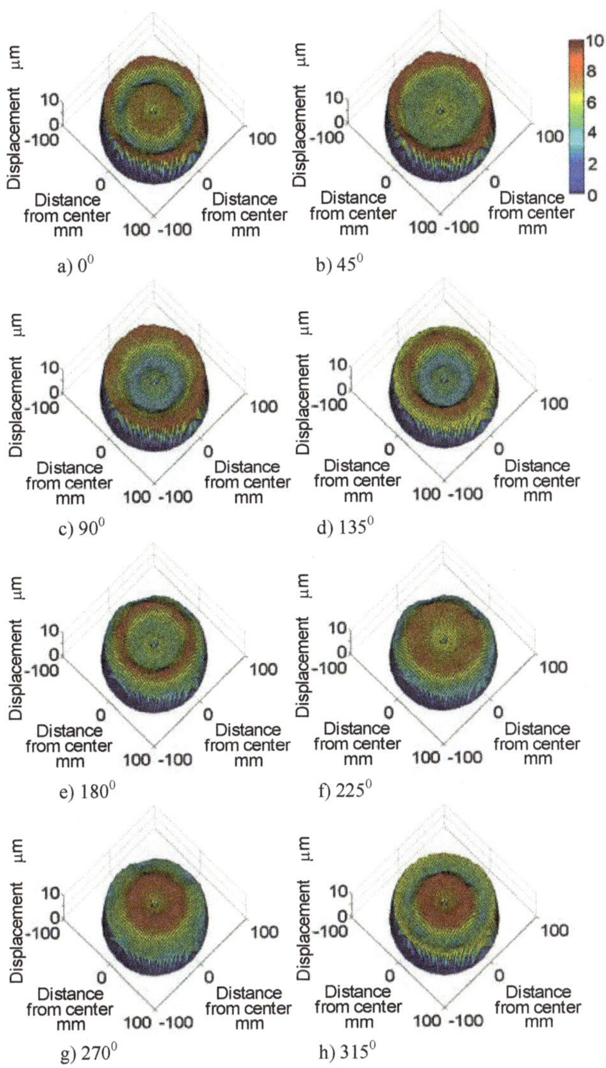

Figure 13. One cycle of traveling wave (Conditions: Freq.: 100Hz, Voltage: 80Vp-p)

3.3.2. *Maximum load capacity*

Utilizing the above traveling wave method, maximum load capacity of the bearing is measured on the prototype bearing. Weights are stacked on the center of the device. Critical weight load between the contact and the non-contact is recorded as the maximum load capacity. Figure 14 shows the maximum load capacity measured at various amplitudes and frequencies of the driving voltage. The maximum load capacity is increased as the amplitude and the frequency are increased. This tendency is the same as that of the floating displacement, but the effect of the frequency is relatively small in this case. The maximum load capacity of about 100 N can be obtained at 80 Vp-p.

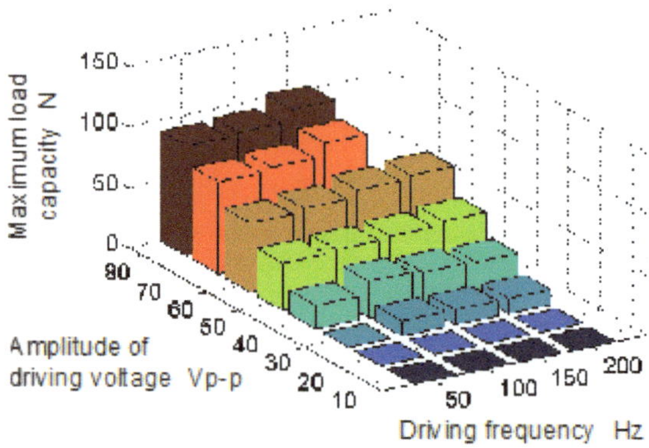

Figure 14. Maximum load capacity under various driving conditions.

3.3.3. *Evaluation of various properties*

Other properties of the bearing system are also studied experimentally. An experiment to produce the attractive force is also carried out by reversing the traveling waves. It is confirmed that the device can be attracted to the guide surface. For example, the negative gauge pressure measured with the pressure sensor is about –3.6 kPa at the center under conditions of 80 Vp-p and 50 Hz.

Moment of force is applied to the device, and it is confirmed that the present device has resistance to the moment of force without using plural bearing pads. For example, the developed device is floated even when a weight load of 30 N is applied on it at an eccentric position, which is 40 mm away from the center. This is considered to caused by viscosity of the air, i.e. the air pressure distribution can be asymmetric to produce the resistance force to the moment.

Vibration of the supported object is evaluated by using the optical sensor. For example, the supported object is vibrated vertically with an amplitude of 0.66 μm at the center, when the device is driven at 80 Vp-p and 100 Hz. It is considered that this vibration is mainly caused by the incompleteness of the traveling wave, whose value is roughly the same as this vibration amplitude.

3.4. Conclusion

A new fluid bearing is developed which utilizes traveling waves. Evaluating the prototype device the following remarks can be concluded:

It is possible to realize a non-contact air bearing by the proposed method.

Floating displacement or load capacity is increased with an increase in the driving voltage and frequency, i.e. the device is electrically controllable. Maximum load capacity is recorded as 100N.

The proposed device can also produce attractive force instead of the floating force by reversing the traveling wave.

The bearing also has resistance to moment of force even with a single pad.

4. Mass-producible rapid mixer based on Baker's transformation

We developed a novel methodology to fabricate three-dimensional passive-type mixer based on the baker's transformation (BT). BT is the best transformation for mixing fluids of laminar flow. We newly designed the BT structure with isovolumetric change without any separation/joining process of two channels. It is a suitable solution for mass-producing BT mold structures by utilizing precision cutting techniques. Two scales of BT mixers with similar structures are introduced herein. The one is for microfluidic analytical systems to accomplish well-mixed solutions in a short channel length, and the other one in miniature scale aims at high performance mixing of high viscosity fluids in food processing or resin blending. An ultraprecision five-axis planing machine and diamond cutting tools were used for a microfluidic BT mixer mold on a oxygen-free copper block, in which the flow passage area was 3.2E-9 m^2. For a miniature BT mixer mold on an aluminium block, a precision machining center and an end mill with a 1 mm radius were used. The flow passage area was 3.2E-5 m^2. We studied their mixing performances by numerical analyses and obtained the BT mixing results showing good similarities with that of numerical analyses. Moreover, the mixing performance of the micro-BT mixer was quantitatively examined to accomplish complete mixing over a wide range of flow rates.

4.1. Background of micromixers for bio-informatics

In the past decades microfluidic systems have been widely used in chemistry, biology, and nanobiotechnology, including DNA [18] or protein analysis [19], cell sorting [20], and chemical reactions [21]. Mixing inside microchannels plays an important role in those microfluidic analysis systems, and many researchers have made efforts to develop innovative mixing techniques inside microchannels [22]. In particular, mixing of solutes with a low diffusion coefficient inside microchannels is important and useful for a variety of applications including immunoassay [23], and it is desirable to get the most efficient and rapid mixing possible inside them.

Most passive-type mixers [24-27] depend on simple mixing techniques without any external power sources, unlike active-type mixers, which use such sources as ultrasonic [28], magnetic stirring [29], and bubble induced acoustic actuation [30]. In the passive-type mixers, only the structural design induces the mixing of fluids affected by the convection flow and large interface of fluids.

Comparison of active type micromixers with the passive type shows that the former can realize excellent mixing performance, but with some disadvantages: 1) difficulty in integrating other microfluidic components; 2) high cost from the standpoint of being a disposable device; 3) complex control units or an external power source. On the other hand, passive (static) type micromixers have three advantages: 1) easy integration with other microfluidic components; 2) low cost; 3) no external power source is needed. These passive structures, however, did not have all the desired features of providing rapid mixing, high mixing efficiency, a wide range of flow rates, and having no dead volume.

4.2. Microfluidic mixer utilizing Baker's transformation

In this paper, we propose a new design for the 3D micromixer, which is based on the baker's transformation (BT in short), which provides the highest mixing performance as demonstrated in chaotic theories [31].

4.2.1. Baker's transformation

The schematic illustration of the BT process (stretch, cut and fuse) is shown in Figure 15. The BT process transform fluid layers from one into two, and therefore the transformation of n times produces 2^n fluid layers. Consequently, the baker's transformation can exponentially shorten the diffusion length.

Figure 16a illustrates the BT mixing process. Two types of fluids are indicated in blue and yellow colors. The successive 3D configuration changes fold the blue fluid onto the yellow fluid gradually. After completing the folding, the combined fluids are stretched and then half of the fluids are bent 90º, which these steps make a great difference between baker's transformation and parallel lamination. The bent part of the fluids are moved to the opposite side. Then the moved part is gradually folded onto the original fluids again.

4.2.2. Numerical CFD analysis

The microfluidic distribution of vertical cross-sectional views was investigated numerically, by using CFD (computational fluid dynamics) software (ANSYS CFX). The boundary conditions of the numerical simulation were: 20 mm/s flow rate for the inlets; 0 Pa static pressure for the outlets; and 0 mm/s flow rate for the channel wall (the wall was regarded as having no roughness and skidding did not occur). The numerical simulation results in Figure 16b.

4.2.3. Fabrication method

We have fabricated baker's transformation structures with isovolumetric change (without dead volume) on an oxygen-free copper block. We selected a planing process by using an ultraprecision planing machine, NIC-300 (Nagase Integrex Co., Ltd.), which consists of three feed tables with double hydrostatic oil guide ways on the XYZ axes, two rotary index tables on the BC axes and a five-axis control system.

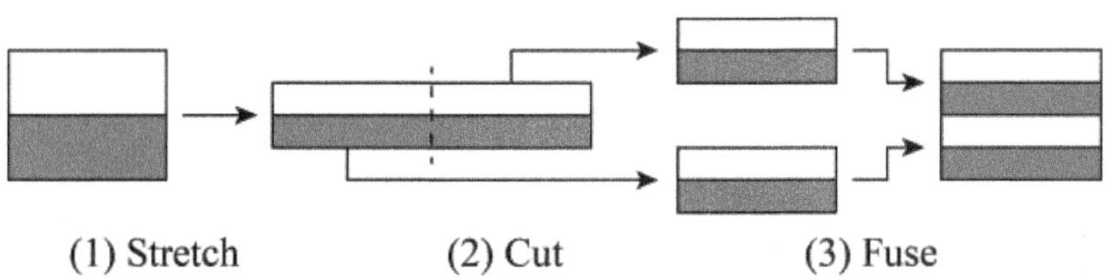

Figure 15. Schema of Baker's transformation (BT)

A custom-ordered ultraprecision diamond cutting tool (A.L.M.T. Diamond Corp.), which had a 1 mm nose radius, was used to finish the top surface of the mold. The proposed microchannels were then machined with another ultraprecision diamond tool (A.L.M.T. Diamond Corp.). A schematic of the mold is shown in Figure 17a. On an oxygen-free copper block (30×60 mm), the BT device is designed to have two inlets, injection microchannels with 10 mm length and 20 μm height, 10.4 mm length BT structures, and outlet channel with 33 mm length and 20 μm height.

PDMS type microfluidic BT mixer was then developed. PDMS (Dow Corning Inc.) and curing agent (Dow Corning Inc.) were mixed at a ratio of 10 to 1, and then the mixture was poured onto the mold, cured at 65 ºC for 2 h, peeled from the mold, and baked at 120 ºC for 30 min. Because the baked PDMS was sequentially used for the second mold, it was soaked in a commercial detergent solution (5%) for 5 min. Then, it was rinsed with double distilled water and dried in a vacuum chamber. After drying, the PDMS mixture was poured onto the PDMS mold, cured, and peeled from the mold as above. Before bonding of the second replication PDMS and a commercial slide glass, access holes were punched into the PDMS and then the PDMS and slide glass were both treated under oxygen plasma.

The 3D configuration of the structures is shown in Figure 17b. These structures were designed to have transverse (x-y) and longitudinal (x-z) movement of solutions at the same time to realize the BT like that of Figure 16. Vertical cross-sections of the BT device along the y axis are illustrated in Figure 17c. The dotted lines indicate the replication point in Figure 17b. One cycle of the BT device is 1040 μm and there are 10 cycles in all. SEM image of the BT device is given as a cross-sectional view (Figure 17d). The scale bar is 100 μm.

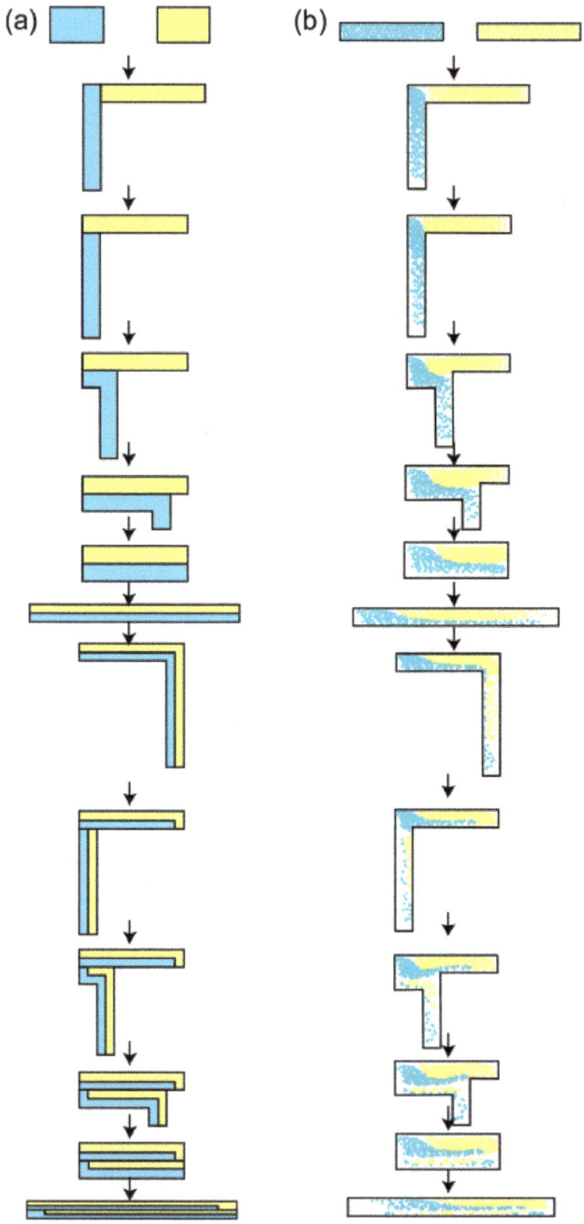

Figure 16. Schematics of (a)mixing process of BT device and (b)microfluidic distributions derived from numerical simulations.

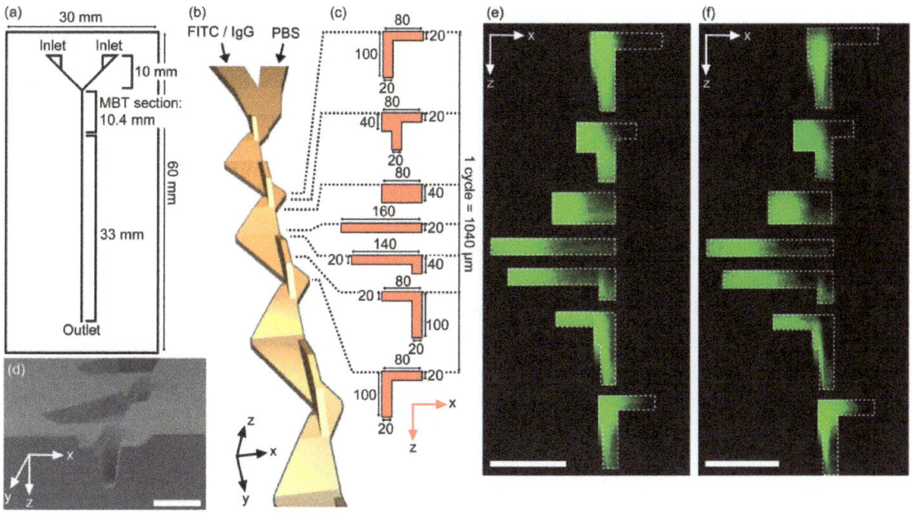

Figure 17. (a) Schematic of the BT device. (b) Schematic 3D diagram of the mold for the BT device. (c) Schematic illustrations of vertical cross-sections of the BT device. (d) SEM image of the BT device. (e)&(f) Series of confocal micrographs for one cycle in a microchannel.

4.2.4. Analytical evaluation of BT mixing performance[Yasui et al., 2011]

We quantitatively studied its mixing performance to attain complete mixing over a wide range of flow rates. The mixing performance of the BT device is evaluated using FITC (fluorescein isothiocyanate; diffusion coefficient: 2.6×10^{-10} m^2s^{-1}) [32] and goat anti-mouse IgG (immunoglobulin G; diffusion coefficient: 4.6×10^{-11} m^2s^{-1}) [33] dissolved in phosphate buffered saline (PBS).

4.2.4.1. Confocal microscopy

A confocal microscope (FV1000, Olympus) was used to observe the mixing behaviour through a focus lens 40x/0.90 (UPLSAPO, Olympus). Excitation laser was 488 nm. Confocal images were captured at a data acquisition rate of 0.90 frames per second. Stacks of each confocal x-y scan of 512×512 pixels were collected with a step of 0.5 μm in the z direction. Scan speed was 200 μs/pixel. Z-series images were merged into vertical cross-sectional images and analyzed. The flames in Figure 17e and Figure 17f show confocal micrographs of the vertical cross-sections of a channel similar to those in Figure 17c for one cycle; the mixing of FITC solution and PBS (Figure 17e) and the mixing of IgG solution and PBS (Figure 17f). Flow rate was 100 mm/s. The scale bar is 100 μm and the white dotted lines indicate channel outlines. The mixing behaviour showed a good agreement with the fluid distributions of numerical simulations.

To evaluate the mixing performance, the mixing ratio was calculated using the following formula [34];

$$\left(1 - \sqrt{\frac{1}{N}\sum_{i=1}^{N}\left(I_i - I_i^{perf.mix}\right)^2 \Big/ \frac{1}{N}\sum_{i=1}^{N}\left(I_i^0 - I_i^{perf.mix}\right)^2}\right) \times 100 \tag{1}$$

where N, I_i, I_i^0, and $I_i^{perf.mix}$ are the total number of pixels, the fluorescence intensity at pixel i, the fluorescence intensity at pixel i without mixing or diffusion, and the fluorescence intensity of the completely mixed solution at pixel i, respectively. Generally, the 90% mixing ratio was regarded as complete mixing.

4.2.4.2. Analysis of mixing ratio

The mixing ratios of FITC solution vs. PBS and IgG solution vs. PBS are shown in Figures 18a and 18b, respectively. The mixing ratio was calculated using formula (1). Circles, squares, and triangles show Δy=1.04, 5.02, and 10.4 mm (i.e. after 1, 5, and 10 cycles of mixing), respectively. The dotted line indicates 90% mixing ratio. The BT device provided complete mixing of FITC solution and PBS at flow velocities up to 400 mm/s for 10 cycles, and complete mixing of IgG solution and PBS was attained at velocities up to 50 mm/s for 10 cycles; these were characterized by the values of low Reynolds number ($Re = Ul/v < 100$, where U is the average flow velocity, l is the typical cross-sectional dimension, and v is the kinematic viscosity of the fluid). The difference in maximum flow rate to attain complete mixing between FITC (400 mm/s) and IgG solution (50 mm/s) was attributed to the different diffusion coefficients. From these figures, we saw the mixing performance increased as the number of cycles increased, e.g. when mixing IgG solution and PBS at 4 mm/s flow velocity, the mixing ratios were 44, 72, and 97% after 1, 5 and 10 cycles, respectively.

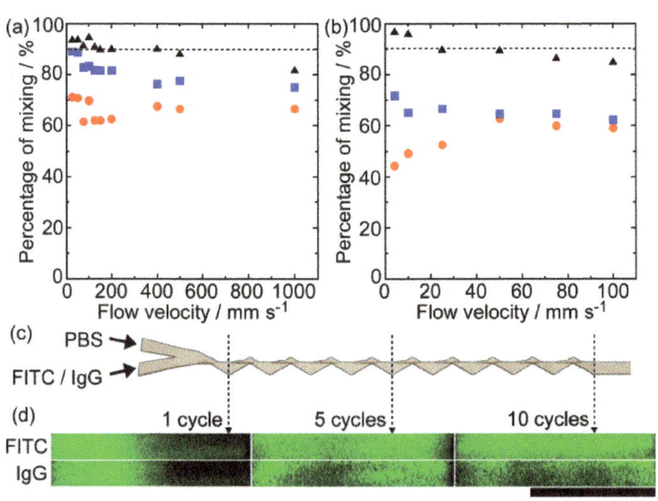

Figure 18. (a) Mixing ratio of FITC solution and PBS vs. flow rate. (b) Mixing ratio of IgG solution and PBS vs. flow rate. (c) Schematic diagrams of the BT device. (d) Confocal micrographs of vertical cross-sections of a microchannel.

Figure 18d shows confocal micrographs of the vertical cross-sections of the most stretched channel (160 μm × 20 μm, width and height, respectively) for different cycles as indicated in Figure 18c. The scale bar is 100 μm.

In these micrographs, FITC solution and PBS in the upper part of Figure 18d or IgG solution and PBS in the lower part of Figure 18d were introduced to observe the mixing behaviour inside the BT device; a syringe pump operated at constant flow velocity of 100 mm/s was used. As the simulation results indicated, the right edge of the channel was not fully mixed for both of FITC and IgG after 5 cycles. At this flow velocity, FITC solution and PBS were completely mixed after 10 cycles, but IgG solution and PBS were not. This insufficient mixing comes from the 10-fold smaller diffusion coefficient of IgG compared to FITC.

To determine residence time to attain complete mixing, we calculated the time to achieve complete mixing from Figure 18a and Figure 18b by dividing the length by flow velocity. Figure 19 shows the mixing ratio vs. residence time after 10 cycles with the BT device. Filled and open symbols show FITC and IgG mixing, respectively. The dotted line indicates 90% mixing ratio. The logarithmic fitting curves are expressed as Y=98.35+6.49logX (red filled circles), Y=94.50+8.74logX (red open circles), Y=52.59+62.71logX (black filled triangles), and Y=6.12+34.12logX (black open triangles); where Y is percentage of mixing and X is residence time.

The residence time for FITC in the BT device was 51 ms and for IgG 306 ms. Considering the residence times in the microchannel without BT structures of 4.0 s for FITC and 297 s for IgG (that could not be attained in this mixing length but was calculated from the fitting curve), our BT device showed significant potential for mixing FITC solution and IgG solution more efficiently and rapidly in a 10.4 mm mixing length microchannel than in a microchannel without BT structures, and mixing rate was more than 70-fold faster for FITC solution and 900-fold faster for IgG solution.

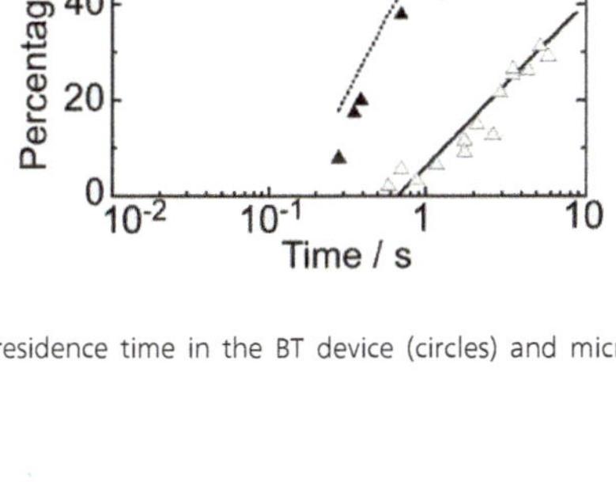

Figure 19. The mixing ratio vs. residence time in the BT device (circles) and microchannel without BT structures (triangles).

4.3. Miniature BT device for high viscosity fluid mixing

We have also developed the miniature BT mixer, which was a scale-up model of the mass-producible micro BT mixer, to make it possible that the mixing methodology we proposed was available for multi-scale structures [35]. With the mass-producible mold design, its short length and high mixing ratio characteristics, the miniature BT mixer would feasible for commercial use of treating high viscosity solutions in mixing/heating/chemistry. Herein, we developed the miniature fluid channel structure along with the basic guideline just like as the development of micro BT mixer.

High viscosity solution forms the laminar flow when flowing inside the miniature fluid channel with tens of mm^2 passage area. In the food industry, oils or flavouring materials need to be mixed smoothly and homogeneously. The same need is for material industries like mixing resins or other chemicals. The viscosity of foods has the distribution of 0.1 – 100 Pa·s, which is 100 – 100,000 times bigger than that of water. Just like the water solutions in a simple micro channel, a simple macro channel cannot mix food solutions, therefore baker's transformation (BT) is feasible.

4.3.1. Prototype

A miniature BT device for mixing high viscosity liquids was prototyped as a mixer itself, not as a mold herein. The aluminium alloy plate was fabricated by using the machining process with a 2 mm diameter end mill, along with the design illustrated in Figure 20. The fabricated structure is shown in Figure 21.

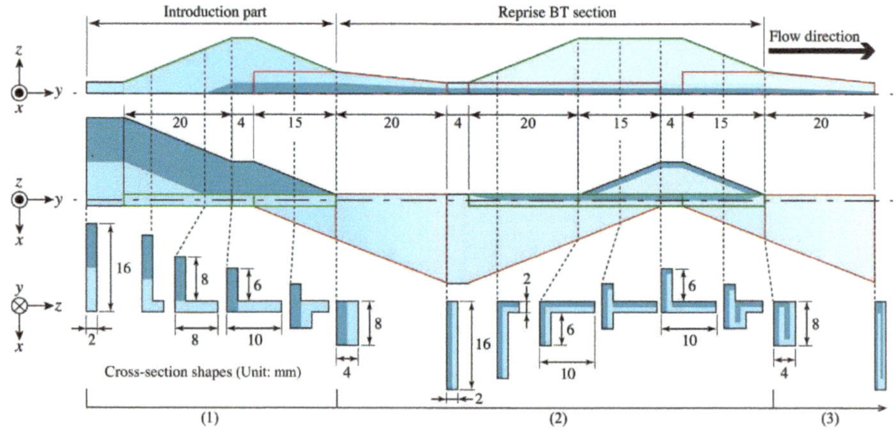

Figure 20. Illustration of three-dimensional BT mixer:

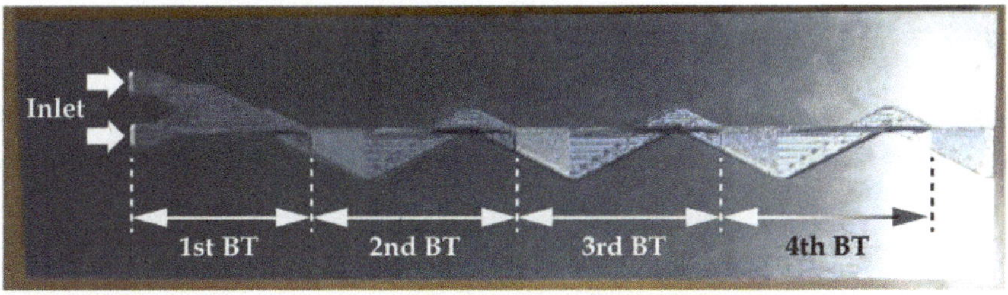

Figure 21. Prototype BT mixer made of hard aluminium alloy

In Figure 20, the side, top, and cross-section view are shown in upper, middle, and lower level, respectively. The dotted lines indicate the replication points in the flow channel. The far left of the channel illustration is connected to two separate inlets. Two colors represent the different fluids mixed by BT ideally: (1) 1st BT forming 2 folds; (2) 2nd BT forming 4 folds; (3) reprise of the 2nd BT. The n times of BT forms 2^n folds, meaning the diffusion distance among every layer results in $1/2^n$ to be the exponential acceleration of fluid mixing.

4.3.2. Preliminary experiment

In order to confirm the flow distribution directly, we introduced cure liquid silicone rubbers to cut them when they totally cured in 8 hours, and observed cross sections. The fabricated BT structure was sealed with the flat aluminum plate, and two inlets were connected to syringes, providing the silicone rubbers. Rotating a handle moves syringes translational to push the silicone rubbers out. We used the white silicone rubber TSE3504 (Momentive Performance Materials Inc.; viscosity: 10 Pa s at 23°C, specific gravity: 1.22 at 23°C, demold time: 8 hours at 25°C) for one inlet, and for the other, the blue colored master ME50-M (Momentive Performance Materials Inc.; viscosity: 800 Pa s at 23°C) was blended into TSE3504 with 1 wt% for distinction.

The distribution of vertical cross-sectional views was also investigated numerically by using ANSYS CFX. A control volume was the tetrahedral shape and the length of each side was 0.2 mm at the maximum. The boundary conditions were: 6 mm/s flow rate for the inlets; 0 Pa static pressure for the outlets; and 0 mm/s flow rate for the channel wall. The solution viscosity and specific gravity were set to 10 Pa s and 1.0, respectively.

4.3.3. Results

Both the experimental results and numerical simulation results are displayed in Figure 22. The top line is the cured silicone rubber. The middle and the bottom line are the experimental and the simulation results as cross-sectional views in the flow direction, respectively.

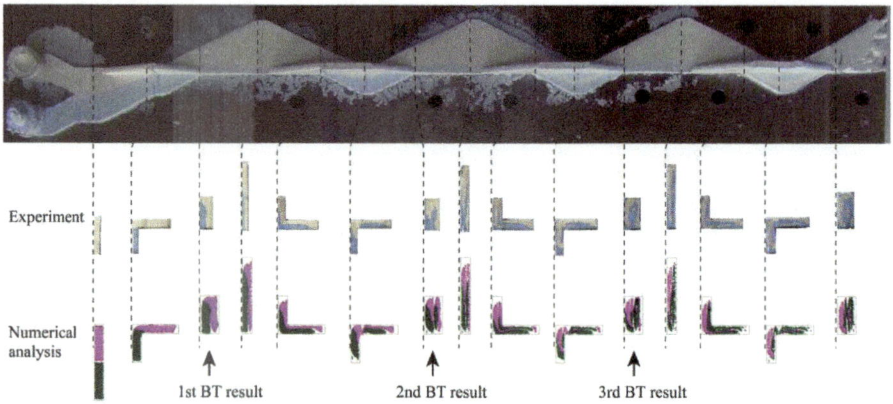

Figure 22. Results of four cycles of BT procedures

In Figure 23, the experimental folding results in the first three cycles (center column) were compared with the ideal BT illustration (left column) and the numerical analysis (right column).

The experimental results show relatively good similarities with the numerical analytical results of two, four, and eight layers folded inside flow passages. Meanwhile, compared with the ideal BT illustration, there are not small differences, especially in its marginal regions. The every folded edge wasn't processed perfectly because of the shear resistance decreasing the flow rate. The mixing performance, however, relies mainly on interlayer distances near the center of flow path. Therefore, the proposed miniature BT mixer is able to achieve high mixing performance even though it doesn't show the ideal BT mixing.

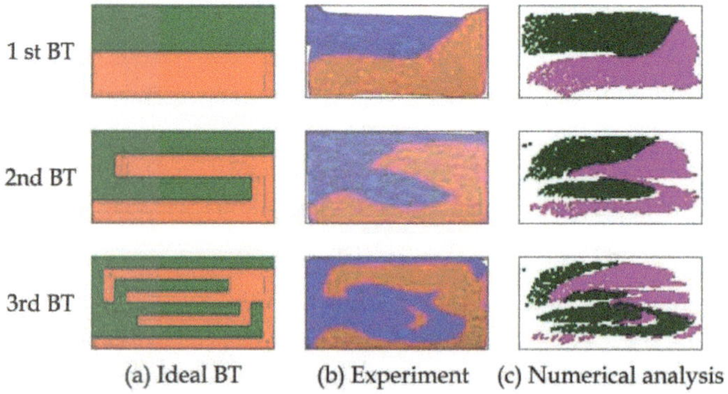

Figure 23. Comparison among cross-sections after each BT cycle

4.4. Conclusion

In order to meet the demand of mixing fluids for multi-scale mixers, We have developed a novel methodology to fabricate three-dimensional passive-type mixers based on the baker's transformation. We aimed at fabricating BT structure with isovolumetric change without any separation/joining process of two channels for the sake of mass-procucing BT mold structures by utilizing precision cutting techniques. Two scales of BT mixers, one for microfluidic analytical system and the other for mixing high viscosity fluids in food processing or resin blending, were developed. We performed laboratory experiments for evaluating the mixing performances of two prototypes. The confocal microscopy for 10 cycles (10.4 mm in length) of microfluidic BT showed that the significant potential for mixing FITC solution and IgG solution more efficiently and rapidly than in a microchannel without BT structures, and mixing rate was more than 70-fold faster for FITC solution and 900-fold faster for IgG solution. The comparison of numerical analysis and the experimental result of mixing two colored silicone rubbers by using prototyped miniature BT mixer (3 cycles) showed good similarities in their cross-sectional phases.

Author details

Eiji Shamoto, Norikazu Suzuki, Takashi Kato and Burak Sencer

Department of Mechanical Science and Engineering, Nagoya University, Nagoya, Japan

References

[1] Suzuki N, Yokoi H, Shamoto E. Micro/nano sculpturing of hardened steel by controlling vibration amplitude in elliptical vibration cutting. Precision engineering 2011;35 44-50.

[2] Shamoto E, Moriwaki T. Study on Elliptical Vibration Cutting. Annals of the CIRP 1994;43(1) 35-38.

[3] Shamoto E, Moriwaki T. Ultraprecision Diamond Cutting of Hardened Steel by Applying Elliptical Vibration Cutting. Annals of the CIRP 1999;48(1) 441-444.

[4] Shamoto E, Suzuki N, Tsuchiya E, Hori Y, Inagaki H, Yoshino K. Development of 3 DOF ultrasonic vibration tool for elliptical vibration cutting of sculptured surfaces. Annals of the CIRP 2005;54(1) 321-324.

[5] Dow T. A, Miller M. H, Falter P. J. Application of a Fast Tool Servo for Diamond Turning of Nonrotationally Symmetric Surfaces. Precision Engineering 1991;13(4) 243-250.

[6] Suzuki N, Haritani M, Yang J, Hino R, Shamoto E. Elliptical Vibration Cutting of Tungsten Alloy Molds for Optical Glass Parts. Annals of the CIRP 2007;56(1) 127-130.

[7] Shamoto E, Akazawa K. Analytical Prediction of Chatter Stability in Ball End Milling with Tool Inclination. Annals of the CIRP 2009;58(1) 351-354.

[8] Altintas Y, Budak E. Analytical Prediction of Stability Lobes in Milling. Annals of the CIRP 1995;44(1) 357-362.

[9] Altintas Y, Shamoto E, Lee P, Budak E. Analytical Prediction of Stability Lobes in Ball End Milling. ASME Journal of Manufacturing Science and Engineering 1999;121 586-592.

[10] Shamoto E, Altintas Y. Prediction of Shear Angle in Oblique Cutting with Maximum Shear Stress and Minimum Energy Principles. ASME Journal of Manufacturing Science and Engineering 1999;121 399-407.

[11] Shamoto E, Suzuki N, Hamaguchi A. A New Fluid Bearing Utilizing Traveling Waves. Annals of the CIRP 2006;55(1) 411-414.

[12] Kawai T, Ebihara K, Takeuchi Y. Improvement of Machining Accuracy of 5-Axis Control Ultraprecision Machining by Means of Laminarization and Mirror Surface Finishing. Annals of the CIRP 2005;54(1) 329-332.

[13] Shamoto E, Murase H, Moriwaki T. Ultraprecision 6-Axis Table Driven by Means of Walking Drive. Annals of the CIRP 2000;49(1) 299-302.

[14] Shamoto E, Park C, Moriwaki T. Analysis and Improvement of Motion Accuracy of Hydrostatic Feed Table. Annals of the CIRP 2001;50(1) 285-290.

[15] Cameron A. Basic Lubrication Theory. Ellis Horwood Limited; 1976.

[16] Habermann H, Liard G. L. Practical Magnetic Bearings. IEEE Spectrum 1979, September: 26-30.

[17] Salbu E. O. J. Compressible Squeeze Films and Squeeze Bearings. Journal of Basic Engineering. Transactions of the ASME, Series D 1964;86(2) 355-364.

[18] Burns M. A, Johnson B. N, Brahmasandra S. N, Handique K, Webster J. R, Krishnan M, Sammarco T. S, Man P. M, Jones D, Heldsinger D, Mastrangelo C. H, Burke D. T. An Integrated Nanoliter DNA Analysis Device. Science 1998;282(5388) 484-487.

[19] Ikami M, Kawakami A, Kakuta M, Okamoto Y, Kaji N, Tokeshi M, Baba Y. Immunopiller Chip: A New Platform for Rapid and Easy-to-use Immunoassay. Lab Chip 2010;10(24) 3335-3340.

[20] Ozaki K, Sugino H, Shirasaki Y, Aoki T, Arakawa T, Funatsu T, Shoji S. Microfluidic Cell Sorter with Flow Switching Triggered by a Sol-gel Transition of a Thermo-reversible Gelation Polymer. , Sensors and Actuators B 2010;150(1) 449-455.

[21] Losey M. W, Schmidt M. A, Jensen K. F. Microfabricated Multiphase Packed-bed Reactors: Characterization of Mass Transfer and Reactions. Industrial & Engineering Chemistry Research 2001;40(12) 2555-2562.

[22] Nguyen N. T. Micromixers; Fundamentals, Design and Fabrication. Norwich, NY: William Andrew Inc.; 2008.

[23] Tachi T, Kaji N, Tokeshi M, Baba Y. Microchip-based Homogeneous Immunoassay using Fluorescence Polarization Spectroscopy. Lab Chip 2009;9(7) 966-971.

[24] Liu R. H, Stemler M. A, Sharp K. V, Olsen M. G, Santiago J. G, Adrian R. J, Aref H, Beebe D. J. Passive Mixing in a Three-dimensional Serpentine Microchannel. Journal of Microelectromechanical Systems 2000;9(2) 190-197.

[25] Stroock A. D, Dertinger S. K, Ajdari A, Mezic I, Stone H. A, Whitesides G. M. Chaotic Mixer for Microchannels. Science 2002;295(5555) 647-651.

[26] Hong C. C, Choi J. W, Ahn C. H. A novel in-plane passive microfluidic mixer with modified Tesla structures. Lab Chip 2004;4(2) 109-113.

[27] Kim D. S, Lee S. H, Kwon T. H, Ahn C. H. A Serpentine Laminating Micromixer Combining Splitting/Recombination and Advection. Lab Chip 2005;5(7) 739-747.

[28] Yang Z, Goto H, Matsumoto M, Maeda R. Active Micromixer for Microfluidic Systems using Lead-zirconate-titanate (PZT)-generated Ultrasonic Vibration. Electrophoresis 2000;21(1) 116-119.

[29] Lu L. H, Ryu K. S, Liu C. A Magnetic Microstirrer and Array for Microfluidic Mixing. Journal of Microelectromechanical Systems;11(5) 462-469.

[30] Liu R. H, Yang J, Pindera M. Z, Athavale M, Grodzinski P. Bubble-induced Acoustic Micromixing. Lab Chip 2002;2(3) 151-157.

[31] Ottino J. M. The Kinematics of Mixing: Stretching, Chaos, and Transport. Cambridge Texts in Applied Mathematics, Cambridge University Press; 1989.

[32] Periasamy N, Verkman A. S. Analysis of Fluorophore Diffusion by Continuous Distributions of Diffusion Coefficients: Application to Photobleaching Measurements of Multicomponent and Anomalous Diffusion. Biophysical Journal 1998;75(1) 557-567.

[33] de Beer D, Stoodley P, Lewandowski Z. Measurement of Local Diffusion Coefficients in Biofilms by Microinjection and Confocal Microscopy. Biotechnology and Bioengineering 1997;53(2) 151-158.

[34] Jhonson T. J, Ross D, Locascio L. E. Rapid Microfluidic Mixing. Analytical Chemistry 2002;74(1) 45-51.

[35] Yasui T, Omoto Y, Osato K, Kaji N, Suzuki N, Naito T, Watanabe M, Okamoto Y, Tokeshi M, Shamoto E, Baba Y. Microfluidic Baker's Transformation Device for Three-dimensional Rapid Mixing. Lab Chip 2011;11(19) 3356-3360.

Chapter 4

Micro-Nano Robotics and Mechatronics for Biomedical Applications

Tomohiro Kawahara and Fumihito Arai

1. Introduction

In medical field, one recent trend in clinical operation is minimally invasive surgery which is used an endoscope and small instruments to avoid a damage for patients [1, 2]. This surgical technique has been well introduced to many diagnosis and treatment department such as internal medicine, ophthalmology, urology, especially surgical department. As well known, surgical robots have been introduced and operated in this field to support surgical operations [3, 4]. Generally, effect of minimally invasive surgery for patient is defined as follows:

$$PatientValue = \frac{Effective}{Invasiveness^2}$$

Therefore, in order to reduce invasiveness such as scars, it is important to miniaturize the endoscope and surgical tools, while maintaining the performance of devices. In addition, doctors would like to add some functions for improvement of safety and usability of surgical tools during treatments. Therefore, small and thin instrument with high-function are highly required and micro mechatronics makes a substantial contribution to fabricate and integrate such novel tools and systems.

On the other hand, target of scientists in bioscience filed has been changed from tissue to single cell level for specific investigation and understand of biological function. Especially, since investigation of mechanical characteristics of cells is really important to know functions and growth factors of cell, micro-nano actuators are highly required to stimulate and measure for an investigation of single cell [5, 6]. Since conventional commercial manipulators are quite large and slow, it is desired to develop high-performance tools which include micro-nanorobots capable of manipulation and sensing of cells with a high-speed and a high-throughput. By applying micro fabrication technique such as MEMS technology, we can make and integrate tiny mechanical parts as a robot about similar size as cells with diameter of 10-100 μm. Therefore, for bio application, micro-nano fabrication and robotics are also important in order to realize the novel robots and systems.

In this chapter, we introduce how to design/fabricate the micro- device and robot using for actual medical and bioscience applications. Current problem and future direction of micro-nano mechatronics for biomedical applications are also discussed to encourage vigorous research activity in this field.

2. Medical application

2.1. Background

Minimally invasive surgery is a well-established method in modern medicine [7]. In particular, Endoscopic Sub-mucosal Dissection (ESD) is well known for availability of 10 mm-order cancer excision [8]. In ESD, a doctor inserts an oral endoscope into a stomach, and performs surgical procedures by using narrow tools for removing tumor tissues. Therefore, since large incisions are not required, this operation is a minimally invasive procedure to apply a quick recovery for patients.

However, the disadvantages of ESD are the narrow operative field provided by an endoscope and the low operability by a simple mechanism forceps. It uses a hook knife and an insulation-tipped electro-surgical knife (IT knife) to cut the lesion. IT knife is designed to prevent the penetration of the gastric wall by covering the tip of the electrical knife with a ceramic ball. However, perforation is still encountered, leading to prolonged resection time. Therefore, the ESD for gastric cancer requires a doctor with technical skills higher than those required for the other endoscopic procedures. A basic technical principle of the surgical resection is the resection of appropriate tissues, which is forced to stand-out by injecting saline under the tumor. However, this endoscopic surgery is "one-arm surgery" that it can only insert one endoscope to cut and exfoliate the lesion. Ideally, one arm is pulling the lesion and another arm to cut would improve the efficiency and reduce the risk of complications, such as bleeding and perforation from cutting the unconfirmed blood vessels. The exfoliating procedure with a hook or IT knife is especially difficult during in the surgery. It takes the most of the surgery time. A couple of studies have attempted to address this issue. In clinical field, several techniques for assisting endoscopic surgery are proposed and conducted [9, 10]. However, these approaches are not enough to solve the time-consuming and the skill-dependent problems on ESD, therefore, a new approach has been desired for years.

There have been many works discussing surgical assist microarms [11]. Catheter is a good tool for diagnosing inner surface of stomach, large intestine, and so forth [12]. Most catheters have a gripper, so that they can treat the surface of internal organs. However, when a catheter contacts to the stomach wall, it easily has deformation. It is difficult to keep the lifting posture of the catheter for surgical assist. Harada et al. developed the 2-DOF micromanipulator with the diameter of 2.4 mm for intrauterine fetal surgery, even though it has no gripper [13]. On the other hand, the robotic tools included endoscope systems are developed by a couple of companies [14, 15]. Since these robots are embedded in the specially-developed endoscope, we cannot apply these robots for a normal oral endoscope.

Considering this background, we have proposed a surgical concept for ESD by utilizing a couple of microarms, as shown in Figure 1 [16, 17]. In this approach, a doctor can easily operate the surgical procedures without time-consuming manner by the wire driven microarms, which have multi-DOF for the wide-range movement. These conventional microarm, however, have the coupling problem between the wire for the arm joint and the wire for the gripper, as shown in Figure 2. When we open the gripper, the joint is also moved by the wire coupling. Therefore, we could not precisely control the movement of the microarm, and we may apply damage to tissue of organs by this problem. Most popular method for decoupling mechanism in robotics is composed of several gears. In this method, it is difficult to design a compact microarm for endoscope insertion. Ikuta et al. have developed the 5-DOF wire driven microarm with the diameter of 3.0 mm for micro surgery in deep area of organs [18]. In this device, a number of wires are required to control the microarm, even though the decoupled wire drive was achieved by the fine micro-joint design.

We have also proposed the decoupling wire driven exoskeletal microarm with gripper enabling the surgical operation of a cancer in tissues during ESD. Based on both a photolithography and a wire-electric discharge technique, we fabricate the decoupling structure for the metal frame of the microarm [19]. By using this approach, the developed extra-thin 2-DOF microarm can be inserted to the endoscope channel with the diameter of less than 3 mm. The developed microarm is capable of lifting up a tissue layer of organs during ESD.

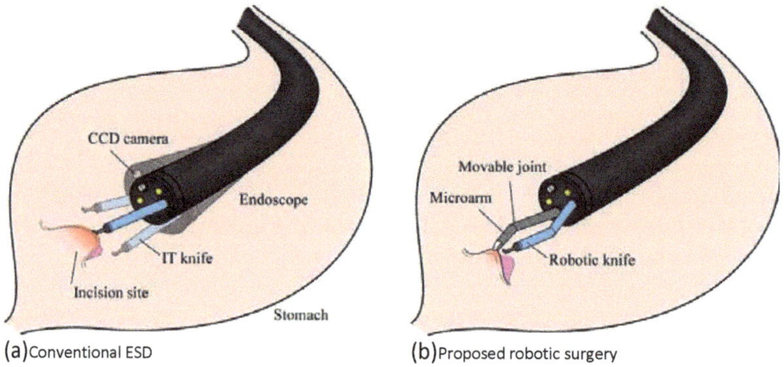

Figure 1. Conceptual image of robotic endoscope surgery

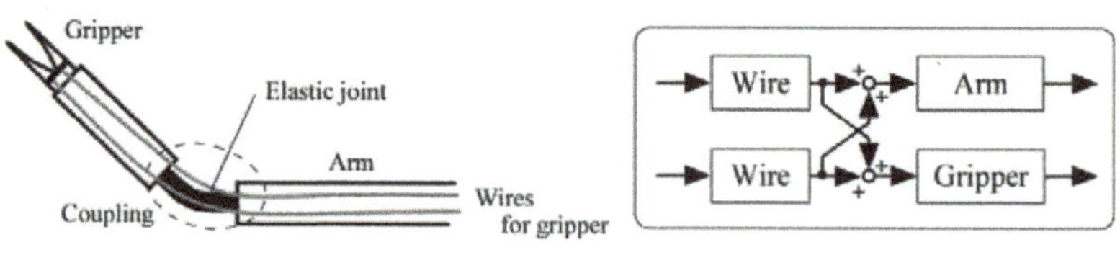

Figure 2. Wires coupling problem to drive a microarm

2.2. Basic concept

In actual ESD, the required specifications for a microarm mounted to an oral endoscope are as follows:

i. Diameter of microarm: less than 3.0 mm

ii. Movable angle of microarm: +/- 60-80 degree

iii. Thickness of tissue layer to handle: 2.0-5.0 mm

iv. Weight of tissue layer to handle: approx. 30 g.

Figure 3(a) shows a conceptual image of the proposed microarm which is composed of a base link, a fore link, an arm joint, and a gripper. Since the size limitation of endoscope channel is approximately 3.0 mm, we have to design the microarm with the diameter of less than 3.0 mm. We apply an exoskeletal structure for the arm design to reduce the size of the microarm. The exoskeletal is well known for a miniaturized structure such as an insect arm. Moreover, since an exoskeletal structure has a rotational joint, we can estimate the exact posture of the microarm by measuring the joint angle from the outside of the body.

At the first, we design the arm joint for bending the microarm and the gripper for grasping/lifting of stomach tissues based on an analytical approach. From these results, we also design the decoupling structure to assembling the mechanical elements of the microarm.

2.3. Design

2.3.1. Joint Design

Figure 3(b) shows the basic structure of the arm joint, which has the shape of trapezoid frame. To increase the movable area of the arm joint, we actively apply the elastic deformation of the frame to the joint design. Figure 4 shows the trapezoid frame model for the arm joint where W, I_i, l_i, b, φ, M_C, R_B, and R_C are a tension by wire, a second moment of area, length of beams, a half-length of bottom of trapezoid, angle from bottom of trapezoid, a moment at a point C, a normal force at point B, and a normal force at point D, respectively (i=1, 2). From this model, we calculate the appropriate parameters of l_i and φ for arm design. Here,

$$b = l_1 \cos\varphi + l_2 \tag{1}$$

Since the microarm has the size limitation, we use b=0.49 mm on this analysis. We can obtain the equation about the deformation of the beam BD as follows:

$$EI\frac{d^2y}{dx^2} = R_B x - M_C H\left(x - l_2\right) - W\left(x - l_2\right)H\left(x - l_2\right) \tag{2}$$

$$R_B = \frac{W(l_2 - d)}{2l_2} \tag{3}$$

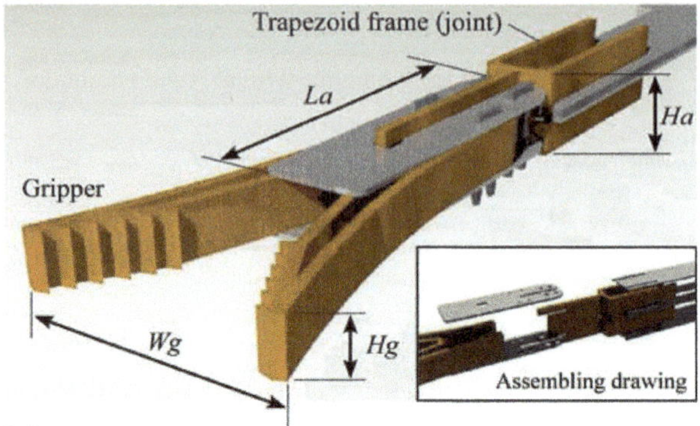

(a)

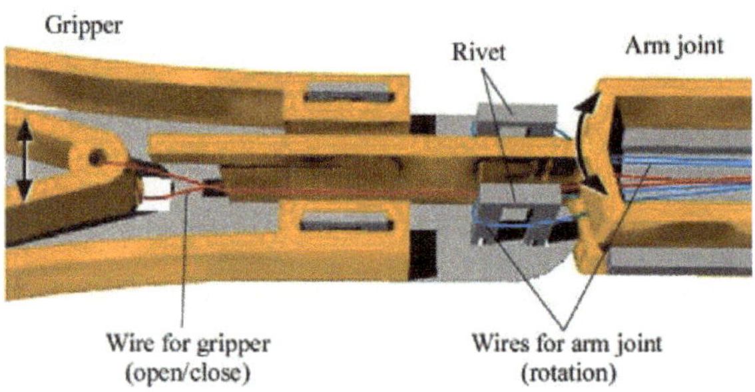

(b)

Figure 3. Design of a robotic microarm with a microgripper

where E, $H(x)$, and d are Young's modulus, Heaviside function, and a distance between the point C and the point of application of tensile force by wire, respectively. By using the equations (1)-(3), we simulate the deformation angle θc at C as follows:

$$\theta_C = \frac{W(b - l_1\cos\varphi)(b - l_1\cos\varphi + d)}{6EI} + \frac{1}{2(b - l_1\cos\varphi)}\{\delta_B\cos(\varphi - \theta_B) + \delta_D\cos(\varphi - \theta_D)\} \quad (4)$$

where δ_B, θ_B, δ_D, and θ_D are the displacement of the beam AB at point B, the angle of the beam AB at point B, the displacement of the beam DF at point D, and the angle of the beam DF at point D, respectively.

For the numerical simulation, we use W=1.0 N and d=0.63 mm. From the result of the simulation as shown in Figure 5, we determine the parameters such as l_1=0.48 mm, l_2=0.00 mm, and $\varphi = 0$ deg. To confirm the availability of the designed parameters, we carried out the FEM analysis by using the COMSOL Multiphysics software. Figure 6 shows an example of the simulation result. From this figure, we can see that the rotation center exists in the center of the joint arm and the arm joint has large deformation by the wire tension. This means that we can realize the extra-thin microarm which has both the wire decoupling and the large movable angle of joint. The required accuracy of the designed parameters (50 μm order) is enough to fabricate the actual microarm by a wire-electric discharge machining.

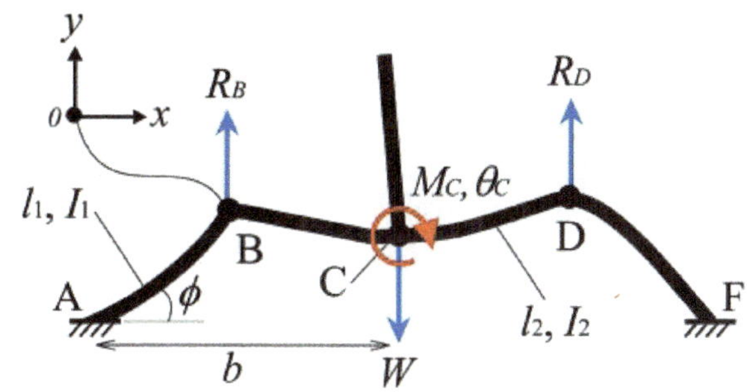

Figure 4. Mechanical model of a trapezoid frame for the arm joint

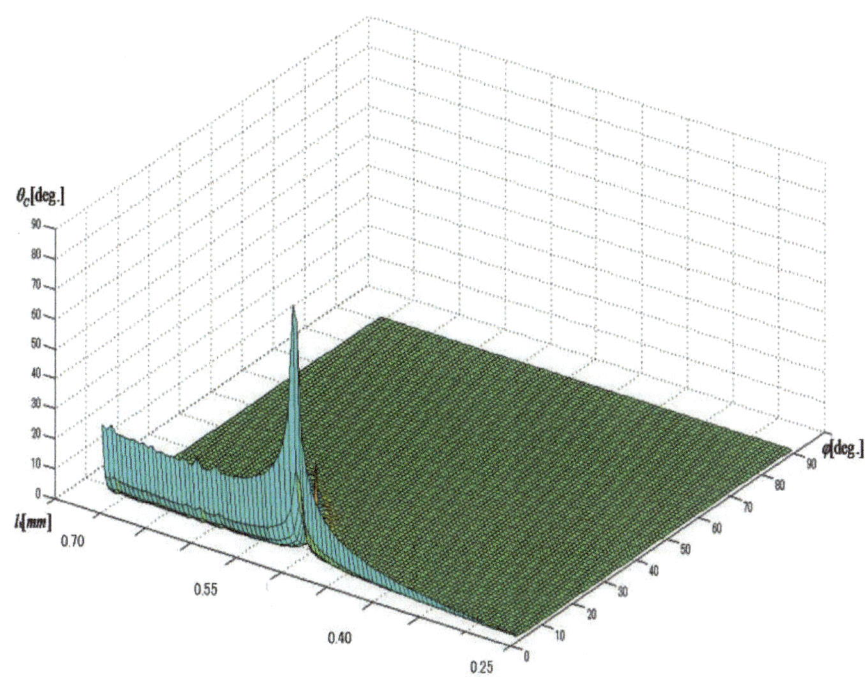

Figure 5. Evaluation of the joint flexibility

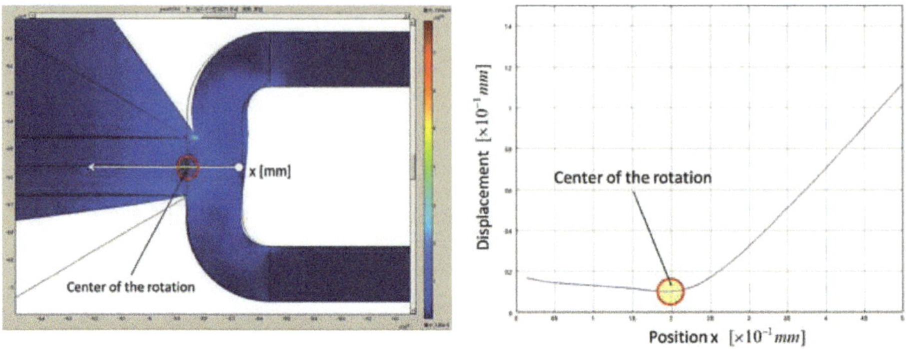

Figure 6. FEM analysis of the designed joint

2.3.2. Gripper design

In order to implement a gripper to the microarm, it is important that gripper has a simple mechanism and a large grip force. From these points, we apply Muramatsu et al. proposed principle which is based on a behavior of long pillar buckling to the gripper design, as shown in Figure 7 [20]. By utilizing this principle, we can develop the simple gripper without large mechanical elements. To determine the shape of the gripper, we carried out FEM analysis by the COMSOL simulator. As the result of trial and error, we decided the shape of the gripper which is normally-open type, as shown in Figure 8. The appropriateness of shape design is confirmed through basic experiments, in section 2.5.

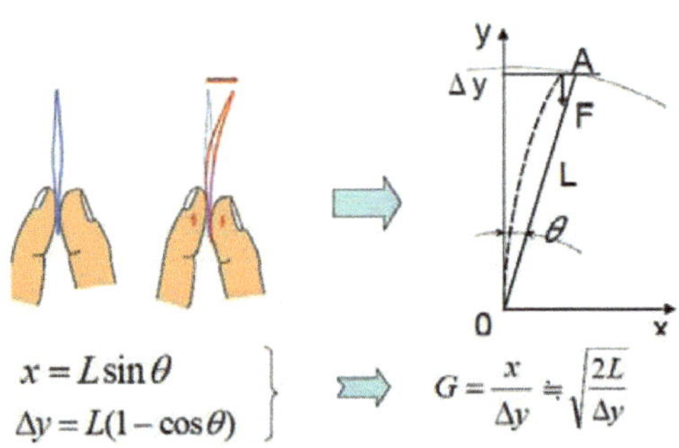

Figure 7. Concept of a buckling mechanism for the gripper design

2.3.3. Decoupling design

From the discussed designs above, we describe the wire decoupling design for microarm and gripper. Figure 3(b) shows the wire implemented microarm, where the red line shows the 1 wire for the gripper and the blue line shows the 2 wires for the arm joint. To achieve the wire decoupling, the wire for gripper is precisely passed through the center position of the arm joint. Therefore, even if the arm joint is bent by wires, the gripper is completely unaffected with the decoupling design. As a result of the design, we decide the parameters of the microarm such as the width of the opened gripper W_g=4.5 mm, the height of the gripper H_g=1.0 mm, the length of the fore link L_a=10.0 mm, and the height of the joint H_a=1.8 mm, respectively.

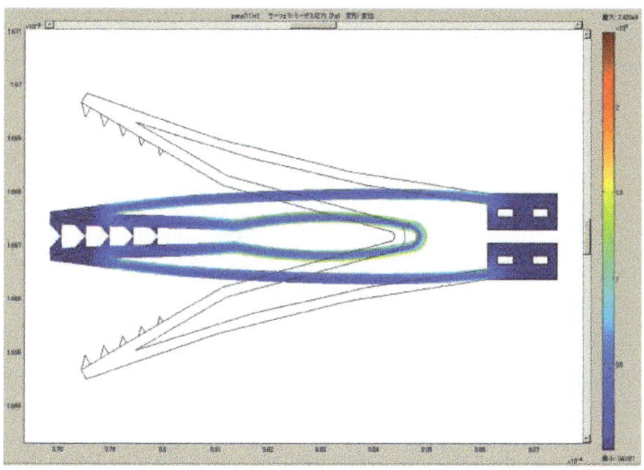

Figure 8. Stress distribution of the gripper

2.4. Fabrication

Figure 9 shows the mechanical parts of microarm which should be fabricated by microfabrication. In order to fabricate and assemble the microparts, we used the combination of a wire-electric discharge machining technique and a photolithography based electroplating technique.

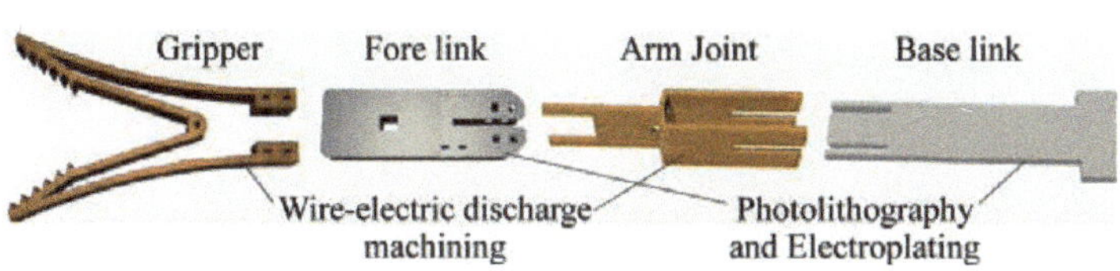

Figure 9. Mechanical parts to construct the microarm

At first, to increase the stiffness of the microarm, the arm joint and gripper were fabricated by using a wire-electric discharge, as shown in Figure 10, where the material of each part is phosphor bronze, and the thickness is 0.2 mm.

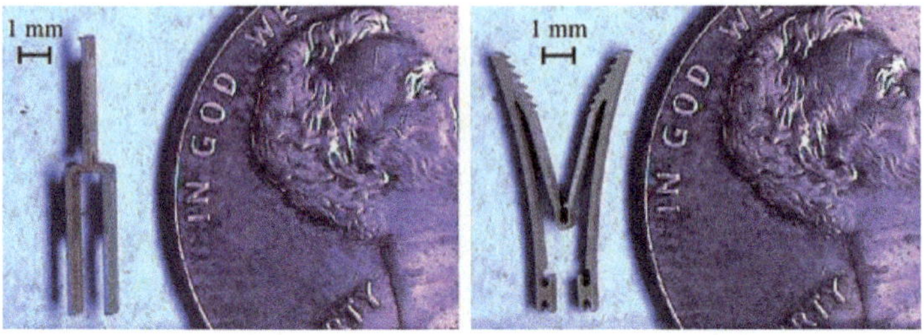

Figure 10. Fabricated arm joint and gripper machined by electrical discharge machining

On the other hand, to assemble of the joint and gripper, we require the link parts with the accuracy of 10 μm order. By applying a photolithography to this fabrication as shown in Figure 11, we developed the extra thin mechanical parts for the microarm. For a first step of the fabrication, the sacrificial layer (LOR 5B, Tokyo Ohka Kogyo Co. Ltd.) was coated on Si wafer. Then Au-Cr was sputtered on the wafer (thickness=300 nm). Next, the photoresist (KMPR 3035, Kayaku Micro Chem Co. Ltd.) was coated on the substrate. After removing photoresist and sacrificial layer (Remover PG, Kayaku Micro Chem Co. Ltd.), ultrasonic cleaning was conducted to peel off the remaining Au layer. Furthermore, by introducing the proposed Stacking Microassembly Process called STAMP as shown in Figure 12 [16], we fabricated the layered link parts to increase the parts thickness. Figure13 shows the fabricated and assembled part (base link) for the microarm.

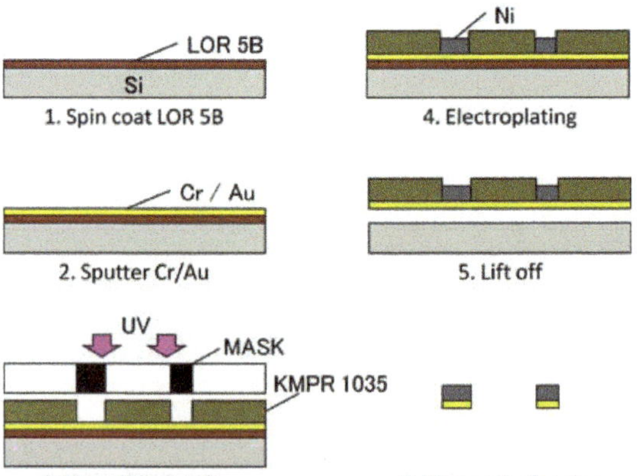

Figure 11. Electroplating based on photolithographic micropattern

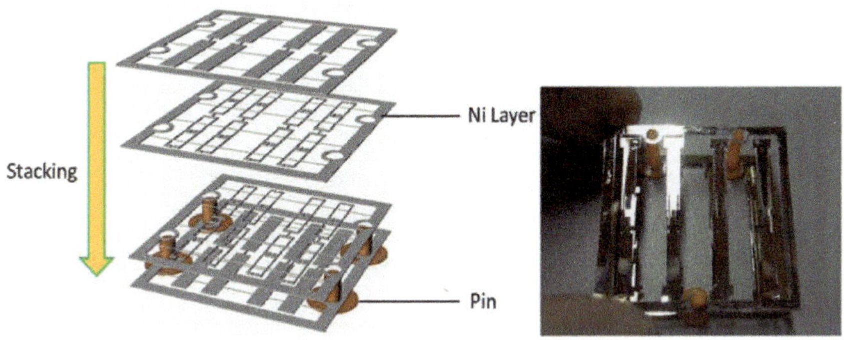

Figure 12. Fabrication method of base links using STAMP (Stacked Micro-assembly Process)

Finally, we completely developed the microarm with the diameter of 2.62 mm (maximum height: 1.9 mm, maximum width: 1.8 mm), as shown in Figure 14.

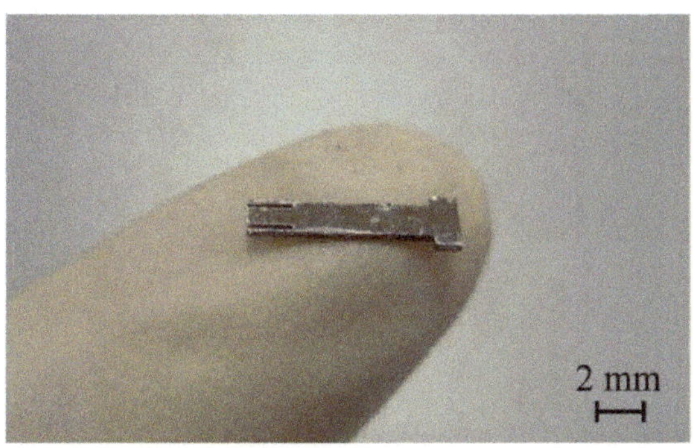

Figure 13. Fabricated and assembled single link part for the microarm

2.5. Experiments

2.5.1. Basic experiments

To confirm the performance of the developed microarm, we performed the basic experiments.

Effect of the decoupling design: The microarm was pulled by Bolfur wires (diameter=100 μm) with manual operation. Then, we measured the rotation angle of the microarm from the captured image by camera (Figure 15(a)-(d)). From this experiment, we confirmed that the maximum rotation angle is 60 deg. with 27.0 N tension force. Furthermore, we also confirmed

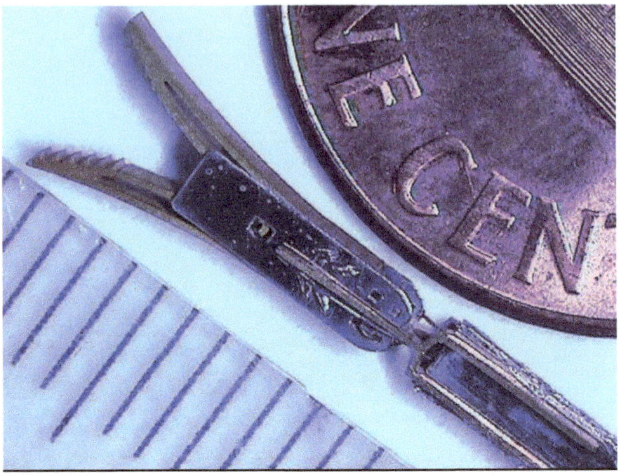

Figure 14. Fabricated and assembled 2DOF microarm which is driven by wires

that the arm joint movement is not mostly affected by that of the gripper, as shown in Figure 16(a) and (b).

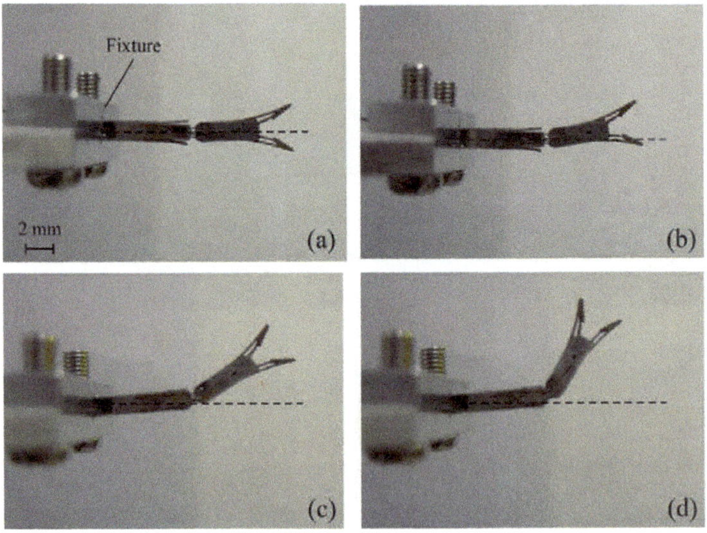

Figure 15. Bending demonstration of the developed microarm

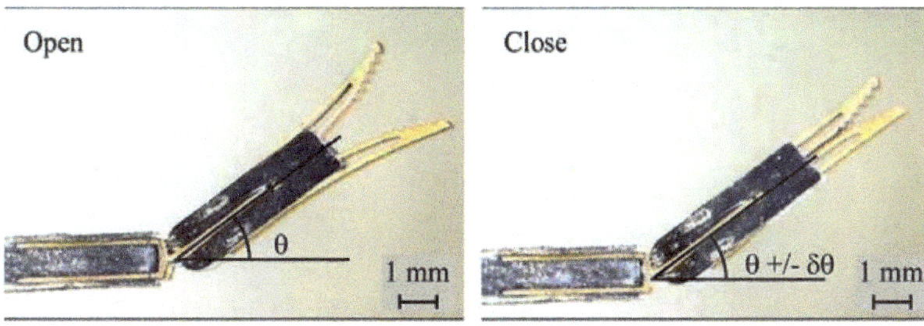

(a) Verification experiment to confirm the effectiveness of the decoupling design.

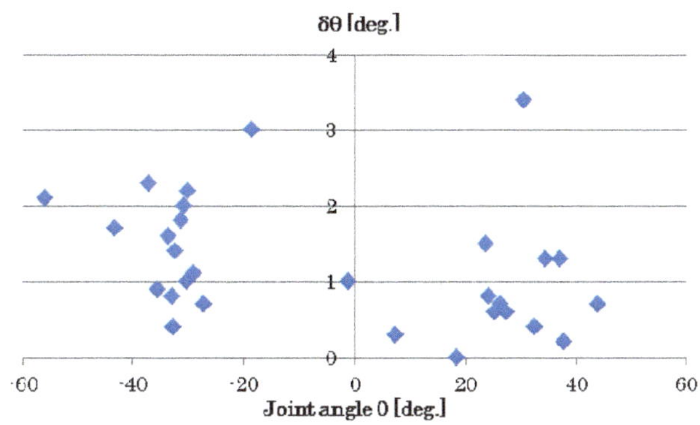

(b) When the joint was bent by wires with θ deg., the gripper was closed by pulling the another wire. Then, we measured the $\delta\theta$ to confirm the effect of the decoupling mechanism.

Figure 16. Evaluation of the decoupling design

Grip force of the microarm: First, the grip force of the gripper was measured by the sensor made by a metal flat plate with a strain gauge. As a result, we confirmed that the grip force of the gripper is 0.52 N at a maximum, as shown in Figure 17. Next, a silicone rubber imitated a human tissue was gripped and lifted by the developed microarm, as shown in Figure 18. In this experiment, we changed the thickness of the 100.0 x 50.0 mm size silicone rubber. From this experiment, it was confirmed that the microarm can lift up the rubber sheet with the thickness of 3.7 mm (21.0 g). In actual ESD, we have to handle a mucous membrane with a thickness of 2.0-5.0 mm and a weight of approximately 30 g. Through this experiment, the usability of the developed microarm was confirmed.

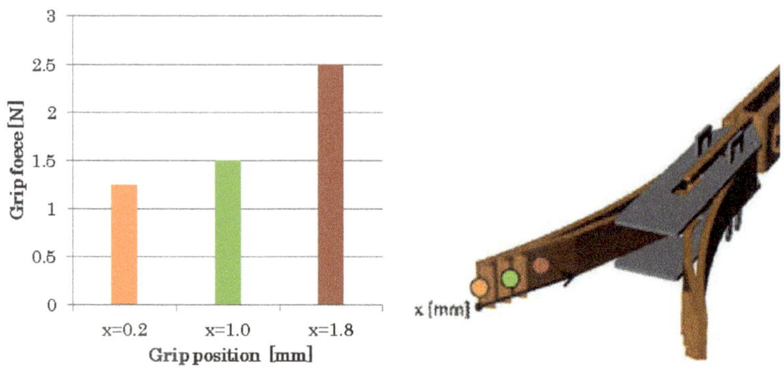

Figure 17. Grip force according to grip position

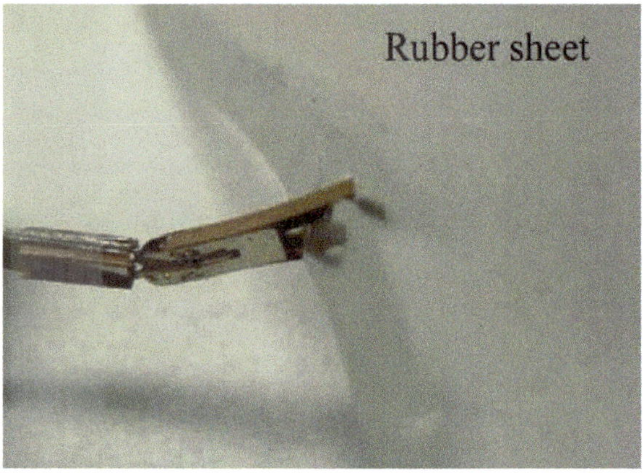

Figure 18. Overview of the lifting experiment

2.5.2. Animal experiments

Based on the experimental results, we applied the developed microarm to the animal experiments. The microarm could insert the actual endoscope channel (2.7 mm), as shown in Figure 19(a), and Figure 19(b) shows the overview of the experiment conducted in the pig stomach. In this experiment, we succeeded in inserting and driving the microarm in actual stomach. However, since the limitation of the wire-electric discharge machining, the gripper could not close completely. Therefore, we think that the grip force of the microarm (0.52 N) is not enough to grasp the tissue without slipping during ESD. In the future work, we should redesign and fabricate the gripper to improve the grip force. The specific evaluation of the grip force will be conducted (e.g. the relationship between the gripper angle and the grip force). On the other hand, the developed microarm was made from phosphor bronze and nickel which may cause allergic reaction. For a clinical application, the microarm should be coated by biocompatible material.

We believe that the developed microarm can be applied to other endoscopic surgeries, for example NOTES (natural orifice translumenal endoscopic surgery), SPS (single port endoscopic surgery) and so on [21, 22].

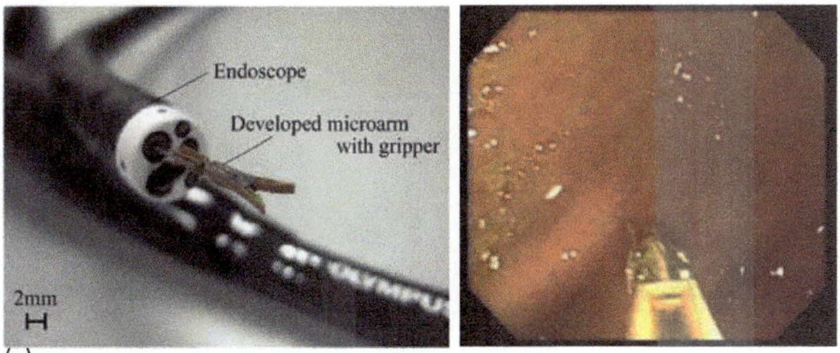

(a) (b)

Figure 19. Overview of the animal experiment

2.6. Discussion and summary

Here, we described a developed decoupling wire driven exoskeletal microarm that can be applied to ESD. The main results of our study are as follows:

1. The 1.8 mm thickness microarm is composed of a gripper with 0.52 N grip force, an arm joint with +/- 60 deg. rotatable angle, and two link parts. These mechanical elements are fabricated by a wire-electric discharge and a photolithography technique.

2. Through our basic experiments, we have confirmed the developed microarm can be installed to an endoscope channel with the diameter of the 2.7 mm and the wire decoupling of the microarm is achieved.

3. Available handling size of the developed microarm is 100.0 x 50.0 x 3.7 mm (21.0 g), we apply the microarm to actual endoscope during ESD.

For one of the future direction of microsurgical robot, new design approach is also required to drastically improve the performance of the microarm. Currently, since these conventional instruments are consisted of rigid materials, it is difficult to achieve both the flexibility and the biocompatibility to avoid damage of tissues. Therefore, more versatile instruments are needed for application of actual endoscopic surgery. In contrast, biological structure such as legs of insect or crustaceans can move flexibly and exert good strength because of their exoskeletal structure. Exoskeletal structure has the advantage to miniaturize the whole structure with a rotational joint, which make it easy to estimate the exact posture of the joint by measuring the joint angle from the outside of the structure. Moreover, joint has an anisotropic stiffness. Considered by these mechanisms, bionic design is the mechanism and the material introduced to realize such as a living structure. By using this approach, bionic joint has a potential to solve the conventional problems on mechanical joint made by rigid materials. For instance, we have developed a bionic joint which is made by different stiffness materials that are silicone and polymer [23]. The joint is small (< 2 mm), and it can insert to the channel of oral endoscope to assist of the surgical procedure, as shown in Fig. 20. By using this approach, we believe that a doctor will be able to easily access and operate affected areas during endoscopic surgery.

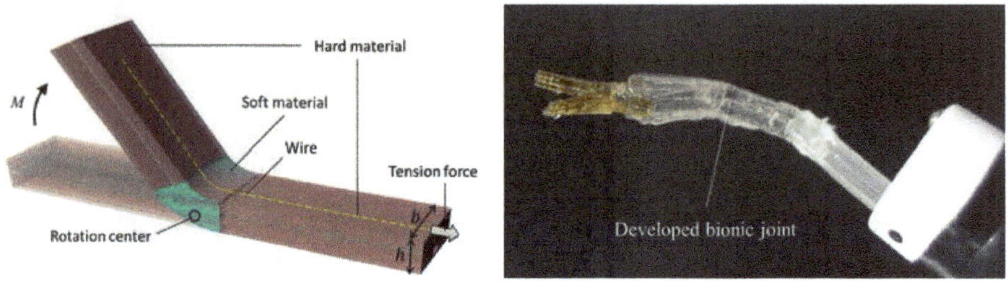

Figure 20. Design concept of bionic joint and fabricated microarm

3. Bioscience application

3.1. Background

Recently, in bioscience field, mechanical stimulation to a single cell is highly required to figure out functions and mechanical characteristics of cells. Especially, this approach is very useful for understanding the mechanism of aquatic microorganisms in terms of neurology and biofuel technology [24]. Since the mechanical characteristics of a number of aquatic microorganisms have not been fully understood yet, mechanical approach is really important to understand the cell interaction, growth factor of cells, and function of mechanoreceptor. In order to cope with this issue, conventional investigation for microorganisms was carried out manually or by using micromanipulators under microscope. However, throughput and dexterity of measurement were very low, because the manipulator had to handle the slender glass tools. Additionally, since petri dish was used these operation, we cannot maintain the stable condition of measurement system without contamination and disturbance.

There are two important points to evaluate the relationship between the mechanical stimulation and the response of microorganism. The first thing is that we should measure the amount of applied stimulation to cell as an applied force. The other thing is that, we should keep the experimental condition to observe the chemical reaction after stimulation of cell precisely.

On the other hand, there are many works of the micro-nano-robots and sensors based on MEMS/NEMS technologies to manipulate and/or measure the microorganisms [25-29]. Sun et al. have been developed the SOI based 2DOF force sensor which is consisted of the comb structures to detect the change of the capacitance [30, 31]. This sensor was applied to the evaluation of the mechanical property of oocyte. Nakajima et al. have been developed the nanoprobes to measure the stiffness of the *C. elegans* by the Atomic Force Microscope (AFM), which is the cantilever based force sensing [32]. Cappelleri et al. have proposed the μN force sensor made by Polydimethylsiloxane (PDMS) for microrobotics [33, 34]. This sensor can measure the force of two directions by using the vision sensor. These sensors, however, require the operation by micromanipulators to measure the objects in the narrow space. This means that it is difficult to close the total measurement system to prevent a contamination of cell and liquid.

In contrast, Abbott et al. have been continuously researching magnetic actuators operated by pairs of Helmholtz coils. They fabricated spiral shape microrobot and the microrobot swim in liquid environment and it can achieve 6-DOF precise actuation [35]. Pawashe et al. also applied pairs of Helmholtz coils for microrobot actuation. They actively employed stick-slip effects on the sliding magnetic object by switching the magnetic fields of vertical direction and horizontal direction [36]. Sakar et al. also developed a Helmholtz coils driven microtransporter [37]. They integrated the magnetic actuation with vision feedback control, and they achieved to deliver the microgel to neuron. We think that these robots with both the sub-micron-order positioning accuracy and the millimeter-order drive speed are very useful to investigate microorganisms. However, they did not have any sensors to evaluate the characteristics of cell in a closed space. Recently, optical force driven nanorobots with nm positioning accuracy have been developed. Ikuta et al. developed 6DOF-nanorobot made by light curing resin. They succeeded in the measurement of applied force to Yeast cells with approx. 5 μm diameters in a fluidic environment [38]. Arai et al. also developed the force feedback system by using the multi-beam optical tweezers [39]. The disadvantage of the optical drive method is from nN to pN order generative force. It is not enough to push 100 μm size microorganisms which have a stiff structure. Therefore, on-chip manipulation/sensing of microorganisms by functionalized microrobots are highly requited.

3.2. Basic concept

In the conventional works of our group, the magnetically driven microtools (MMT) have been proposed for automation of cell manipulation, as shown in Figure 21 [40]. The MMT, which is placed in a microfluidic chip, is composed of a magnetic material (Ni), and can be actuated by a permanent magnet mounted on a motorized stage (an XY-linear stage and a rotation motor) outside the microfluidic chip. Using four magnets to drive a single MMT, we can achieve stable movement of the MMT with 3DOF (X-direction, Y-direction and rotation). This allows the manipulation of cells in a completely closed biochip with generative forces of the order of mN. Since a closed microchip is used, we can avoid any contamination and maintain stable conditions during the measurements. This system contributes to our ability to observe continuously the response of the chemicals of microorganisms after stimulation. Further, our approach has the advantages of being a downsized actuation system and having a disposable chip structure. Conventionally, we succeeded in increasing the positioning accuracy of a 3DOF MMT from several hundred μm to 1 μm by using both the new method of arranging the permanent magnets [41] and a vibration-based method of reducing friction between the MMT and the substrate [42].

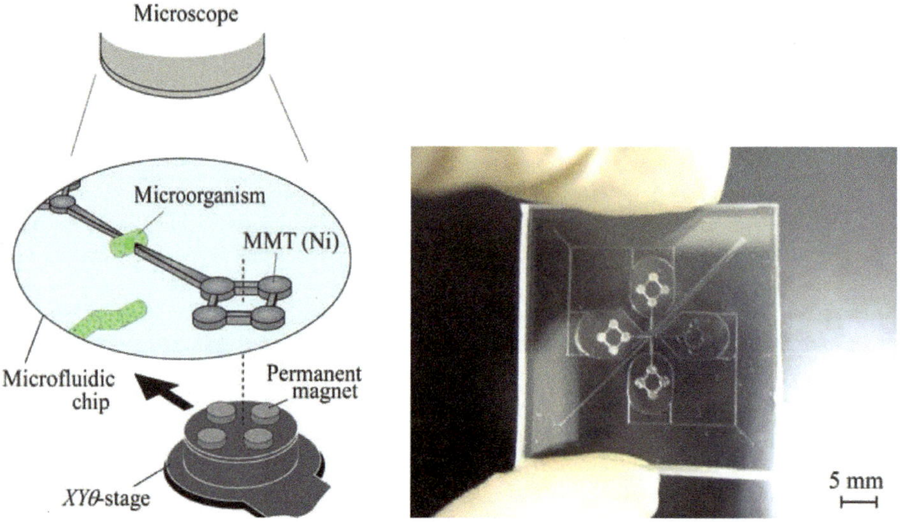

Figure 21. Basic concept of on-chip microtools and fabricated microfluidic chip with four microtools made of Ni.

Here, we consider that the beam structure is mounted on the tip of the MMT and the deformation of the beam is measured by the microscope to achieve the on-chip force sensing, as shown in Figure 22(a). However, there is a large friction between the MMT and the substrate of the biochip (normally glass) because the MMT is attracted by the magnet. It is big issue for on-chip accurate force sensing because the effect of the friction is included to the measured force data. Furthermore, since the beam structure becomes very thin, we concern that it is difficult to assemble the MMT to the microchip without any damage. To avoid these issues, we newly propose the layer type MMT which includes the MMT with the beam structure supported by the spring, the magnetic material, and the microchannel, as shown in Figure 22(b). The space between the MMT and the substrate is maintained by the microspacer. Therefore, there is no friction on the force sensing structure of the MMT. By using the layer structure, the micropattern of the MMT is protected and easy to assemble the several microparts

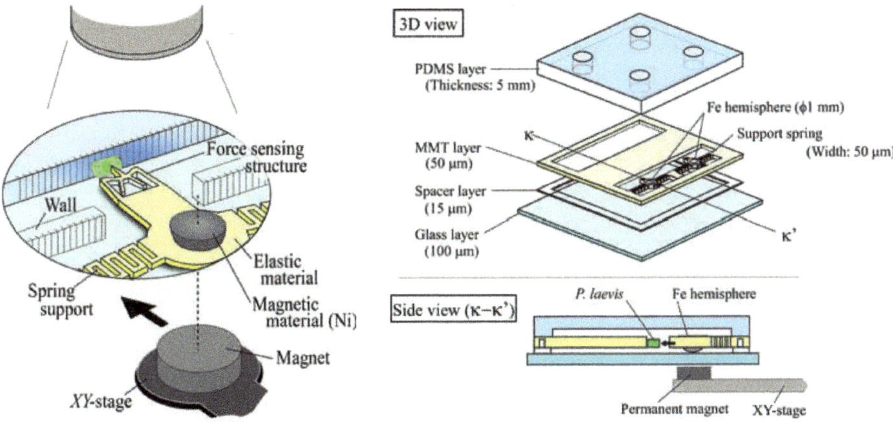

(a) 2DOF MMT with a force sensing structure (b) Layer fabrication method to avoid friction and damage of MMT

Figure 22. Concept of force sensing microtool and layer fabrication method

3.3. Basic sensing principle

In the first step of the on-chip force sensing, we design the frame-shaped beam to the MMT. This structure not a cantilever structure is important to keep straight the posture of the MMT when we push cells. In addition, to increase the sensitivity of the force sensing by the camera, the magnification mechanism of the beam deformation is placed into the frame, as shown in Figure 23.

To design the parameter of the force sensing structure, we set the mechanical model of the frame, as shown in Figure 23, where F, δ, l, h, d, r, $I_1=b_1h_1^3/12$, $I_2=b_2h_2^3/12$, and E are the force to be measured, the displacement of the beam, the width of the frame, the height of the frame, the width of the magnification mechanism, the height of the magnification mechanism, Young's modulus, the second moment of area of the beam BC, and the second moment of area of the beam CD, respectively. Here, the displacement of the magnification mechanism is

$$X = \alpha\delta \tag{5}$$

where α is the magnification ratio. Therefore, from this static model, we can obtain the applied force as follows [43]:

$$\begin{aligned} F &= kX \\ &= \frac{4EI_1}{r}\left[\frac{d(L-l)(I_1H+I_2l)}{(I_1H+2I_2l)}\right]^{-1}X \end{aligned} \tag{6}$$

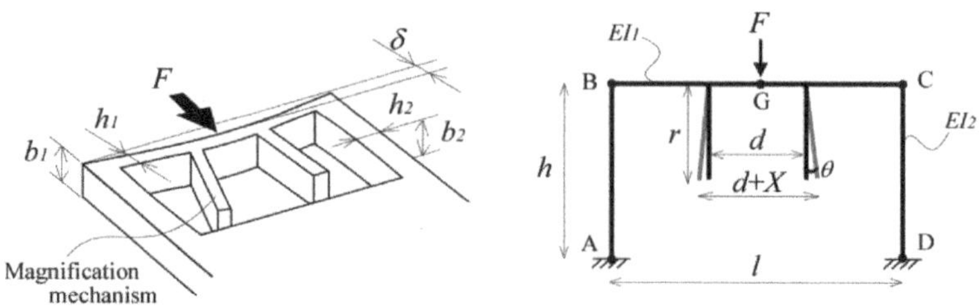

Parameter	Description	Value
h_1	width of frame BC	0.02 mm
h_2	width of frame AB and CD	0.04 mm
b_1	height of frame BC	0.05 mm
b_2	height of frame AB and CD	0.05 mm
h	length of frame AB and CD	0.60 mm
l	length of frame BC	0.92 mm
r	length of magnification mechanism	0.50 mm
d	Distance between magnification mechanism	0.50 mm
E	Young's modulus of frame	3.2 GPa

Figure 23. Frame model of force sensing structure and designed parameters

3.4. Fabrication

By considering the results of the previous works which applied the force to the cells, we selected an SU-8 negative photoresists for a material of the MMT to prevent the damage to cells or microorganisms. As well known, SU-8 is flexible (approx. 2 to 10 GPa), and we can easily make the micropattern by using the photolithography technique [44, 45]. The Young's modulus of SU-8 depends on the thickness of the SU-8 and the exposure amount. By considering the diameter of targeted cell with a diameter of approx. 0.05-0.1 mm, we set that the thickness of the MMT was $b_1=b_2=0.05$ mm and the exposure amount (UV Dose) was 500 mJ/cm^2. Then, we used 3.2 GPa Young's modulus for the SU-8 layer design. On the other hand, from the limitation of the field of view of the microscope with the 10X objective lens (1.5 x 1.0 mm), we set $l=0.92$ mm, $h=0.6$ mm, $r=0.5$ mm, and $d=0.5$ mm to observe the whole force sensing structure during measurements. Finally, from the resolution of the photolithography of the SU-8 with the emulsion mask, we decided as $h_1=0.02$ mm and $h_2=0.04$ mm.

Figure 24 shows the results of the FEM analysis based on the designed parameters. To determine the optimal condition of the movable area of the MMT, we varied the width of the support spring. Finally, we determined that a spring width of 50 μm was suitable. From the results of the simulation, we confirmed that the support spring can be manipulated by magnetic actuation, and that the force sensing structure works well, even though there were some differences between the analysis and the simulation because of modeling error.

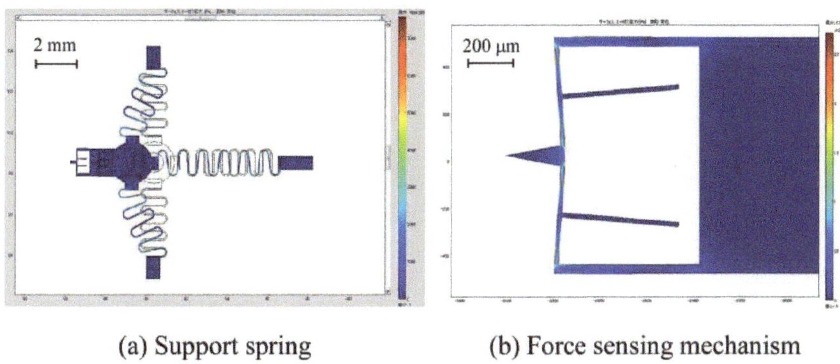

Figure 24. FEM simulation of support springs and force sensing structure

Figure 25 shows the fabrication process for the proposed layer type MMT with the frame structure. The biochip was composed of the glass substrate with the thickness of 100 μm, the microspacer with the thickness of 15 μm, the SU-8 layer with the thickness of 50 μm includes the MMTs and the microchannel, the hemispheres with the diameter of 1 mm to actuate the MMTs, and the PDMS cover with the thickness of 5 mm. Figure 26(a) shows the fabricated SU-8 layer, and Figure 26(b) shows the enlarged view of the fabricated MMT. To push the local point of a small cell, a triangle-shaped probe was attached to the tip of the MMT. Figure 27 shows the side view of the MMT. The space between the force sensing structure and the glass substrate is maintained by the spring structure.

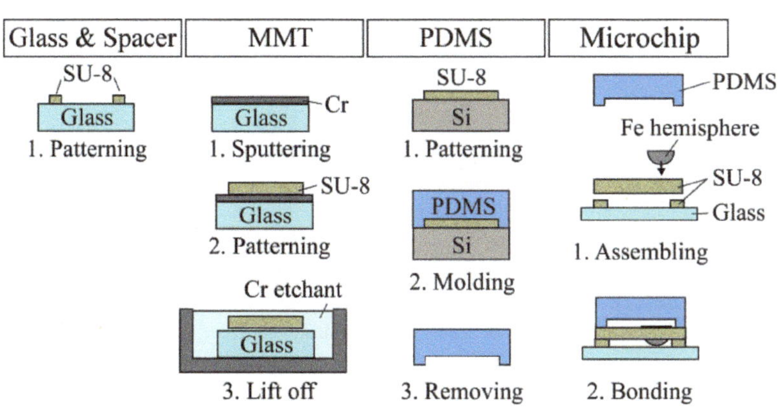

Figure 25. Fabrication process of MMT with a force sensing structure

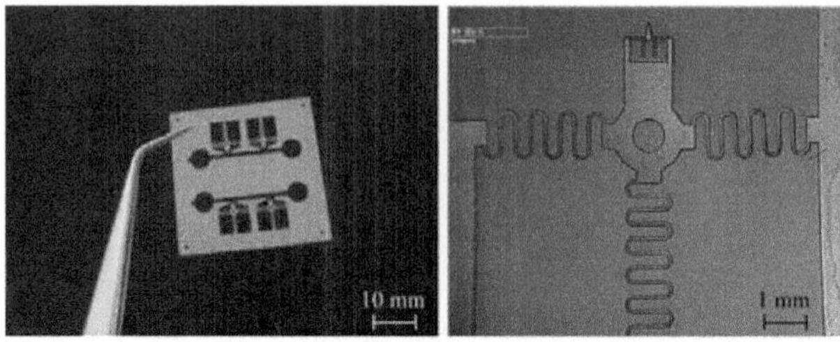

(a) SU-8 layer with 4 MMTs (b) Overview of the MMT

Figure 26. Fabricated MMT with force sensing structure

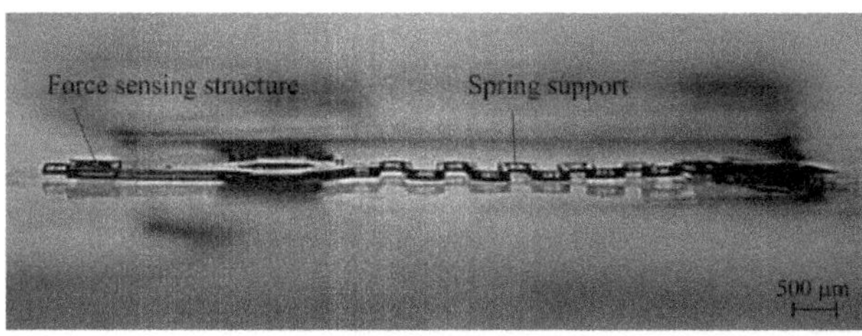

(a) Side view

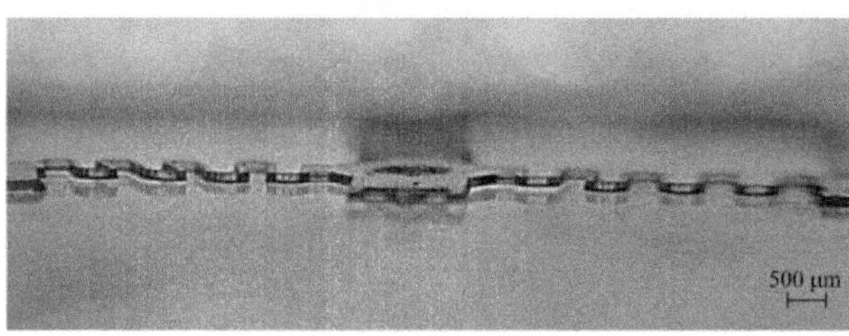

(b) Front view

Figure 27. Assembled MMT layer (without Fe hemisphere) and microspacer layer

3.5. Experiments

Next, we evaluated the force measurement by the developed MMT. In order to measure force data accurately, the MMT was pushed against a commercial force sensor (KYOWA, LVS-5GA) with an accuracy of 10 μN by using a linear stage with a position accuracy of 2 μm, as shown in Figure 28. Then, we measured the deformation using a microscope equipped with a CCD camera (resolution: 1 μm/pixel). In this experiment, we confirmed that there was a high linearity between the displacement of frame δ and the displacement of the magnification mechanism X, and that the magnification ratio δ was 2.72. Therefore, X can be used to estimate the applied force.

Figure 29(a) shows the relationship between the displacement X and the force measured by the commercial sensor. From this result, we confirmed that there is a high linearity between the displacement δ and the X. Therefore, the magnification ratio is approximately 2.7 and X can be used to estimate the applied force.

Figure 29(b) shows the relationship between the X and the force data measured by the commercial sensor. From these results, we can see that there is a hysteresis in different conditions. Furthermore, when the applied force is larger than 3 mN, the shape of the frame does not return to the initial state (offset). Based on these characteristics, we have to only use the displacement X in the pushing phase. And, it is better to limit the force range to less than 3 mN to estimate the force data properly.

Fig. 29(c) left and center show the measured force data by the commercial sensor and the estimated force data calculated from (6). From this result, we can see that there is a difference between the measured and the estimated forces. We think that this was caused by errors in the fabrication and in the numerical analysis. By correcting the estimated data, we considered the fabrication error, and we can measure accurate force data, as shown in the right side of Fig. 29(c).

Next, Fig. 29(d) shows the individual difference between MMTs in the SU-8 layer. From these results, we confirmed that the variation among the MMTs is less than 10 %. This means that by calibrating one of the MMTs from the fabricated layer, we can use other MMTs (included in the layer) without calibration.

Through these experiments, we obtained the minimum resolution of the force sensing as approximately 100 μN. The resolution for this type the force sensor was determined by the resolution of the camera used and the sensitivity of the frame deformation. A large deformable frame caused a large hysteresis and offset as discussed above. Therefore, by using a high-resolution camera, we can increase the sensor performance with comparative ease. As a first step, we believe that this performance is reasonable to determine the range of force required to push a 100 μm size single cell.

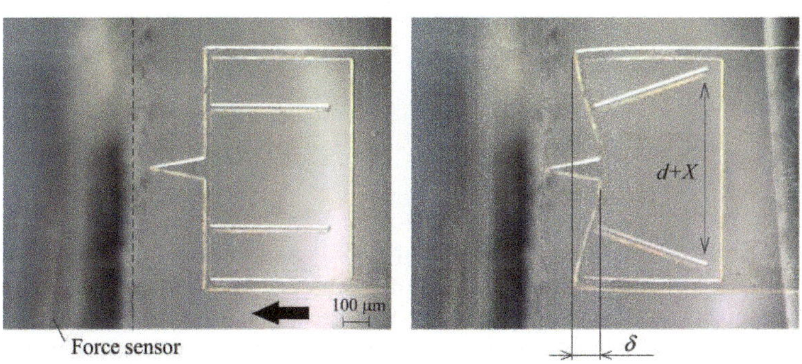

Figure 28. Overview of force calibration experiment

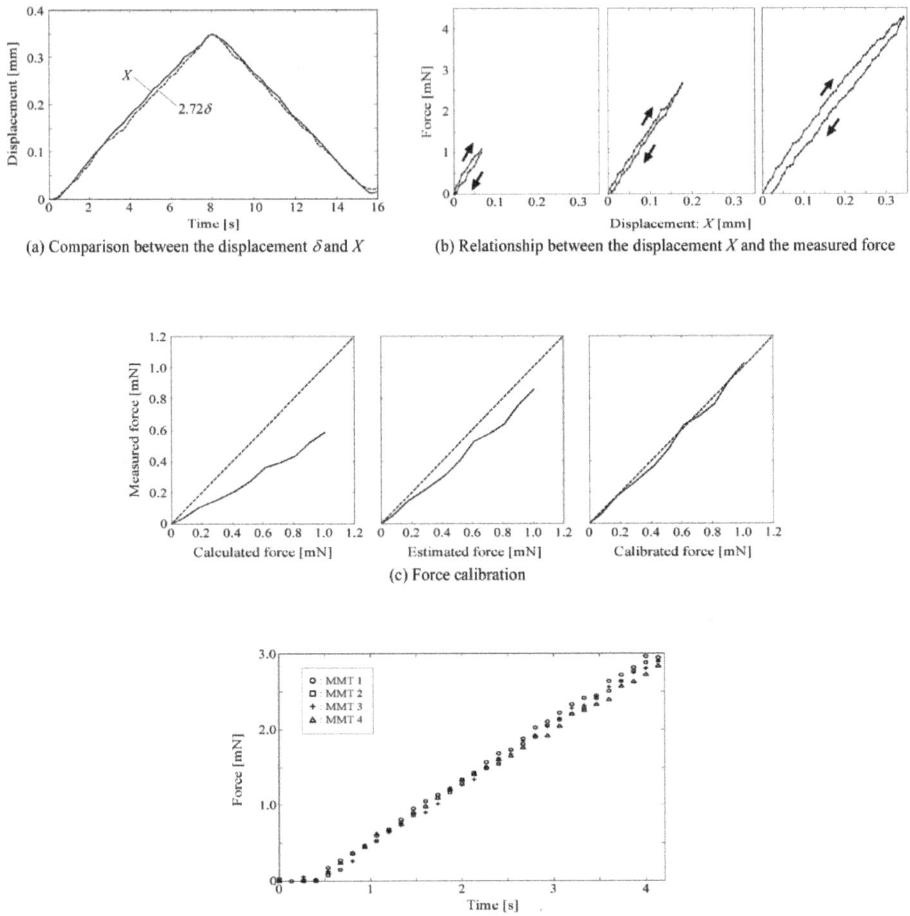

(a) Comparison between the displacement δ and X

(b) Relationship between the displacement X and the measured force

(c) Force calibration

(d) Individual difference

Figure 29. Evaluation of calibration of fabricated force sensing structure

3.6. Application

We applied the developed MMT to actual microorganism. In this experiment, we used one of centric diatom called *Pleurosira laevis*. This cell has a unique behavior, that is, it is stimulated, the chloroplasts within it agglomerate around the nucleus, as shown in Figure 30 [46]. Furthermore, this agglomeration phenomenon is transmitted to other connected cells, as shown in Figure 31. Interestingly, the phenomenon even transmits to unconnected cells. The reason for this spreading of chloroplast agglomeration is to be because of the release of unknown components from the stimulated cell to stimulate other cells to increase growth. This phenomenon is considered to be similar to information transmission in neurons. From the perspective of neurology and algae-based biofuel, the investigation of this phenomenon is will be useful for understanding the mechanism of communication between aquatic microorganism. Although it is well known that these cells have a receptors, and thus, that the agglomeration phenomenon is caused by optical and electrical stimulations, the mechanism of this phenomenon is still not fully understood [47, 48]. Conventional studies with a mechanical approach, however, could not clarify the relationship between the applied force and the response of *P. laevis* due to force sensing problem [49]. Therefore, quantitative investigation of the applied force must be performed in order to elucidate the agglomeration phenomenon of chloroplasts of *P. laevis* by mechanical stimulation.

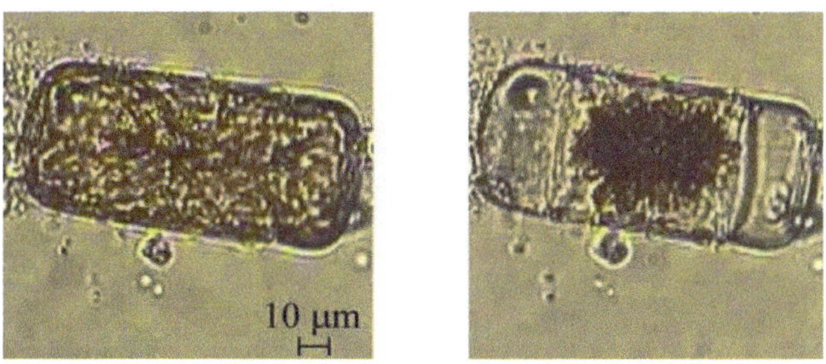

(a) Overview of single cell (b) After optical/electrical/mechanical/chemical stimulation

Figure 30. A single cell of *Pleurosira laevis* (*P. laevis*)

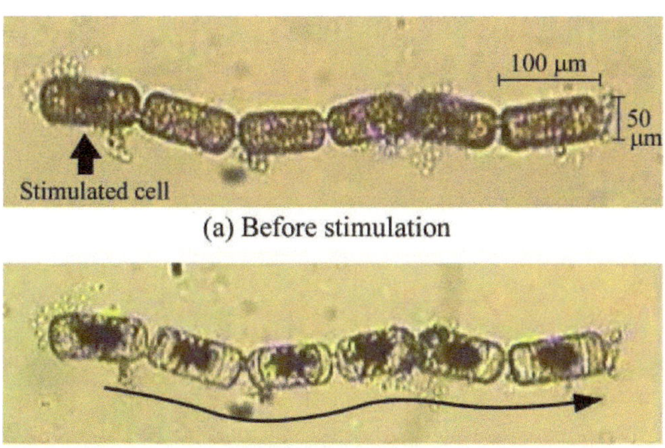

(a) Before stimulation

(b) After stimulation

Figure 31. Propagation of agglomeration in *P. laevis*

Fig. 32 shows the overview of the experiment by developed MMT. As a result, we succeeded in rotating, capturing, stimulation of microorganisms by using dual-MMT with force sensing structure. Then, the applied force was estimated as shown in Figure 33. The agglomeration propagation phenomenon of *P. laevis* was also observed by MMT stimulation, as shown in Figure 34. In future work, the relationship between the applied force and the agglomeration phenomenon is investigated by the proposed system. The optimized microchannel design is also required to reduce the whole measurement time and the observations.

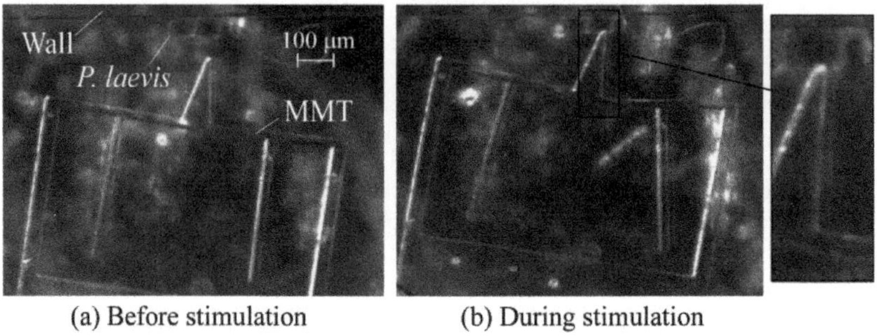

Figure 32. On-chip *P. laevis* stimulation by MMT

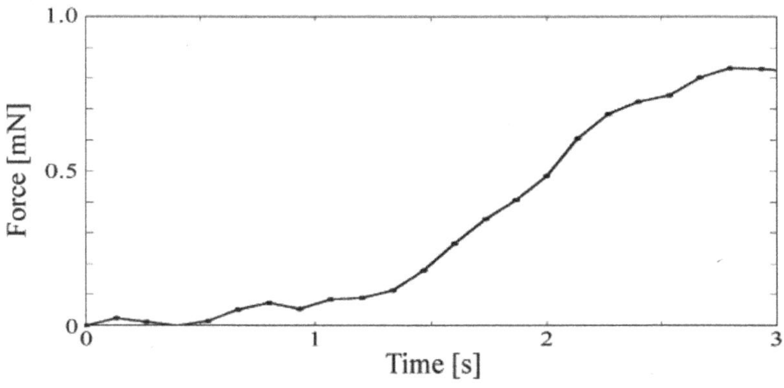

Figure 33. Measured force data by force sensing structure

3.7. Discussion and summary

In this section, we discussed an on-chip force sensing by magnetically driven microtool (MMT) which enables measurement of stimulant property of single cell:

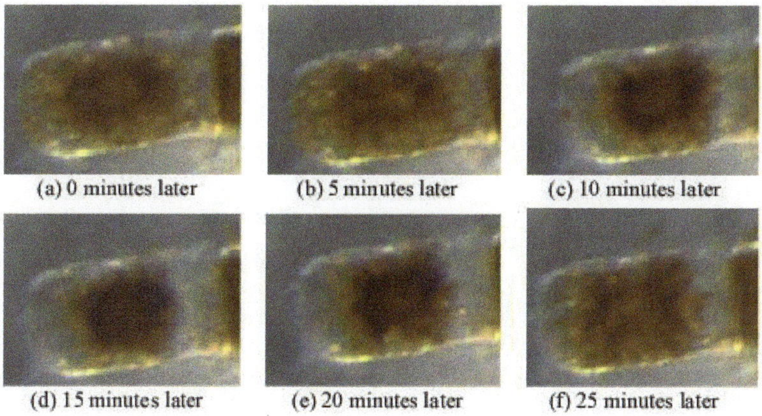

Figure 34. Measured agglomeration propagation phenomenon

1. The microchip is composed of a microspaceer, an MMT layer, an Fe hemisphere and a PDMS cover. These parts are assembled by newly proposed layer fabrication technique.

2. The frame based force sensing structure with the displacement magnification mechanism is designed and fabricated based on analytical approach.

3. Through our basic experiments, we have confirmed the effectiveness of the fabrication method and the performance of the force sensing.

4. Through our experiments for *P. laevis* stimulation, we confirmed that the proposed force sensing mechanism works well with accuracy of 100 μN, and the MMT has a enough power to push *P. laevis*.

Currently, the force sensing structure of the MMT was designed to measure a one-directional force, taking into account both the specifications of the *P. laevis* measurement that were requested and the difficulties involved in the fabrication process. The MMT was found to be satisfactory for investigating the stimulant property of *P. laevis*. For wider applications, however, it would be better to mount a multi-DOF force sensing structure on to the tip of the MMT. Additionally, in order to measure the microorganism precisely, we should consider a fully-automated 3DOF position control of MMT. Automated vision-based force estimation and MMT position control can be introduced into our developed system by exploiting-robotics technologies. We believe that our approach is useful for investigating of living cells and microorganisms in a closed microfluidic chip. The combination of the simple and disposable SU-8 based structure and the camera based measurement achieves very low cost force sensing system.

In our future plan, the developed system will be applied to measure other microorganisms and the MMT has a potential to capture and measure motile microorganisms which contain a large proportion of aquatic microorganisms [50]. In parallel, to improve the force sensing resolution from μN-order to nN-order, electrical sensors should be implemented to tip of the

MMT, although there is a couple of problem on fabrication process. In addition, we are now adding more functions for MMT to expand application of microrobots [51]. The performance of the MMT such as positioning accuracy and driving speed may be increased by introducing a sub-micro and a nano-order fabrication approach [52, 53].

4. Conclusion

We showed two specific approaches based on micro mechatronics and fabrication for medical and bioscience applications. The results indicated that microrobotics is also really important to drive, control and measure the developed devices. Currently, manipulation and sensing of smaller object are strongly requited in biomedical field. In order to realize a nano-scale manipulation and measurement, appropriate integration of micro and nano mechatronics are most important factor, because there are difficulties to seamlessly integrate the functioning of micro and nano objects. In addition, handling and sensing of micro-scale wet-object such cells are also important as mentioned above. Therefore, it is required to establish and develop innovative methodologies regarding control, sensing and fabrication in the fields of micro-nano mechatronics and robotics. For the purpose of ensuring the appropriate and effective integration of the technology and science, we should be continued the ongoing promotion of corroboration in a broad range of areas.

Acknowledgements

This work has been supported by the Ministry of Education, Culture, Sports, Science and Technology Great-in-Aid for Scientific Research (21656064, 22860030) and the Nagoya University Global COE Program, ``COE for Education and Research of Micro-Nano Mechatronics,".

Author details

Tomohiro Kawahara and Fumihito Arai

Department of Micro-Nano Systems Engineering, Nagoya University, Nagoya, Japan

References

[1] Hasegawa H, Kabeshima Y, Watanabe M, Yamamoto S, Kitajima M. Randomized Controlled Trial of Laparoscopic versus Open Colectomy for Advanced Colorectal Cancer. Surgical Endoscopy 2003;17(4) 636-640.

[2] The Clinical Outcomes of Surgical Therapy Study Group. A Comparison of Laparoscopic Assisted and Open Colectomy for Colon Cancer. The New England Journal of Mdicene 2004;350(20) 2050-2059.

[3] ntuitive Surgical Inc. da Vinci Surgical System. http://www.intuitivesurgical.com/ (accessed 13 May 2012)

[4] Spaun G. O, Zheng B, Swanstrom L. L. A Multitasking Platform for Natural Orifice Translumenal Endoscopic Surgery (NOTES): a Benchtop Comparison of a New Device for Flexible Endoscopic Surgery and a Standard Dual-Channel Endoscope. Surgical Endoscopy 2009;23 2720-2727.

[5] Discher D. E, Janmey P, Wang Y. L. Tissue Cells Feel and Respond to the Stiffness of Their Substrate. Science 2005;310 1139-1143.

[6] Engler A. J, Sweeney H. L, Discher D. E. Matrix Elasticity Directs Stem Cell Lineage Specification. Cell 2006;126 677-689.

[7] Kondo H, Gotoda T, Ono H, Oda I, Kozu T, Fujishiro M, Saito D, Yoshida S. Percutaneous Traction-Assisted EMR by using an Insulation-Tipped Electrosurgical Knife for Early Stage Gastric Cancer. Gastrointestinal Endoscopy 2004;59(2) 284-288.

[8] Miyamoto S, Muto M, Hamamoto Y, Boku N, Ohtsu A, Baba S, Yoshida M, Ohkuma M, Hosokawa K, Tajiri H, Yoshida S. A New Technique for Endoscopic Mucosal Resection with an Insulated-Tip Electrosurgical Knife Improves the Completeness of Resection of Intramucosal Gastric Neoplasms. Gastrointestinal Endoscopy 2002;55(4) 576-581.

[9] Minami S, Gotoda T, Ono H, Oda I, Hamanaka H. Complete Endoscopic Closure of Gastric Perforation Induced by Endoscopic Resection of Early Gastric Cancer using Endoclips can Prevent Surgery. Gastrointestinal Endoscopy 2006;63(4) 596-601.

[10] Sakamoto N, Osada T, Shibuya T, Beppu K, Matsumoto K, Shimada Y, Konno A, Kurosawa A, Nagahara A, Ohkusa T, Ogihara T, Watanabe S. The Facilitation of a New Traction Device (S-O clip) Assisting Endoscopic Submucosal Dissection for Superficial Colorectal Neoplasms. Endoscopy 2008;40 E94-95.

[11] Martin C, Chapelle F, Lemaire J. J, Gogu G. Neurosurgical Robot Design and Interactive Motion Planning for Resection Task. In: Proceedings of the IEEE/RSJ International Conference on Intelligent Robots and Systems, 11-15 October 2009, St. Louis, USA; 2009.

[12] Arai F, Fuji R, Fukuda T, Negoro M. Robot-Assisted Active Catheter Insertion: Algorithms and Experiments. The International Journal of Robotics Research 2009;28(9) 1101-1117.

[13] Harada K, Tsubouchi K, Fujie M. G, Chiba T. Micro manipulators for intrauterine fetal surgery in an Open MRI. In: Proceedings of the IEEE International Conference on Robotics and Automation. 18-22 April 2005, Barcelona, Spain; 2005.

[14] Hattori A, Suzuki N, Hayashibe M, Suzuki S, Otake Y, Tajiri H, Kobayashi S. Development of a Navigation Function for an Endosocopic Robot Surgery System. Studies in Health Technology and Informatics 2005;111 167-171.

[15] Rothstein R. I, Swanstrom L. L. Use of the Direct Drive Endoscopic System (DDES) for In-Vivo Mucosal Resection in a Porcine Model. Gastrointestinal Endoscopy2008;67 AB146.

[16] Narumi K, Azuma D, Arai F. Fabrication of Articulated Microarm for Endoscopy by Stacked Microassembly Process (STAMP). Journal of Robotics and Mechatronics 2009;21(3) 396-401.

[17] Azuma D, Lee J, Narumi K, Arai F. Fabrication and Feedback Control of an Articulated Microarm. In: Proceedings of the IEEE International Conference on Robotics and Automation. 12-17 May 2009, Kobe, Japan; 2009.

[18] Ikuta K, Yamamoto K, Sasaki K. Development of Remote Microsurgery Robot and New Surgical Procedure for Deep and Narrow Space. In: Proceedings of the IEEE International Conference on Robotics and Automation. 14-19 September 2003, Taipei, Taiwan; 2003.

[19] Kawahara T, Matsumoto T, Muramatsu T, Osada T, Sakamoto N, Arai F. Development of a Decoupling Wire Driven Exoskeletal Microarm for Endoscopic Submucosal Dissection. In: Proceedings of the 3rd IEEE/RAS-EMBS International Conference on Biomedical Robotics and Biomechatronics, 26-29 September 2010, Tokyo, Japan; 2010.

[20] Muramatsu N, Ando H. Efficiency of Gripping Mechanism using Buckling Phenomenon of Long Column. In: Proceedings of the IEEE International Conference on Robotics and Biomimetics. 18-22 December 2009, Guilin, Guangxi, China; 2009.

[21] Ko C. W, Kallo A. N. Per-Oral Transgastric Abdominal Surgery. Chinese Journal of Digestive Diseases 2006;7 67-70.

[22] Baron T. H. Natural Orifice Transluminal Endoscopic Surgery. British Journal of Surgery 2007;94 1-2.

[23] Owaki H, Kawahara T, Arai F. Bionic Design of Microjoint for Minimally Invasive Surgical Instrument. In: Proceedings of IEEE International Symposium on Micro-Nano Mechatronics and Human Science. 6-9 November 2011, Nagoya, Japan; 2011.

[24] Yadugiri V. T. Milking Diatoms - a New Route to Sustainable Energy. Current Science 2009;97(6) 748-750.

[25] Fauver M. E, Dunaway D. L, Lilienfeld D. H, Craighead H. G, Pollack G. H. Microfabricated Cantilevers for Measurement of Subcellular and Molecular Forces. IEEE Transactions on Biomedical Engineering 1998;45(7) 891-898.

[26] Papi M, Sylla L, Parasassi T, Brunelli R, Monaci M, Maulucci G, Missori M, Arcovito G, Ursini F, Spirito M. D. Evidence of Elastic to Plastic Transition in the Zona Pelluci-

da of Oocytes using Atomic Force Spectroscopy. Applied Physics Letters 2009;95 153902-1-153902-3.

[27] Jeong K. H, Keller C, Lee L. Direct Force Measurements of Biomolecular Interactions by Nanomechanical Force Gauge. Applied Physics Letters 2005;86 193901-1-193901-3.

[28] Koch S, Thayer G, Corwin A, de Boer M. Micromachined Piconewton Force Sensor for Biophysics Investigations. Applied Physics Letters 2006;89 173901-1-173901-3.

[29] Wacogne B, Pieralli C, Roux C, Gharbi T. Measuring the Mechanical Behavior of Human Oocytes with a Very Simple SU-8 Micro-Tool. Biomedical Microdevices 2008;10 411-419.

[30] Sun Y, Nelson B. J, Potasek D. P, Enikov E. A Bulkmicrofabricated Multi-Axis Capacitive Cellular Force Sensor using Transverse Comb Drive. Journal of Micromechanics and Microengineering 2002;12 832-840.

[31] Sun Y, Wan K. T, Roberts K. P, Bischof J. C, Nelson B. J. Mechanical Property Characterization of Mouse Zona Pellucida. IEEE Transactions on Nanobioscience 2003;2(4) 279-286.

[32] Nakajima M, Ahmad M. R, Kojima S, Honmma M, Fukuda T. Local Stiffness Measurements of C. elegans by Buckling Nanoprobes Inside an Environmental SEM. In: Proceedings of the IEEE/RSJ International Conference on Intelligent Robots and Systems, 11-15 October 2009, St. Louis, USA; 2009.

[33] Cappelleri D. J, Piazza G, Kumar V. Two-Dimensional, Vision-based μN force Sensor for Microrobotics. In: Proceedings of the IEEE International Conference on Robotics and Automation, 12-17 May 2009, Kobe, Japan; 2009.

[34] Cappelleri D. J, Krishnan G, Kim C, Kumar V, Kota S. Toward the Design of a Decoupled, Two-Dimensional, Vision-based μN Force Sensor. Journal of Mechanisms and Robotics 2010;2 021010-1-021010-9.

[35] Abbott J. J, Peyer K. E, Lagomarsino M. C, Zhang L, Dong L, Kaliakatsos I. K, Nelson B. J. How Should Microrobots Swim? The International Journal of Robotics Research 2009;28(11-12) 1434-1447.

[36] Pawashe C, Floyd C. S, Sitti M. Modeling and Experimental Characterization of an Untethered Magnetic Micro-Robot. The International Journal of Robotics Research 2009;28(8) 1077-1094.

[37] Sakar M. S, Steager E. B, Kim D. H, Kim M. J, Pappas G. J, Kumar V. Single Cell Manipulation using Ferromagnetic Composite Microtransporters, Applied Physics Letters 2010;96 043705-1-043705-3.

[38] Ikuta K, Sato F, Kadoguchi K, Itoh S. Optical Driven Master-Slave Controllable Nano-Manipulator with Real-Time Force Sensing. In: Proceedings of the IEEE Inter-

national Conference on Micro Electro Mechanical Systems, 13-17 January 2008, Tucson, Arizona, USA; 2008.

[39] Arai F, Ichikawa A, Ogawa M, Fukuda T, Horio K, Itoigawa K. High-Speed Separation System of Randomly Suspended Single Living Cells by Laser Trap and Dielectrophoresis. In: Proceedings of the IEEE International Conference on Robotics and Automation, 21-26 May 2001, Seoul, Korea; 2001.

[40] Yamanishi Y, Sakuma S, Onda K, Arai F. Powerful Actuation of Magnetized Microtools by Focused Magnetic Field for Particle Sorting in a Chip. Biomedical Microdevices 2008;10 411-419.

[41] Hagiwara M, Kawahara T, Yamanishi Y, Arai F. Driving Method of Microtool by Horizontally Arranged Permanent Magnets for Single Cell Manipulation. Applied Physics Letters 2010;97 013701-1-013701-3.

[42] Hagiwara M, Kawahara T, Yamanishi Y, Masuda T, Feng L, Arai F. On-Chip Magnetically Actuated Robot with Ultrasonic Vibration for Single Cell Manipulations. Lab on a Chip 2011;11 2049-2054.

[43] Kawahara T, Sugita M, Hagiwara M, Yamanishi Y, Arai F, Kawano H, Shihira-Ishikawa I, Miyawaki A. On-Chip Force Sensing by Magnetically Driven Microtool for Measurement of Stimulant Property of P. laevis. In: Proceedings of the IEEE International Conference on Robotics and Automation, 9-13 May 2011, Shanghai, China; 2011.

[44] Khoo H. S, Liu K. K, Tseng F. G. Mechanical Strength and Interfacial Failure Analysis of Cantilevered SU-8 Microposts. Journal of Micromechanics and Microengineering 2003;13 822-831.

[45] Bachmann D, Schoberle B, Kuhne S, Leiner Y, Hierold C. Fabrication and Characterization of Folded SU-8 Suspensions for MEMS Applications. Sensors and Actuators A 2006;130-131 379-386.

[46] Makita N, Shihira-Ishikawa I. Chloroplast Assemblage by Mechanical Stimulation and Its Intercellular Transmission in Diatom Cells. Protoplasma 1997;197 86-95.

[47] Furukawa T, Watanabe M, Shihira-Ishikawa I. Green- and blue-light-mediated chloroplast migration in the centric diatom Pleurosira laevis. Protoplasma 1998;203 214-220.

[48] Shihira-Ishikawa I, Nakamura T, Higashi S, Watanabe M. Distinct Responses of Chloroplasts to Blue and Green Laser Microbeam Irradiations in the Centric Diatom Pleurosira laevis. Photochemistry and Photobiology 2007;83 1101-1109.

[49] Hanada Y, Sugioka K, Kawano H, Shihira-Ishikawa I, Miyawaki A, Midorikawa K. Nano-Aquarium for Dynamic Observation of Living Cells Fabricated by Femtosecond Laser Direct Writing of Photostructurable Glass. Biomedical Microdevices 2008;10 403-410.

[50] Kawahara T, Sugita M, Hagiwara M, Yamanishi Y, Arai F. Ultra-High-Speed Robot Hand and Eye for Investigation of Microorganisms in a Chip. In: Proceedings of the 15th International Conference on Miniaturized Systems for Chemistry and Life Sciences, 2-6 October 2011, Seattle, USA; 2011.

[51] Ichikawa A, Arai F. On-Chip Noncontact Actuation of a Micro-Pipette Driven by Permanent Magnets. In: Proceedings of the International Conference on Micro Electro Mechanical Systems, 29 January - 2 February 2012, Paris, France; 2012.

[52] Sakuma S, Arai F. Noncontact Nanometric Positioning of Probe Tip for Measurement of Mechanical Parameters of Cell. In: Proceedings of the International Conference on Micro Electro Mechanical Systems, 29 January - 2 February 2012, Paris, France; 2012.

[53] Hagiwara M, Kawahara T, Iijima T, Arai F. High Speed Microrobot Actuation in a Microfluidic Chip by Levitated Structure with Riblet Surface. In: Proceedings of the IEEE International Conference on Robotics and Automation, 14-18 May 2012, St. Paul, USA; 2012.

Chapter 5

Neural Interfaces: Bilateral Communication Between Peripheral Nerves and Electrical Control Devices

Goro Obinata, Hitoshi Hirata, Chikara Nagai,
Shigeru Kurimoto, Shuichi Kato and
Tomonori Nakano

1. Introduction

Neural Interfaces are data links between human nervous system and an external device, and allow the transmission of information to and from the human nervous system to the external device. Bioelectric signals can be obtained from the nervous system, and/or transmitted to the nervous system via implanted or surface electrodes. One beneficial application of such neural interfaces is to obtain control signals from undamaged sensory-motor areas of patient's brain for controlling neuroprosthetic devices such as artificial limbs or wheelchairs. Paralyzed or amputated people can reconstruct certain motor functions by using such neuroprosthetic devices. Brain-machine interfaces, which are direct data links between human brain and machines, are one kind of neural interfaces [1]. Such interfaces have been proposed during this decade to control prosthetic limbs, or to control machines such as wheelchairs or robotic manipulators [2]. However, our present knowledge on brain functions is so limited that we do not fully understand the coding of information expressing the behavior in motions; specially, we do not know the variation of the coding in individual differences or in related thoughts or emotions.

In order to design practical neuroprosthetic devices, we focus on neural interfaces which link peripheral nervous system and external devices in this paper (PNI, Peripheral Neural Interface). Since peripheral nervous system is much simpler than central nervous system, we may avoid the difficult problems which come from our limited knowledge on brain functions while the neuroprosthetic devices interact just with the peripheral nerves. In Section 2, we describe interfaces to peripheral motor neurons by using reinnervation type electrodes. The electrodes are constructed by implanting embryonic neurons into peripheral motor units. The endplates of the implanted neurons which grow into muscles make biological interfaces for motor commands to the muscles. The other ends of the implanted neurons are connected by special types of electrodes to communicate with electric devices of motor controllers. The diameter of mammalian motor neurons is from 0.5 μm to 20 μm. There is a reason why our research requires bio-compatible micro-nano technologies for achieving such interfaces of new reinnervation type electrodes.

Neuroprosthetic devices with interfaces detecting electromyogram (EMG) are in practical use [3]. Such EMG interfaces require to place the electrodes at the end plate of a motor neuron; thus, the paralyzed or amputated users can not control the paralyzed or amputated muscles with the same passes of motor neurons as before paralyzed or amputated. Moreover, EMG interfaces pickup the signals at end plate of a motor neuron; sensory signals from receptors are not obtainable via EMG. This means that two-way communication by EMG interfaces is impossible. We can assemble sensors and stimulators into a neuroprosthetic device. If we have a way to send the signals from the assembled stimulators to sensory nervous system, ideal neuroprosthetic devices with two-way communication will be achievable. The combination of EMG pickup and electrical stimulation with surface electrodes was proposed for two-way communication [4]. However, there is no clear result on the amount of transmissive information through the afferent passes. In Section 3, we show a preliminary study on the possibility of sensory feedback via axial fibers of peripheral sensory neurons. In other words, we try to achieve an artificial afferent pass for feeding signals back to the brain. A method for improving on the amount of transmissive information has been proposed by using a configuration with multi-channel electrodes.

One application of PNIs is to achieve unconscious muscular movements such as walking, writing, dancing, playing musical instruments, and so on. For an example, in walking motions some neural networks generate the pattern of walking, and the steady behavior is closed within the peripheral nervous system and the pattern generators in spinal cord. The upper central nervous system provides the triggers of walking such as start/stop, speedup/slowdown, turn-right/turn-left. The understanding on pattern generators for walking is now enough to simulate human walking or some animals' walking. In Section 4, we give a walking simulation of "rat" to show the possibility of practical usages of PNIs.

2. Transplantation of embryonic neurons into peripheral nerve forms functional motor units

2.1. Background and purpose

There has been a rapid surge in clinical trials involving stem cell therapies over the last three to four years [5]. Those trials are establishing the clinical pathways for regenerative medicine, especially in nervous system. Since derivation of human embryonic stem cells (hESCs) in 1998 [6], hESCs have been thought a promising source of replacement cells for regenerative medicine. Although Geron Corporation has been no longer enrolling patients for the trial, they conducted the first clinical trial in the United States to evaluate the safety of oligodendrocyte precursor cells derived from hESCs in patients with thoracic spinal cord injuries [7]. The company Advanced Cell Technology has reported promising results from the clinical trials on Stargardt's Macular Dystrophy with retinal pigment epithelium derived from hESCs [8]. However, neuronal replacement in central nervous system is still limited due to a number of cells required to reconstruct complex structures and narrow therapeutic time window [9]. There are still several hurdles to be overcome in order to establish clinical application of neuronal replacement therapies in central nervous systems.

One experimental approach to rescue denervated muscles from damage to lower motor neurons or axonal disconnection is transplantation of motoneurons into a peripheral nerve as a source of neurons for muscle reinnervation. Since Erb reported the reinnervation of denervated muscle by embryonic motoneurons transplanted into peripheral nerve in 1993 [10], several studies have investigated the factors that improve motoneuron survival in peripheral nerve [11-13]. Considering the simplicity of neural network and wide windows of opportunity for the treatment [14], peripheral nerve system can be an ideal target for neuronal replacement therapy. This transplantation strategy may provide the potential to excite these muscles artificially with electrical stimulation. The aim of this study was to evaluate whether transplanted rat embryonic motorneurons into adult rat peripheral nerve would survive, and whether transplanted motorneurons would form functional motor units.

2.2. Methods

The sciatic nerves of adult Fischer 344 rats were transected; 1 week later, dissociated embryonic spinal neurons were transplanted into the distal stump of the tibial and peroneal nerves. Surgical controls underwent the same surgeries but had only medium injected into the peripheral nerves. Tissue analysis and measurement of ankle angles were performed twelve weeks after neural transplantation.

2.2.1. Cell preparation

Ventral spinal cord cell were obtained from Fischer 344 rat embryos (Japan SLC, Inc., Shizuoka, Japan). After Fischer rats on day 14 of pregnancy were anesthetized with isoflurane, ventral spinal cords were resected from the fetuses using a surgical microscope and were cut into small pieces in ice-cold Hanks' balanced salt solution. Ventral spinal neurons were dissociated using papain-containing separation solution (MB-X9901; Sumitomo Bakelite Co. Ltd, Tokyo, Japan). For implantation, dissociated neurons were suspended in Neurobasal medium (Gibco) containing B27 supplement (Gibco), Glutamax (Gibco), and N-2 supplement (Gibco).

2.2.2. Surgical procedures and transplantation

All procedures were performed on 8-week-old male Fischer 344 rats (Japan SLC, Inc.) under isoflurane anesthesia. The sciatic nerves of Fischer rats were completely transected at midthigh. The nerves were ligated on both ends and the proximal nerve stump was sutured into hip muscles to prevent reinnervation. Approximately 1 x 10^6 neurons contained within 10 μl of medium were injected into the distal stumps of the tibial and peroneal nerves after 1 week of nerve transection using a Hamilton syringe. The injection site was 20 mm proximal to those entries into the lateral gastrocnemius and anterior tibialis muscles. Surgical controls underwent the same surgeries but had only medium injected into the peripheral nerves.

2.2.3. Reproduction of ankle motions

The stainless steel wire electrodes were placed on the peroneal and tibial nerves of the neural transplantation group and those were covered with silicone gel for insulation and immobilization. The other ends of wires were passed through the dorsal neck skin and connected to the generator (Neuropack MEB-5504; Nihon Kohden, Tokyo, Japan). Using video technique, ankle angles of the neural transplantation group were measured with electrical stimulation of the peroneal nerves at 60 Hz, 0.4 mA. The ankle angle is measured from the line connecting the knee and ankle joints, and the line connecting the ankle joint and the metatarsal head [15].

2.2.4. Tissue analysis

After the measurement of ankle angles, the rats were perfused through the left ventricle with 4% paraformaldehyde. The tibial nerves, gastrocnemius and anterior tibialis muscles were removed for immunohistochemical and histochemical analysis. The tibial nerve distal to the transplant was then fixed in 4% glutaraldehyde in phosphate buffer. The nerves were embedded in Epon and 1-μm thick sections were stained with Toluidine blue (Sigma-Aldrich, Germany) for light microscopic examination. The total number of myelinated axons was measured. The gastrocnemius muscles were dissected free from the origin and insertion and weighed immediately. Then the gastrocnemius muscles were embedded in paraffin, sectioned into 10-μm cross sections and stained with hematoxylin and eosin. The mean muscle fiber area was measured at the middle of the lateral gastrocnemius muscle belly.

Neuron survival was examined in the cell transplantation group. The transplant sites of tibial nerves were cryoprotected in sucrose, and then frozen in dry-ice-cooled isopentane. The tibial nerve sections of 10-μm thichness were stained with the antibodies against choline acetyltransferase (ChAT, 1:50; Millipore, Billerica, MA, USA) in order to assess the survival of motoneurons and those axons in the tibial nerve. The anterior tibialis muscle were sectioned at 50μm to assess neuromuscular junctions using anti-neurofilament H antibody (1:500; Millipore, Billerica, MA) and α-bungarotoxin (1:200, Molecular Probes, Eugene, OR) which reveals terminal muscle innervation by staining motor end plates. Sections were stained using a full automatic Immunohistochemical staining system: Ventana HX system (Ventana, Yokohama, Japan) according to the manufacturer's instruction.

2.2.5. Statistics

For statistical analysis, Student t test or ANOVA with Tukey post hoc comparisons was used, as appropriate. All statistical analyses were conducted using the Statistical Package for Social Science version 19.0 software (SPSS Inc, Chicago, IL, USA). The significance level was set at 0.05. All data are expressed as mean ± standard deviation.

2.3. Result

2.3.1. Motoneuron survival and neuromuscular junction formations

Transplanted motoneuron survival was observed up to 12 weeks in all of the cell-transplanted tibial nerves. Neuromuscular junction formations were also present in the lateral gastrocnemius muscle of the rats that received cell transplants (Figure 1).

2.3.2. Axon regeneration

No myelinated axons were present in the surgical control group. The mean number of myelinated axons in the tibial nerves of animals that received cell transplantion was less than half of uninjured animals (1198 ± 1218 versus 2510 ± 802, p=0.079) (Figure 2).

2.3.3. Analysis of muscle anatomy

No significant differences were found in muscle fiber cross-sectional area between transplantation and surgical control groups (326μm^2 ± 91 versus 176μm^2 ± 105, p=0.28). The naive group showed the largest gastrocnemius muscle fiber area (2873 ± 233μm^2, p<0.001)

2.3.4. Reproduction of ankle motions

The reinnervated muscle of transplantation group could reproduce ankle dorsiflexion artificially through electrical stimulation. Video-based analysis revealed that the ankle angles in the neural transplantation group significantly decreased with electrical stimulation of the peroneal nerve as compared to surgical control (30.4 ± 7.7 versus 106.7 ± 9.7, p < 0.001) (Figure 3).

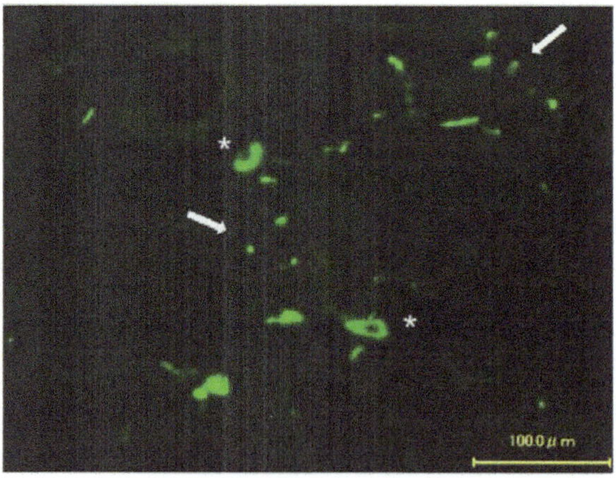

Figure 1. Twelve weeks after transplantation, motoneurons survived in the tibial nerve. The ChAT (choline acetyltransferase) positive axons and motoneurons (arrow) were observed in the tibial nerve.

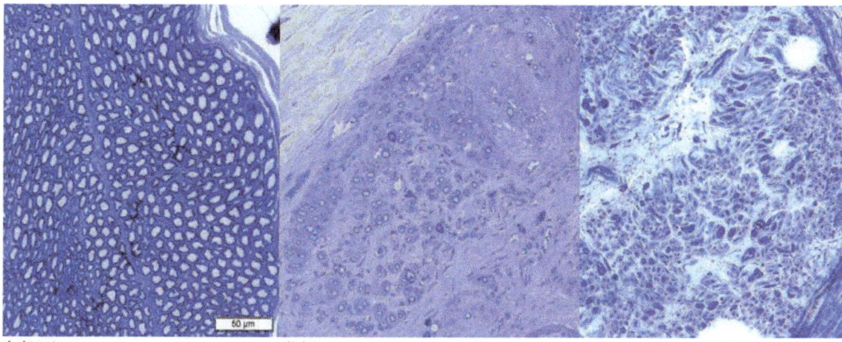

(a)Naive (b)Neuron transplantation (c)Surgical control

Figure 2. Myelinated axons in tibial nerve were assessed with toluidine blue stain.

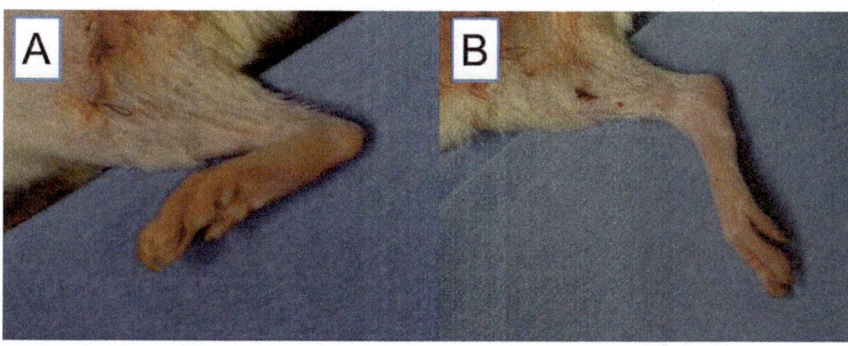

Figure 3. Controlled reproduction of ankle motion and measurements of dorsiflexed angles. The ankle motions were restored with electrical stimulation of the peroneal nerve in neural transplantation group (A) as compared to surgical control group (B).

2.4. Discussion

The transplanted rat embryonic motoneurons could survive in the rat peripheral nerve, and transplanted motoneurons formed functional motor units. Transplantation of motoneurons into peripheral nerve provided the reproduction of ankle dorsiflexion. The transplantation of embryonic neurons into peripheral nerve could restore ankle motions in conjunction with electrical stimulation, even though no neural connection between central nervous system and muscle were present.

Although all rats were able to flex their ankle, the small number of axons was present in the tibial nerves of cell transplantation group as compared with naive group. Endoneurial environment in peripheral nerve may not provide suitable condition for motoneurons. Transplanting motoneuron with glial cells in the central nerve systems, such as astrocytes, may promote the survival of motoneurons. Further studies are needed to clarify the necessity of central nervous system derived cells to improve survival rate of transplanted neurons in peripheral nerve.

No significant differences were found in muscle fiber area between transplantation and surgical control groups. And the muscle fiber area was much smaller for the cell transplantation group than for the naive group. Before neural connection between motoneurons and muscle was established, other treatment strategy had been needed to prevent muscle atrophy. After the neural connection, continuation of electrical stimulation and exercise beyond the period of this study may improve muscle fiber area and the quality of gait.

In clinical setting, several investigators have stated that performing reconstruction of denervated muscle with nerve transfer technique in peripheral nerve within 9 months notably improves functional outcome [16, 17]. Motoneurons can be placed very closely to neuromuscular junction using the neural transplantation into peripheral nerve. Target muscles can be reinnervated in short-term. Considering these facts, useful functional recovery of denervated muscle can be expected with this method even if the treatment was delayed until more than 6 months after Wallerian degeneration. Considering that generating neural cells from induced pluripotent stem cells (iPSCs) takes a few months, the wide windows of opportunity for the treatment can be advantage in propagating safe motoneurons derived from iPSCs made from the injured patient's own cells.

Another advantage of transplantation into peripheral nerve is that much less number of motoneurons will be required, compared with neural transplantation into the central nervous system. The iPSCs are likely to carry a higher risk of tumorigenicity than ES cells, due to the inappropriate reprogramming of these somatic cells, the activation of exogenous transcription factors, or other reasons [18, 19]. A larger number of transplanting cells are needed, the higher risk of tumorigenicity remains an obstacle for neural transplantation therapy. Presumably much less number of motoneurons are supposed to be required for neural transplantation into peripheral nerve, because the number of motor unit is usually about several hundred [20]. Considering these issues, transplantation of motoneurons into peripheral nerve can be an ideal target for regenerative therapy from a clinical viewpoint.

Transplantation of rat embryonic motoneurons into peripheral nerve provides reproduction of simple behaviors in conjunction with electrical stimulation. Recent breakthroughs in stem cell biology have raised possibilities of the clinical application of this treatment strategy. Our results lead us to believe that transplantation of iPS- or ES-derived motoneurons into peripheral nerve changes treatment strategy for the diminished voluntary muscle function.

3. Sensory feedback by electrical stimulation

3.1. Purpose

The purpose of this section is to show the possibility of sensory feedback in nueroprosthetic devices. If we have a way to send the signals from the assembled sensors to sensory nervous system, leading neuroprosthetic devices with two-way communication will be achievable. In this Section, we show a preliminary study on the possibility of sensory feedback via axial fibers of peripheral sensory neurons. We have tested several patterns of electrical signals and the placements of surface electrodes for carrying sensory information to brain via axial fibers of neurons. In a preliminary experiment with one channel electrical stimulation, the amplitude modulation of the stimulation voltage achieves only 2 bit resolution. Then, we increase the channels and take a method of frequency modulation of the stimulation electrical signal.

3.2. Methods

One pair of surface electrodes, one anode and one cathode, is usually used for one channel electrical stimulation. The two pairs of surface electrodes in our experiments are shown in Figure 4a. The placements of the electrodes were well controlled for each experiment to certify the repeatability based on the markers on skin shown in Figure 4b. We set the two pairs of electrodes (A, B) and (C, D), and made the electrical current flow from A to B or from C to D in each pair. The duty ratio of stimulation voltages was 5%, and the frequency was set at 2000 Hz for the channel (A, B). The duty ratio 5% was defined because we found it looks good for all five subjects to identify the stimulation frequency in several trials with one channel stimulation.

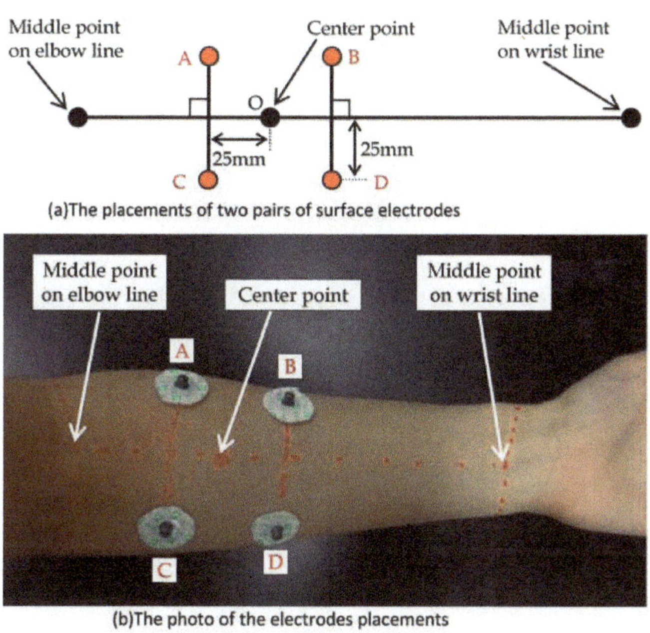

Figure 4. The electrodes placement

The applied voltages were tuned for each subject in similar way for the good identifiability. The electrical stimulation of surface electrodes with about 2000 Hz gives some feeling of vibration around at the electrodes without any pain. The stimulation frequency of the channel (C, D) was set at five frequencies: 2000 Hz, 2001 Hz, 2002 Hz, 2004 Hz and 2010 Hz. The change of stimulation frequency on the channel (C, D) causes different sensations to the subjects. More specifically, the subjects sense the frequency difference between the two channels in addition to the base frequency 2000 Hz. In other words, the subjects sense the interference potential of the pair electrical potentials. For an example, 1 Hz sensation will be felt by the subject if the frequencies of 2000 Hz and 2001 Hz are selected. We applied electrical stimulations of 2000 Hz to a subject by using the pair of the electrodes (A, B), and set the same and different frequencies for the pair of the electrodes (C, D) for the discrimination experiments. The phase shift was zero degree when the both frequency were set at 2000 Hz. After enough trials for each combination of the stimulation frequencies, each subject carried out 40 trials of the discrimination. In the trials, all combinations of frequencies were presented in random order. Five subjects were asked to identify the combinations of the electrical stimulations.

3.3. Results

The average rate of the correct answers over 40 tests per subject was 90%. All errors were occurred in the cases of nearest frequencies difference of the two pairs electrodes. This results show that the difference of the two channel stimulation frequencies is key point for the identifiability of stimulations. The bigger difference in stimulation frequencies, the higher rate of the correct answers will be achieved.

3.4. Discussion

The simple amplitude modulation of one channel electrical stimulation achieves only 2 bits resolution for transferring sensory information to brain. On the other hand, frequency modulation with interference potential of the pair electrical potentials has achieved a high resolution which corresponds to the number of combination of given different stimulation frequencies. For an example, we can obtain 3.5 bits (1 digit in decadal system) resolution if we use five different frequencies. We can increase number of the channels; thus, interference potential of the pair electrical signals with more than two pairs of electrodes holds forth the possibility that we can achieve a high resolution for sensory feedback via axial fibers of peripheral sensory neurons.

Further research on multi-channel electrodes, specially more than three channels, will be required to clarify the effect of multiple interference of stimulation frequencies on the sensory resolution of transferring the information. The frequency range which is most appropriate for transferring sensory information should be also clarified. It is common that electrical stimulations with a certain frequency band involve a pain of the subject. Appropriate frequency bands should be identified not only to achieve a high resolution, but to avoid subject's pain.

4. Restoration of walking by artificial peripheral nerve system — Simulation study with neuro-musuculo-skeletal model of rat

4.1. Purpose

One application of PNIs is to replace the part of neural networks for a particular movement by artificial electrical networks or artificial digital networks. In this section, we propose a method of PNIs for partially impaired neural networks. When we look at injured persons, it is common that some muscles do not work because the corresponding motor neurons are damaged but the muscles do not receive any damage themselves. In such cases, the muscles can work if the motor commands are provided.

In this section, we make the following assumptions as conditions for our simulation of injured rat's walking. First, motor commands to muscles for walking are generated by specified neural networks. The networks is called central pattern generator (CPG), which has been identified in spinal cords of mammalian. The upper central nervous system is assumed to give only the trigger of walking. Second, some sensory systems, which may be muscle spindles or another somatosensory organs, are damaged but we are able to obtain the necessary signals for constructing the motor commands by using some artificial sensors instead of the damaged sensory systems. Third, we are able to provide the constructed motor commands for the target muscles through reinnervation type electrodes, which are described in Section 2. Based on these assumptions, we give a walking simulation of rat to show the possibility of practical usages of PNIs. Simulations of human bipedal walking have been successfully proposed [21, 22]; therefore, we may restore the walking motion for patients with spiral cord injury if our proposed method works well for rat walking.

An example of the experimental setup corresponding to the simulation is given in Figure 5. The camera system including image processing is to obtain the necessary signals such as joint angles, joint angular velocities, and other signals for the positions and orientations of body

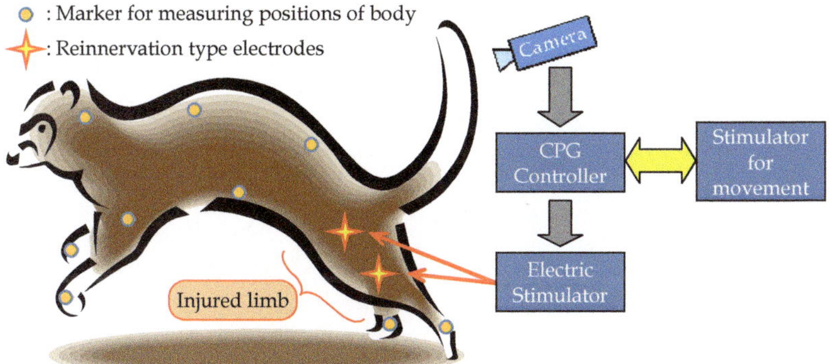

Figure 5. Proposed experimental setup for reconstructing walking motion.

segments. It is possible to extend this type of neuroprosthetic system with portable sensors and other electrical devices to more practical one for disabled persons.

The purpose of this section is to conduct walking simulation which will play an important role of our succedent neuroprosthetic system. The basic concept of the neuroprosthetic system is shown in Figure 5. The role is to check whether the designed artificial CPG controller works adequately or not, and whether the sensing system is able to provide necessary information for the controller's adapting to the environmental changes. Thus, the walking simulation given in this section means the preliminary study of motion planning and control for our succedent neuroprosthetic system.

4.2. Neuro-musuculo-skeletal model

The neuro-musculo-skeletal model of rat is composed of a rigid link system, joint torque generators, neural controllers of CPGs with the upper level triggers. The inertial properties of the entire body are represented by a three dimensional 17-rigid-link system with 16 joints. All the joints are one degree-of-freedom, and the axes of the joints are perpendicular to sagittal plane. The schematic views in sagittal and horizontal planes are given in Figure 6. The dappled circles indicate the center of mass of each segment.

A viscoelastic passive moment induced by soft tissues and a passive moment of nonlinear elasticity induced by tendons and bones assumed to act on each joint. The characteristic is given by

$$passive_i\left(\theta_i,\dot{\theta}_i\right)=k_{i1}^J\exp\left[k_{i2}^J\left\{k_{i3}^J-\left(\theta_i-\bar{\theta}_i\right)\right\}\right]+k_{i4}^J\exp\left[k_{i5}^J\left\{\left(\theta_i-\bar{\theta}_i\right)-k_{i6}^J\right\}\right]+c_i^J\,\dot{\theta}_i \tag{1}$$

where θ_i is the i-th joint angle variable, $\dot{\theta}_i$ is the i-th joint angular velocity variable, and another parameters are constants representing the joint characteristics. These parameters were determined from the estimations for human joints [23]. The inertia properties of each body segment, such as the mass and the moment of inertia, were estimated from the size of anatomized each body segment with the average density of rat body [24].

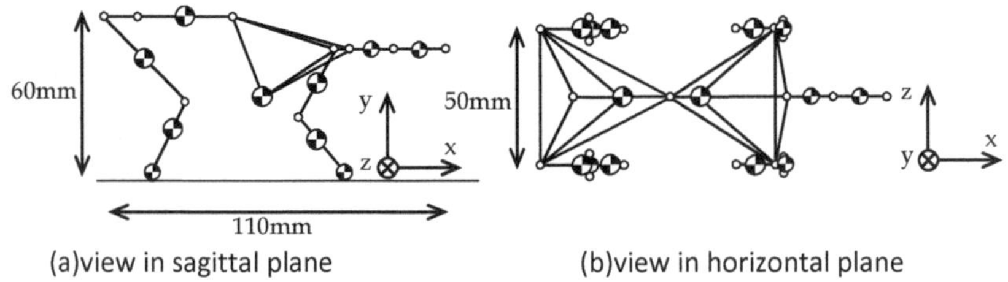

Figure 6. Assumed skeletal system of rat.

The functions of the neural control system are divided into two types. The first function is a rhythm pattern generator corresponding to the spinal cord level. This subsystem composes CPGs which work as command generators for the joints. The command generation of each CPG starts by receiving a stimulus from the higher center. Each unit of CPG is a second-order nonlinear dynamical system with the inputs from the higher center and the sensory systems. During movement of the link system, somatic sensory signals are fed back to the CPGs so that the musculo-skeletal system is able to interact with the environment. This is the second function of the neural control system.

The neural control system of CPG is a nonlinear feedback controller :

$$\frac{1}{T_r}\dot{x}_i + x_i = -\sum_{j=1}^{n} a_{ij} y_j - b z_i + u_i + s_i \tag{2}$$

$$\frac{1}{T_a}\dot{z}_i + z_i = y_i \tag{3}$$

$$y_i = \max(0, x_i) \tag{4}$$

Here, i and j are the oscillator numbers that are coupled in the antagonistic relation, x_i is the membrane potential of the neuron, T_r, T_d and b are the coefficients corresponding to aging variation, y_i is the firing rate of the neuron, a_{ij} is the weight coefficient, z_i is a variable corresponding to the adaptation level, s_i is the stimulation input from the brain, and u_i is the feedback signal from the somatic sensation. The oscillator neurons represented by equations (2), (3) and (4) are coupled to one another in a depressive manner, and the entire walking behavior undergoes an adaptation process by causing an oscillation in each joint movement [25]. Movement of a joint is coupled with that of other joints. In this simulation, only the four joints at shoulders and hips were assumed to generate actively the torques. One pair of the oscillator unit is attached at one of the four joints. The extra-connections between CPGs of left and right shoulders, and also between CPGs of left and right hips are assumed for coordinating the motions of left and right limbs. These extra-connections cause mutually inhibitions between left and right CPGs and then generate alternate motions between the left and right limbs. The structure of CPGs is given in Figure 7.

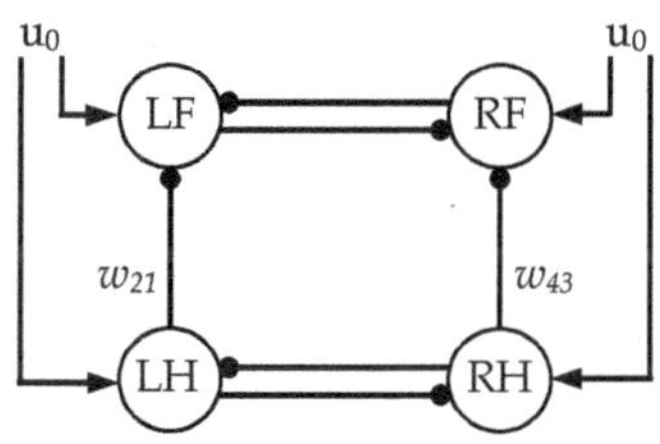

Figure 7. Structure of CPGs.

The joint torque is determined as follow:

$$\tau = p_F y_F - p_E y_E \tag{5}$$

where y_F and y_E are the outputs of one pair of CPGs in which one CPG generates flexion torque and the other CPG generates extension torque at each active joint. For another joints except these active joints, the rotational motions will be induced according to the passive characteristics defined by (1).

The interaction between the foot and the ground was modelled as a combination of springs and dampers. The ground reaction forces produced by the springs and dampers were assumed to act on four points of each foot: two in the heel and two in the toe.

In this simulation, we did not take the muscle model into considerations. However, the generated muscle forces corresponding to the generated torques by this neural controller can be estimated with an appropriate procedure by the method in [23].

We searched for suitable parameters of the CPG-based controller such as T_r, T_d, a_{ij}, b, etc. in each joint to achieve a realistic walking pattern, using a performance index that is a weighted linear combination of energy consumption per unit walking distance, muscle fatigue index and foot clearance.

4.3. Simulation results

Figure 8 shows the walking simulation results in sagittal plane which are calculated by 0.1 second. Mammalians with four limbs are usually able to move in some patterns, such as walk, trot, and gallop. Each pattern is appeared around the corresponding optimal velocity. In mammalians' moving, "optimal" means energy efficient way in gate patterns for getting unit moving distance. From Figure 8, the pattern of "walk" is observed and the moving velocity is matched approximately to the optimal velocity of rat walking. Figure 9 shows that the outputs of CPGs successfully generated rhythmic patterns as functions of time and indicated appropriate phase shifts each other for coordinated motions in four limbs. The corresponding generated torques are given in Figure 10. The similar patterns as the CPG outputs are observed. We can define the period of walking from Figure 10, and it is 0.2 second. The phase shift is -180 degree from the left fore-limb to the right fore-limb, and also -180 degree from the left fore-limb to the left hind-limb. On the other hand, the phase shift between the left fore-limb and hind-limb is -97 degree and in the right side it is the same. These phase shifts match to the pattern of "walk" [26]. Figure 10 shows the active joints angles controlled by the CPG controllers. It seems that the difference between left and right angels both in fore-limbs and hind-limbs comes from the effect of initial conditions on the joint angels and joint angular velocities.

4.4. Discussion

We observed a large rolling motion of the trunk in the walking simulation. This can be explained by the fact that no joint of rolling motion is introduced in the skeletal model. At this stage, we do not have any knowledge about rolling joints of rat. The placements, passive characteristics such as range of joint angle and viscoelastic property should be clarified before the rolling joints are introduced in the skeletal model. The proper introduction such rolling joints reduces the magnitude of trunk rolling motion and may improve the simulation quality in comparison with the actual walking motion of rat.

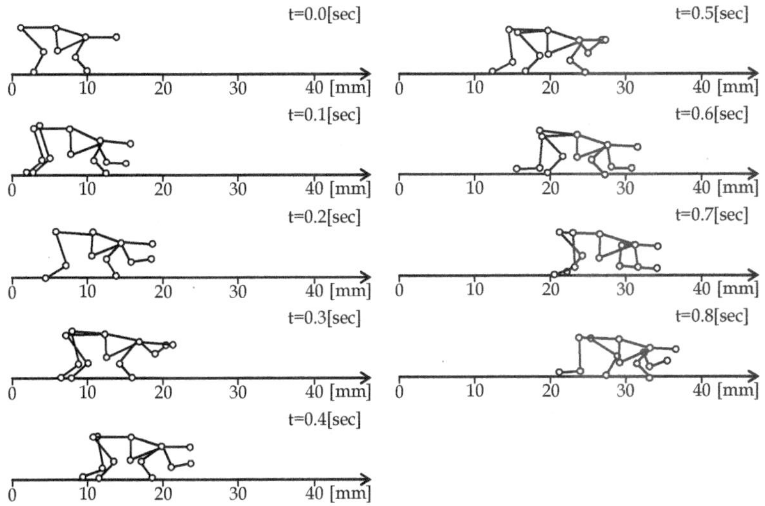

Figure 8. Walking simulation results.

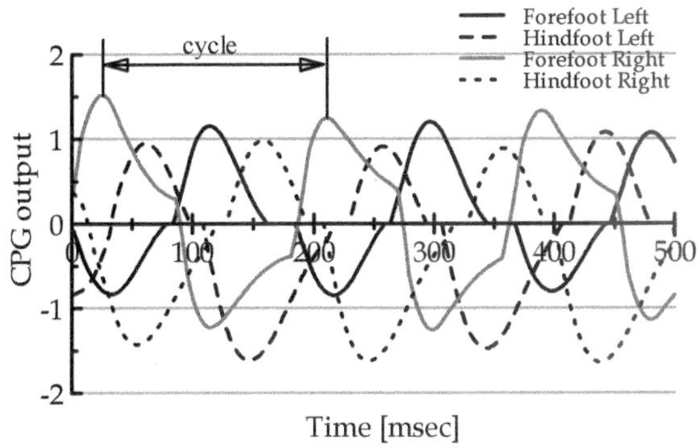

Figure 9. CPGs' outputs as commands for joint torques in steady state.

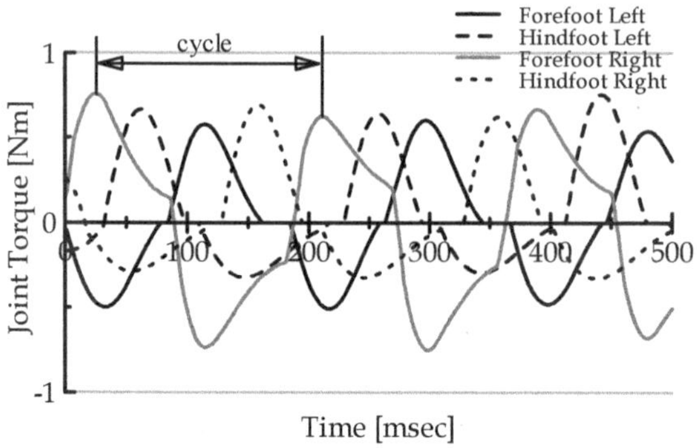

Figure 10. Generated joint torques.

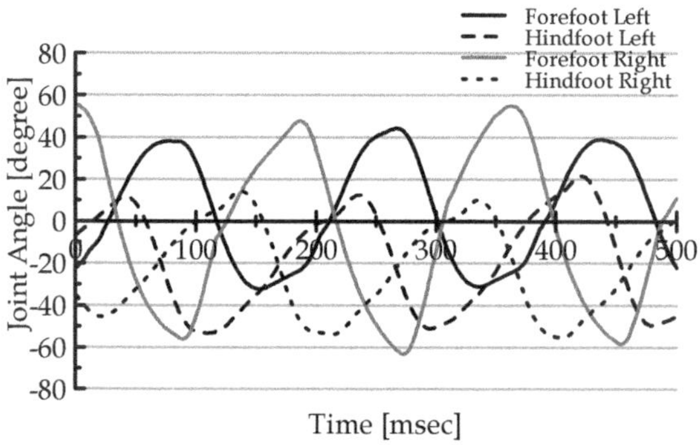

Figure 11. Joint angles at shoulders and hips.

We extended the model to the case of more active joints controlled by the neural controllers. Other four active joints were introduced at elbows in fore-limbs and at knees in hind-limbs. The simulation results showed that the generated motion was a pattern of "gallop". From this simulation, it will be required that we consider the method for distinguish patterns of moving when we search the parameters in the neural controllers.

Although there exist several problems to be solved, the simulation results presented here indicate the possibility of functional restoration for the cases of partially impaired neural networks. We will be able to use this simulation technology for checking whether the designed artificial CPG controller works adequately or not, and for providing necessary information to the controllers adapting to the environmental changes. The usage of this kind of online simulator is also illustrated in Figure 5.

5. Conclusion

In this chapter, we presented three technologies for achieving PNIs. The PNIs will be designed for patients of partially impaired neural networks. In Section 2, we show that restoration of motoneurons is possible by using reinnervation type electrodes which are constructed by implanting embryonic neurons into peripheral motor units. Several methods for restoring motoneuron functions have been proposed during these three decades because the muscles can generate the forces if some electric potentials are provided at the motor points. These electric potentials may be provided by artificial ones with some electrodes. Functional electrical stimulation (FES) has been applied to human patients with spinal cord injury and has shown higher potential for restoring functional movements [27]. Usually in FES, some metal wires are used for transmitting the electrical signals as the motor commands for the muscles. However, the mechanical properties of the metal wires such as elasticity are different from biological tissues: then, the mismatching between the wires and the tissues will appear and result in malfunctions of the FES systems. If we use reinnervation type electrodes constructed from embryonic neurons, which are almost same as the original motor neurons, the demanding problem of mechanical property mismatching with metal wire transmission lines will be solved. This is the reason why we seek to establish the technology around reinnervation type electrodes.

Sensory feedback in control of functional movements plays an important role for the robustness and the adaptation to environmental changes in the task executions. A few studies have been carried out to introduce sensory feedback for the restoration of motor functions [28]. We can improve greatly the quality of functional restoration by sensory feedback for the cases of partially impaired neural networks. In Section 3, we showed the possibility of sensory feedback via axial fibers of neurons in nueroprosthetic devices. A method for passing the feedback signals via axial fibers of neurons with surface electrodes has been tested. It is shown that three or four bits resolution for sensory feedback can be achieved with the proposed method.

In Section 4, we showed the possibility of practical usages of PNIs by taking walking simulation of rat. In the concept of restoring motor functions, controlling joint torques with sensory feedback is a key point of the technology. The reinnervation type electrodes described in Section 2 and the sensory feedback via axial fibers of neurons described in Section 3 will be included in such a new type of nueroprosthetic device.

Author details

Goro Obinata[1], Hitoshi Hirata[2], Chikara Nagai[3], Shigeru Kurimoto[2], Shuichi Kato[2] and Tomonori Nakano[2]

1 EcoTopia Science Institute, Nagoya University, Nagoya, Japan

2 Graduate School of Medicine, Nagoya University, Nagoya, Japan

3 Graduate School of Engineering, Nagoya University, Nagoya, Japan

References

[1] Wolpaw J. R, Birbaumer N, McFarland D. J, Pfurtscheller G, Vaughan T. M. Brain-computer interfaces for communication and control. Clinical Neurophysiology 2002;113(6) 767-791.

[2] Patil P. G, Turner D. A. The development of brain-machine interface neuroprosthetic devices. The Journal of the American Society for Experimental NeuroTherapeutics 2008;5 137-146.

[3] Patterson P. E, Katz J. A. Design and evaluation of a sensory feedback system that provides grasping pressure in a myoelectric hand. Journal of Rehabilitation Research and Development 1992;29(1) 1-8.

[4] Scott R. N, Brittain R. H, Caldwell R. R, Cameron A. B, Dunfield V. A. Sensory-feedback system compatible with myoelectric control. Medical and Biological Engineering and Computing 1980;18(1) 65-69.

[5] Trounson A, Thakar R.G, Lomax G, Gibbons D. Clinical trials for stem cell therapies. BMC medicine 2011;9 52.

[6] Thomson J.A, Itskovitz-Eldor J, Shapiro S.S, Waknitz M.A, Swiergiel J.J, Marshall V.S, Jones J.M. Embryonic stem cell lines derived from human blastocysts. Science 1998;282 1145-1147.

[7] Alper J. Geron gets green light for human trial of ES cell-derived product. Nature biotechnology 2009;27 213-214.

[8] Schwartz S.D, Hubschman J.P, Heilwell G, Franco-Cardenas V, Pan C. K, Ostrick R. M, Mickunas E, Gay R, Klimanskaya I, Lanza R. Embryonic stem cell trials for macular degeneration: a preliminary report. Lancet 2012;379(9817) 713-720.

[9] Okano H. Stem cell biology of the central nervous system. Journal of Neuroscience Research 2002;69 698-707.

[10] Erb D. E, Mora R. J, Bunge R. P. Reinnervation of adult rat gastrocnemius muscle by embryonic motoneurons transplanted into the axotomized tibial nerve. Experimental Neurology1993;124 372-376.

[11] Grumbles R. M, Casella G. T, Rudinsky M. J, Godfrey S, Wood P. M, Thomas C. K. The immunophilin ligand FK506, but not the P38 kinase inhibitor SB203580, improves function of adult rat muscle reinnervated from transplants of embryonic neurons. Neuroscience 2005;130 619-630.

[12] Grumbles R. M, Sesodia S, Wood P. M, Thomas C. K. Neurotrophic factors improve motoneuron survival and function of muscle reinnervated by embryonic neurons. Journal of Neuropathology & Experimental Neurology 2009;68 736-746.

[13] Grumbles R. M, Wood P, Rudinsky M, Gomez A. M, Thomas C. K. Muscle reinnervation with delayed or immediate transplant of embryonic ventral spinal cord cells into adult rat peripheral nerve. Cell Transplantation 2002;11 241-250.

[14] Merrell G. A, Barrie K. A, Katz D. L, Wolfe S. W. Results of nerve transfer techniques for restoration of shoulder and elbow function in the context of a meta-analysis of the English literature. Journal of Hand Surgery American Volume 2011;26 303-314.

[15] Yu P, Matloub H.S, Sanger J.R, Narini P. Gait analysis in rats with peripheral nerve injury. Muscle & Nerve 2001;24 231-239.

[16] Millesi H. Surgical management of brachial plexus injuries. Journal of Hand Surgery American Volume 1977;2 367-378.

[17] Nagano A, Tsuyama N, Ochiai N, Hara T, Takahashi M. Direct nerve crossing with the intercostal nerve to treat avulsion injuries of the brachial plexus. Journal of Hand Surgery American Volume 1989;14 980-985.

[18] Miura K, Okada Y, Aoi T, Okada A, Takahashi K, Okita K, Nakagawa M, Koyanagi M, Tanabe K, Ohnuki M, Ogawa D, Ikeda E, Okano H, Yamanaka S. Variation in the safety of induced pluripotent stem cell lines. Nature biotechnology 2009;27 743-745.

[19] Yamanaka S. Strategies and new developments in the generation of patient-specific pluripotent stem cells. Cell stem cell 2007;1 39-49.

[20] Bromberg M. B. Updating motor unit number estimation (MUNE). Clinical neurophysiology : official journal of the International Federation of Clinical Neurophysiology 2007;118 1-8.

[21] Taga G, Yamaguchi Y, Shimizu H. Self-organized control of bipedal locomotion by neural oscillators in unpredictable environment. Biological Cybernetics 1991;65(3) 147-159.

[22] Hase K, Yamazaki N. Computer simulation study of human locomotion with a three-dimentional entire-body neuro-musculo-skeletal model. JSME International Journal. Series C 2002;45(4) 1040-1050.

[23] Hase K, Yamazaki N. Development of three-dimensional musculoskeletal model for various motion analyses. Transactions of the Japan Society of Mechanical Engineers. C 1995;61(591) 4417-4412.

[24] Hayakawa T, Iwaki T. A color atlas of sectional anatomy of the rat. Tokyo: Adthree; 2008.

[25] Matsuoka K. Sustained oscillations generated by mutually inhibiting neurons with adaptation. Biological Cybernetics 1985;52(6) 367-376.

[26] Nanua P, Waldron K. J. Energy comparison between trot, bound, and gallop using a simple model. Journal of Biomechanical Engineering 1995;117(4) 466-473.

[27] Shimada Y, Sato K, Kagaya H, et al. Closed-loop control for functional electrical stimulation with percutaneous electrodes in paraplegics. In: proceedings of the 5th Vienna International Workshop on FES, 17-19 August 1995, Vienna, Austria. Austrian Society for Artificial Organs, Biomaterials and Medical Replacement Devices; 1995.

[28] Hoffer J. A, Baru M, Bedard S, Calderon E, Desmoulin G, Dhawan P, Jenne G, Kerr J, Whittaker M, Zwimpfer T. J. Initial results with fully implanted NeurostepTM FES system for foot drop. In: proceedings of 10th Annual Conference of the International FES Society, Montreal, Canada: 2005.

Chapter 6

Synthesis of Nanomaterials by Solution Plasma Processing

Osamu Takai, Maria Antoaneta Bratescu,
Tomonaga Ueno and Nagahiro Saito

1. Introduction

Recently, plasmas in the liquid phase have attracted a great attention because of its applications to industrial materials processing [1]. In particular, glow discharge in the liquid phase ("solution plasma" - SP) is a useful tool for the synthesis of nanomaterials. We named the synthesis process using plasma in liquids as Solution Plasma Processing (SPP).

Plasma in water has been produced in 1899 by Dr. Wilsing between different types of metal electrodes, in order to study the spectral properties in connection with the astronomical observations [2]. The earliest works on discharge in liquids have studied the arc and spark discharge in water and salt solutions [3].

The detailed structure of the solution plasma is still unclear at present. Currently, SPs are generated by nanoseconds pulsed or dc voltages. Also, ac excitations with frequencies raging from 50 Hz up to MHz were used [4-7].

In our research group, we focused on the fundamental aspects of the solution plasma diagnostics related with the synthesis of novel nanomaterisls processed by SP. Figure 1 shows a model of the solution plasma. The characteristic regions of SP are the plasma gas phase, the liquid phase, the interface between plasma gas and liquid, and the interface between electrode surface and gas plasma. The emission center of plasma is located in the gas phase which is surrounded by the liquid phase. Near the gas/liquid interface an ion sheath is formed. The plasma is confined by a condensed phase, which produces unique features of the solution plasma. This solution plasma provides extremely rapid reactions using activated chemical species and radicals under high pressure [1].

SPP is a new useful and simple method for the metal nanoparticles (NPs) synthesis since this non-equilibrium plasma can provide extremely rapid reactions due to the reactive chemical species, radicals, and UV radiation produced in an atmospheric pressure plasma [1]. The most important merits of the SPP for the NPs synthesis, as compared with chemical methods, consist in the short processing time (in the range from few minutes to several tens of minutes), preparation in room temperature and pressure conditions, and low energy of plasma. The novelty of the SPP method used in our laboratory consists in the fact that plasma operates in glow discharge limits, offering a suitable medium to control the chemical reactions inside the solutions [10, 11]. This is possible because plasma offers a new reaction medium, where hydrogen, hydroxyl, and oxygen radicals are produced, where the hydrogen radical is the most responsible for the reduction reaction of the gold ion to the neutral atom, and therefore a reduction agent is not necessary. The SPP method seems well-suited for the NPs synthesis offering the possibility to control the size by controlling the surrounding chemistry of the gold NPs, adding thus another level of utility of this procedure to material science [12, 13].

SPP was successfully used for loading metal NPs on carbon materials to prepare composite structures which can improve the catalytic activity of fuel cells [14].

Many other applications of SPP in nanomaterial technology have been performed in our group related with template removal in mesoporous silica synthesis process [15], decomposition of organic dyes or compounds [16, 17], surface modification of metals [19], and sterilization of water [1].

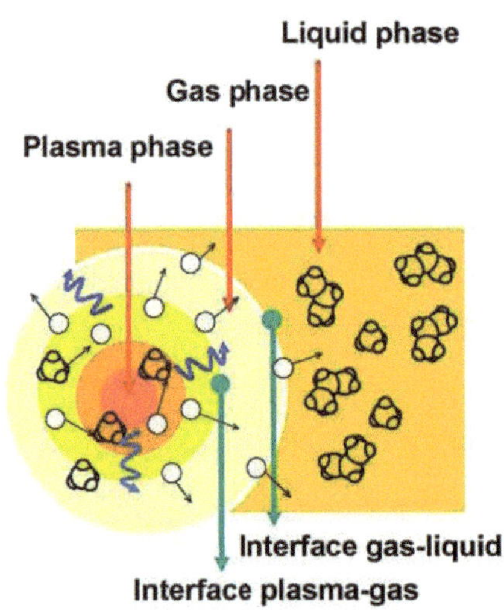

Figure 1. Model of Solution Plasma. The main regions are: plasma gas phase, gas phase, liquid phase and the interfaces between plasma and gas, and gas and liquid.

In the followings we will discuss about the conditions for SP generation [16, 19-21], the role of radicals, ions and surrounding chemistry in NPs synthesis [11, 12, 17, 22], and synthesis of nanomaterials with enhanced catalytic activity [1, 14].

2. Conditions for SP generation

A typical experimental set-up for the generation of SP, with time-dependent electrical and optical diagnostics of plasma and synthesis of nanomaterials is shown schematically in Figure 2. A typical power supply for SP generation has the following characteristics: bipolar pulse type with a maximum voltage and current of 5 kV and 10 A, respectively, a variable repetition frequency in the range from 5 to 60 kHz, and a variable pulse width from 500 ns to 3 μs [10, 16, 18-20].

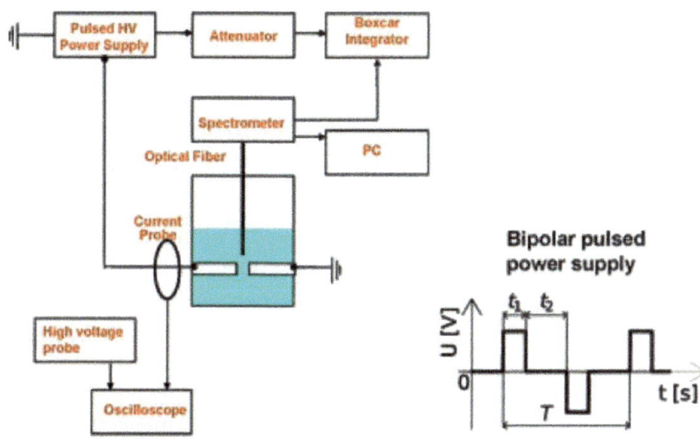

Figure 2. Typical experimental set-up for SP generation, with time-dependent electrical and optical diagnostics of plasma, and synthesis of nanomaterials.

We investigated the dependence of SP characteristics on the inter-electrode distance. In SP, the formation of different radicals, the excited atoms and molecules are strongly influenced by the geometry and the input electrical power in the system. Figure 3 shows various regions of SP depending on the inter-electrode distance and the applied pulsed high voltage. Typical regions are the glow discharge regime, when the inter-electrode space is less than 2 mm and the peak voltage is more than 2 kV, the corona discharge regime, when the inter-electrode distance increases and the peak high voltage is also high, and the pre-breakdown regime where electrochemical reactions dominate [20].

In nanomaterials processing, an important factor consists in controlling of the solution plasma stability. The value of pH and the conductivity of the solution also determine the operation regime of SP. In the diagram from Figure 4 various conditions of plasma determined by solution conductivity, bipolar pulse width and frequency, generate a glow or an arc discharge. If the solution conductivity is high, more than 1 mS/cm, the ionic current through the liquid is high, and at the same input electrical power, plasma is instable, if the pulse width is smaller than 2 μs. If the pulse width increases, the input electrical power increases and even the solution conductivity is high, a stable glow discharge can be obtained.

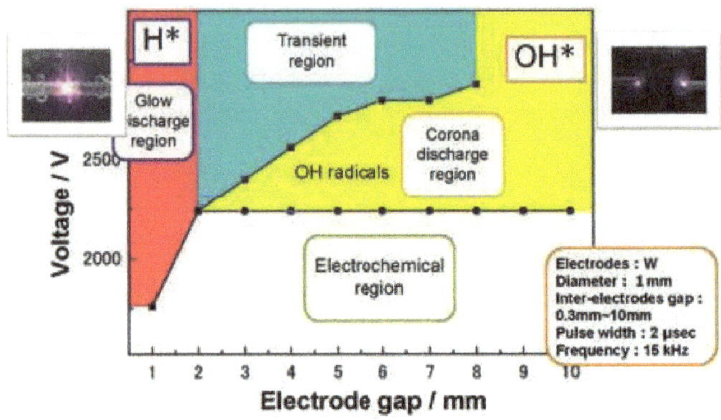

Figure 3. Dependence of SP characteristics on the inter-electrode distance and the applied peak high voltage.

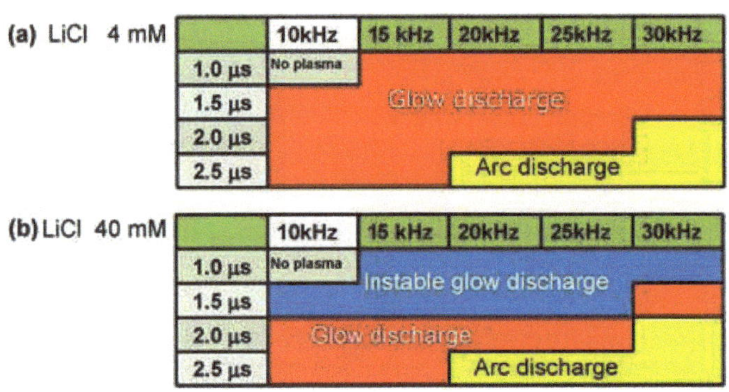

Figure 4. Influence of solution conductivity and input electrical parameters of the pulsed power supply on SP stability. Different operation regimes of SP: glow discharge and arc discharge. (a) Stability conditions of SPP in a solution of LiCl with 4 mM concentration. (b) Stability conditions of SPP in a solution of LiCl with 40 mM concentration.

Independently of the solution conductivity, if the pulse width is higher than 2.5 μs and the pulsed power supply frequency is 30 kHz, the plasma generated in solution switches fast to arc discharge regime.

The optical emission spectra strongly depend on SP regime working. Corona discharge is characterized by a strong emission of the OH radical as compared with the glow discharge regime, when the inter-electrode distance is around 1 mm [20].

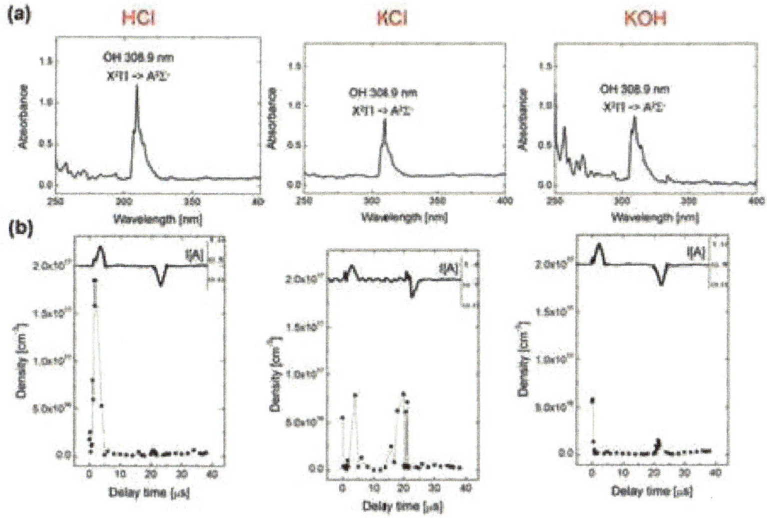

Figure 5. (a) Optical absorption spectra of the OH radical measured in HCl, KCl, and KOH solutions, on the positive applied voltage, under the same experimental conditions (discharge voltage of 700 V, frequency of 25 kHz, and pulse width of 4 μs). (b) The time evolution of the OH radical density measured in HCl, KCl, and KOH solutions, under the same experimental conditions above mentioned. Lines are used as guides for the eyes.

The optical absorption spectra of the OH radical measured in different solution with HCl, KCl, and KOH, under a positive value of the high voltage, and the corresponding time-dependent signals of the emission line of the OH radical, are represented in Figure 5 (a) and (b), respectively. The OH radical number density measured by broad band absorption spectroscopy was highest for the HCl solution plasma ($2x10^{17}$ cm^{-3}) when positive voltage pulses were applied to the electrodes. KOH is highly basic and can be an important source of hydroxyl radicals, but in this experiment the density was the lowest for this type of solution ($\sim 5x10^{16}$ cm^{-3}) [24].

The main chemical reactions responsible for the generation of the reactive species in SPP are [4, 5]:

$$\begin{gathered} H_2O \rightarrow \cdot H + \cdot OH \\ 2H_2O \rightarrow O_2 + 2H_2 \\ O_2 \rightarrow 2\cdot O \\ O_2 \rightarrow 2\cdot H \\ H + e \rightarrow H^* + e \\ O + e \rightarrow O^* + e \end{gathered} \tag{1}$$

3. Role of radicals, ions and surrounding chemistry in nanomaterials synthesis by SPP

In metal NPs synthesis, SPP is used without any chemical reduction agent, since hydrogen radical produced in plasma gas phase is transferred in liquid phase where the metal ion is reduced to the neutral form. The role of hydrogen radical in the reduction process of Au ions from HAuCl4 to Au neutral atoms, which nucleate to produce gold NPs, is illustrated in Figure 6.

In this experiment, solutions with different ion concentrations and conductivities were used to synthesize gold NPs. From the optical emission spectra, we can observe that in a solution with high concentration of ions (40 mM LiCl), the relative number density of hydrogen radical was smaller than in the solution with 4 mM LiCl concentration (Figure 6(a)). Therefore in these conditions, a smaller number of Au ions were reduced to the neutral form and the relative number density of gold NPs was less, as can be observed from the UV-vis spectra (Figure 6 (b)) and photos (Figure 6 (c)) [10].

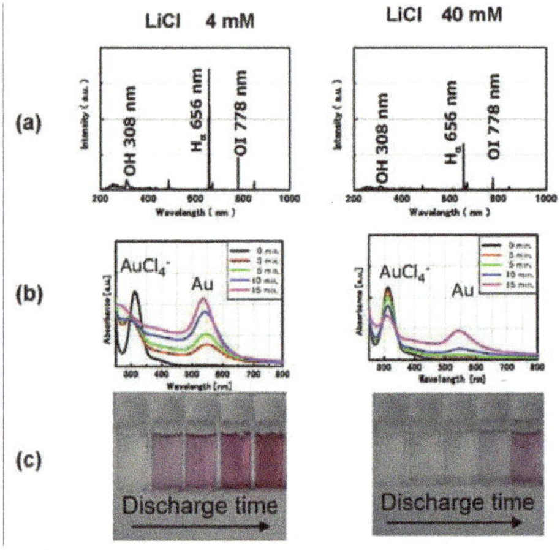

Figure 6. Role of hydrogen radical in the reduction reaction of Au ion to neutral form, in the synthesis process of gold NPs. (a) Optical emission spectra of SP generated in LiCl solution with different molar concentrations. (b) Time evolution of the UV-vis spectra of solutions resulted after SPP, showing gold NPs formation. (c) Photos of the solutions containing gold NPs.

Plasma has the role to provide the reactive species, especially the H radicals necessarily for the reduction process. In order to confirm more the role of H radical, we attempted to synthesize the gold nanoparticles from the solution containing PBN (n-tert-buthyl-α-phenylnitrone), which works as a spin trap agent.

When PBN reacts with H and OH, PBN-H and PBN-OH adducts are produced, respectively. In an ESR measurement, these adducts can be detected. During SPP, the PBN-H adduct was

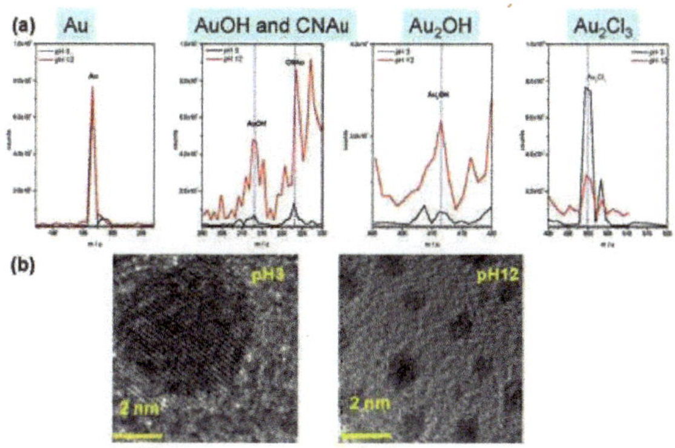

Figure 7. Dependence of chemical surrounding on Au NPs prepared in solutions with different pH values. (a) ToF-SIMS mass spectra of the negative ions fragments Au, AuOH and CNAu, Au_2OH, and Au_2Cl_3. (b) HRTEM images of Au NPs synthesized in solutions with pH 3 and 12.

more detected than PBN-OH adduct. This indicated that H radicals are supplied from plasma to the solution. Moreover, the formation of gold nanoparticles was not observed in the solution containing PBN since the H radicals produced in the solution which represent the reducing agent for the gold ions, were trapped by PBN molecules [11].

Size-controlled gold NPs have been synthesized using SPP. The gold NPs exhibit sizes from 1-2 nm to 10 nm when the solution pH was adjusted in the range from 12 to 3, respectively. The chemical environment surrounding the gold NPs depends on the preparation conditions and determines the electrostatic interaction among the nanoparticles, which alters their final size. Information obtained from XPS analysis, ToF-SIMS mass spectra, and UV-visible absorption spectroscopy were consistent and demonstrate that the gold NPs are partially oxidized on the surface, when synthesized in a pH 12 solution, and remain surrounded by gold chloride compounds when synthesized in a pH 3 solution. Plasma diagnostics shows that a high electron density contributes to generate a larger number of hydrogen radicals, which represent the main component in the reduction process of the gold ion into the neutral form [18].

Figure 7 shows the dependence of chemical surrounding on Au NPs prepared in solutions with different pH values. The ToF-SIMS mass spectra of the negative ions fragments were Au, AuOH and CNAu, Au_2OH, and Au_2Cl_3. The chemical surrounding of gold NPs synthesized in pH 3 solution is mainly formed by AuCl species, while the chemical surrounding of gold NPs produced in pH 12 solution is composed by oxidized form of AuO. Figure 7 (b) displays the HRTEM images of Au NPs synthesized in solutions with pH 3 and 12. We can clearly observe that small size gold NPs were synthesized in solution with pH 12.

The main role of SPP during the synthesis of the gold NPs consists in the production of the H radicals in the plasma gas phase which are necessary to reduce the gold ion Au^{3+} to atomic Au^0, in the liquid phase.

Before plasma, in solution, the hydrolysis of $HAuCl_4$ occurs:

$$AuCl_4^- + jOH^- \leftrightarrow AuCl_{4-j}(OH)_j^- + jCl^- \quad (2)$$

where 0 < j < 4 and the replacement of Cl^- by OH^- depends on solution pH, as was found in the UV-visible results.

In the liquid phase of plasma, the reduction of gold ions occurs in different ways:

$$AuCl_4^- + 3H\cdot \rightarrow 3HCl + Au^0 + Cl^- \quad (3)$$

$$Au(OH)_4^- + 3H\cdot \rightarrow 3H_2O + Au^0 + OH^- \quad (4)$$

where the hydrogen radicals (H•) are provided by the gas phase plasma and the reaction (3) and (4) takes place in the pH 3 and pH 12 solution, respectively. The pH 6 solution is an intermediate case to the other two solutions, where the hydroxo-complexes as $AuCl_3(OH)$- and $AuCl_2(OH)_2$- represent a molar fraction of about 0.6 in the solution [17].

By changing the solution pH in the preparation condition, the size of the gold NPs can be controlled. The connection between the gold NPs size and the solution pH can be understood by: (i) the effect of the redox standard potential in the reactions (3) and (4), (ii) the electrostatic repulsion force between AuO^- ions, and (iii) the protective layer of the surfactant (Figure 7).

We also attempted to regulate the size of the gold NPs by a method based on discharge in reverse micelle solutions [12] (Figure 8). The reactive species generated by the discharge reduced $[AuCl_4]^-$ only inside water droplets in the reverse micelle solutions. At the lower values of water to surfactant ratio (W), the average diameter is smaller and the size distribution is narrower.

The size of gold nanoparticles varied from 4.0 to 11.4 nm. Size of gold NPs formed inside the water droplets was regulated by the size of reverse micelles. This suggests that SPP in glow discharge regime in reverse micelle solutions can be applied as a plasma nanoreactor for nanomaterial fabrication.

We analyzed the gold NPs synthesized by chemical reduction process and SPP and we investigate the microstructural characteristics of these in SPP [24]. Microstructural characteristics of gold nanoparticles (Au NPs) fabricated by SPP in reverse micelle solutions have been studied by high-resolution transmission electron microscopy (HRTEM).

The synthesized Au NPs, with an average size of 6.3 ± 1.4 nm, have different crystal characteristics: fcc single-crystalline particles, multiply-twinned particles (MTPs), and incomplete MTPs (single-nanotwinned fcc configuration). The crystal structure characteristics of the Au NPs synthesized by SPP method were analyzed and compared with similar-size Au NPs obtained by the conventional chemical reduction synthesis (CRS) method. The TEM analysis results show that the Au NPs synthesized by the CRS method have shapes and crystal

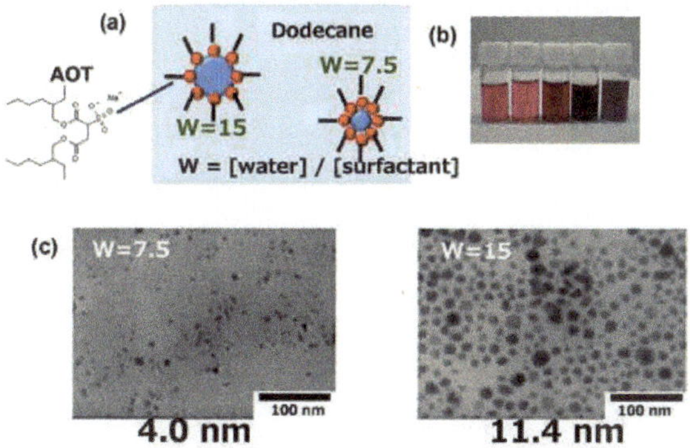

Figure 8. Synthesis of gold NPs in reverse micelle solutions in SPP. (a) Mechanism of reverse micelle solution with dodecane as solvent and sodium bis(2-ethylhexyl)sulfosuccinate (AOT), as surfactant. (b) Photos of solutions obtained by SPP, with different processing times. (c) TEM images of gold NPs synthesized in reverse micelle solutions for two ratios W.

structures similar to those nanoparticles obtained by the SPP method. However, from the detailed HRTEM analysis, the relative amount of the Au MTPs and incomplete MTPs to the total amount of the Au NPs synthesized by the SPP method was observed to be around 94 %, whereas the relative amount of these kinds of crystal structures fabricated by the CRS method was about 63 %. It is most likely that the enhanced formation of the Au MTPs is due to the fact that the SPP method generates highly reaction-activated species under low environmental temperature conditions. Figure 9 shows microstructural characteristics of gold NPs synthesized in SPP.

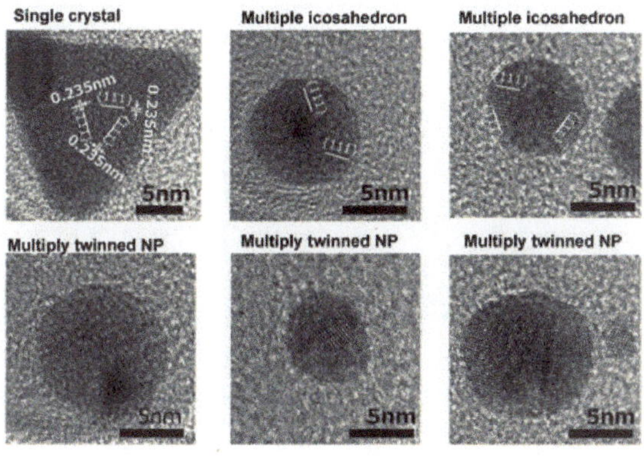

Figure 9. Microstructural characteristics of gold NPs synthesized in SPP. The synthesized Au NPs, with an average size of 6.3 ± 1.4 nm, have different crystal characteristics; fcc single-crystalline particles, multiply-twinned particles (MTPs), and incomplete MTPs (single-nanotwinned fcc configuration).

4. Synthesis of nanomaterials with enhanced catalytic activity

In order to improve the energy-conversion efficiency in fuel cells, we loaded Pt nanoparticles on carbon nanoballs (CNBs) by SPP by reducing Pt ion on CNB surface via a surfactant molecule. In this study, we employed poly (vinylpyrrolidone) (PVP) or sodium dodecyl sulfate (SDS) to prepare Pt nanoparticles supported on CNB (Pt/CNB) by the SPP, and the electrochemical properties as catalyst were evaluated by cyclic voltammetry (CV) [14].

CNBs were prepared by thermal decomposition process of ethylene and hydrogen gases. During the synthesis process, the color of the solution changed from yellow to dark brown indicating the improvement of the dispersibility of CNB in solution. Moreover, TEM images and elemental mapping images showed the Pt NPs supported on CNB.

The catalytic activity of the Pt/CNB using SDS as surfactant was shown to be higher than the Pt/CNB prepared with PVP system. The SDS-containing Pt/CNB also showed the higher activity than that obtained by the conventional chemical method.

We also studied the influence of the solution pH on the size of Pt NPs supported on CNBs and the catalytic activity of the composite nanomaterial. Figure 10 shows the influence of solution pH on the size and catalytic activity of synthesized Pt NPs supported on CNBs. From TEM images of Pt NPs on CNBs for solutions with various pH we can observe that the average size of Pt NPs is decreasing as the solution pH increases. The HRTEM image of Pt NP on carbon shows the crystalline structure of Pt NP and the organic surrounding surfactant PVP. The cyclic voltammetry measurements of Pt@CNBs show that small size Pt NPs have high catalytic activity (Figure 10(d)).

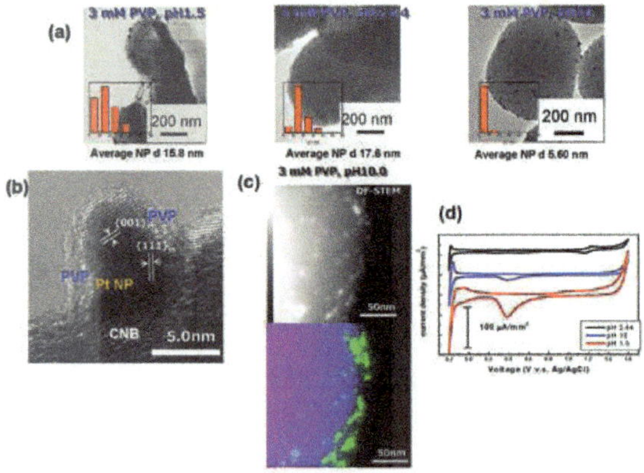

Figure 10. Influence of solution pH on the size and catalytic activity of synthesized Pt NPs supported on CNBs. (a) TEM images of Pt NPs on CNBs for solutions with various pH, indicating the average size of Pt NPs. (b) HRTEM image of Pt NP on carbon, with the organic surrounding PVP. (c) EDS mapping of Pt NPs. (d) Cyclic voltammetry measurements of Pt@CNBs.

5. Conclusions

SPP is a rapid, useful method for nanomaterials synthesis. The complexity of the phenomena makes this method a field of fundamental investigations. A better knowledge of the chemical reactions inside the plasma gas phase, liquid phase, and interfaces can make SPP a more powerful tool in nanotechnology.

Acknowledgements

This work was partially supported by "Tokai Region Nanotechnology Manufacturing Cluster" sponsored by Ministry of Education, Culture, Sports, Science, and Technology (MEXT), Core Research for Evolutional Science and Technology (CREST) of Japan Science and Technology (JST) Agency, and Global COE Project, sponsored by Ministry of Education, Culture, Sports, Science, and Technology (MEXT)

Author details

Osamu Takai[1], Maria Antoaneta Bratescu[1], Tomonaga Ueno[2] and Nagahiro Saito[1]

1 EcoTopia Science Institute, Nagoya University, Nagoya, Japan

2 Green Mobility Collaborative Research Center, Nagoya University, Nagoya, Japan

References

[1] Takai O. Solution plasma processing (SPP). Pure and Applied Chemistry 2008;80(9) 2003-2011.

[2] ilsing J. Astrophysical Journal 1899;10 113-125.

[3] ockyer N. Astrophysical Journal 1902;15(3) 190-198

[4] oshi A. A, Locke B. R, Arce P, Finney W. C. Formation of hydroxyl radicals, hydrogen peroxide and aqueous electrons by pulsed streamer corona discharge in aqueous solution. Journal of Hazardous Mater 1995;41(1) 3-30.

[5] un B, Sato M, Clements J. S. Optical study of active species produced by a pulsed streamer corona discharge in water. Journal of Electrostatics 1997;39(3) 189-202.

[6] un B, Sato M, Harano A, Clements J. S. Non-uniform pulse discharge-induced radical production in distilled water. Journal of Electrostatics 1998;43(2) 115-126.

[7] unka P. Pulse electrical discharges in water and their applications. Physics of Plasmas 2001;8(5) 2587.

[8] olb J. F, Joshi R. P, Xiao S, Schoenbach K. H. Streamers in water and other dielectric liquids. Journal of Physics D: Applied Physics 2008;41(23) 234007

[9] ruggeman P, Leys C. Non-thermal plasmas in and in contact with liquids. Journal of Physics D: Applied Physics 2009;42(5) 053001.

[10] Hieda J, Saito N, Takai O. Size-regulated gold nanoparticles fabricated by a discharge in reverse micelle solutions. Surface & Coatings Technology 2008;202(22-23) 5343-5346.

[11] Saito N, Hieda J, Takai O. Synthesis process of gold nanoparticles in solution plasma. Thin Solid Films 2009;518(3) 912-917.

[12] Hieda J, Saito N, Takai O. Exotic shapes of gold nanoparticles synthesized using plasma in aqueous solution. Journal of Vacuum Science and Technology A 2008;26(4) 854-856.

[13] Bratescu M. A, Saito N, Takai O. Redox reactions in liquid plasma during iron oxide and oxide-hydroxide nanoparticles synthesis. Current Applied Physics 2011;11(5) S30-S34.

[14] Ichin Y, Mitamura K, Saito N, Takai O. Characterization of platinum catalyst supported on carbon nanoballs prepared by solution plasma processing. Journal of Vacuum Science and Technology A 2009;27(4) 826-830.

[15] Pootawang P, Saito N, Takai O. Solution Plasma Process for Template Removal in Mesoporous Silica Synthesis. Japanese Journal of Applied Physics 2010;49(12) 126202-1.

[16] Baroch P, Anita V, Saito N, Takai O. Bipolar pulsed electrical discharge for decomposition of organic compounds in water. Journal of Electrostatics 2008;66(5-6) 294-299.

[17] Bratescu M. A, Cho S. P, Takai O, Saito N. Size-Controlled Gold Nanoparticles Synthesized in Solution Plasma. Journal of Physical Chemistry C 2011;115(50) 24569-24576.

[18] Bratescu M. A, Hieda J, Umemura J, Saito N, Takai O. Analysis of organic pollutant degradation in pulsed plasma by coherent anti-Stokes Raman spectroscopy. Journal of Vacuum Science and Technology A 2011;29(3) 031302-1-7.

[19] Anita V, Saito N, Takai O. Microarc plasma treatment of titanium and aluminum surfaces in electrolytes. Thin Solid Films 2006;506-507 364-368.

[20] Potocky S, Saito N, Takai O. Needle electrode erosion in water plasma discharge. Thin Solid Films;518(3) 918-923.

[21] Miron C, Bratescu M. A, Saito N, Takai O. Time-resolved Optical Emission Spectroscopy in Water Electrical Discharges. Plasma Chemistry Plasma Processing 2010;30(5) 619-631.

[22] Miron C, Bratescu M. A, Saito N, Takai O. Optical diagnostic of bipolar electrical discharges in HCl, KCl, and KOH solutions, Journal of Applied Physics 2011;109(12) 123301.

[23] Cho S. P, Bratescu M. A, Saito N, Takai O. Microstructural characterization of gold nanoparticles synthesized by solution plasma processing. Nanotechnology 2011;22(45) 455701-7.

[24] Miron C, Bratescu M. A, Saito N, Takai O. Effect of the electrode work function on the water plasma breakdown voltage. Current Applied Physics 2011;11(5) S154-S158.

Chapter 7

Electronic Structure Calculations for Nano Materials

Nobutada Ohno, Dai Okumura and
Yusuke Kinoshita

1. Introduction

This is a chapter on electronic structure calculations for nano materials based on first-principles density functional theory (DFT) [1, 2]. The DFT has become the primary tool for electronic structure calculations for solids and has also become popular for atoms and molecules. There are many reviews and books on the DFT [3-5]. In this chapter, the works of Ohno and coworkers on electronic structure calculations for deformed boron nitride nanotubes (BNNTs) using the DFT are described [6, 7].

The existence of BNNTs was theoretically predicted by Rubio et al. [8, 9] and then multi-walled (MW) BNNTs were first synthesized by Chopra et al. [10]. Since then, BNNTs have attracted the attention of many researchers owing to their important properties [11]. The mechanical strength [12, 13] and thermochemical stability [14] of BNNTs are comparable to those of carbon nanotubes (CNTs) [15]. For instance, experiments (using a thermal vibrational amplitude technique [12] and an electric field-induced resonance method [16]) and atomistic simulations (first-principles [17-20], tight-binding [21, 22], and classical molecular mechanics [23, 27] calculations) measured the Young's modulus of BNNTs to be in the range 0.7-1.2 TPa, which is close to that of CNTs (e.g., the average value is 1.8 TPa [28] and 1.25 TPa [29]). In contrast, the electrical conductivity of BNNTs is completely dissimilar to that of CNTs. While CNTs become either metallic or semiconductive depending on the chirality, BNNTs are electrically insulating regardless of the diameter and chirality [11]. This is a notable characteristic of BNNTs that is different from CNTs. Therefore, BNNTs are expected to be used as electrical insulation coatings for conducting or semiconducting nanochains, nanowires, and nanotubes in severeconditions such as high temperatures and chemically hazardous environments.

However, a recent experimental study indicated that a bent MWBNNT was electrically conductive [30], and a theoretical study showed that flattening decreased the energy gap of a zigzag single-walled (SW) BNNT [31]. These results indicate that the usefulness of BNNTs as nanocoatings might be lost under certain conditions (e.g., deformation caused by thermal

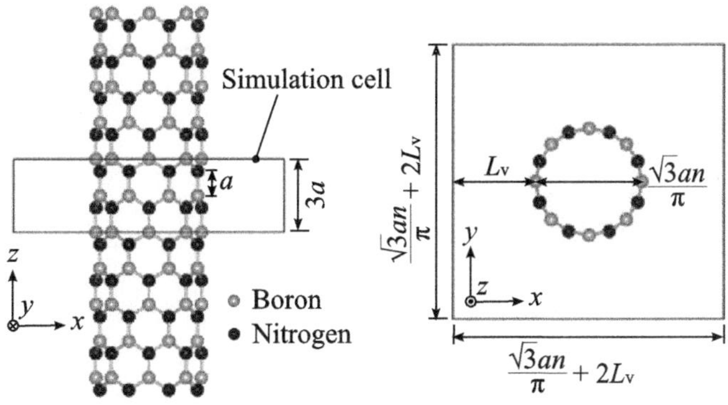

Figure 1. Simulation model of (8,0) SWBNNT.

stress or by a substrate constraint). Alternatively, BNNTs can be used in nanoelectronic devices by introducing deformation. In any case, our aim is to elucidate the electronic structures of deformed BNNTs. In the following sections, electronic structure calculations for SWBNNTs under tension, torsion, and flattening (Section 2) and for MWBNNTs under flattening (Section 3) will be discussed.

2. SWBNNTs under tension, torsion and flattening

2.1. Simulation procedure

This section focuses on $(n,0)$ SWBNNTs with n= 6, 8, 10, where (n, m) is the chiral index.The (n, m) tube has a diameter of $\sqrt{3}a\sqrt{n^2+nm+n^2}/\pi$, where a is the nearest interatomic distance between boron and nitrogen atoms. Figure 1 shows the simulation model of the (8, 0) SWBNNT. The BNNT is located at the center of the unit cell so that the axial direction is parallel to the z direction. The cell size is $\left(\sqrt{3}an/\pi+2L_V\right)\times\left(\sqrt{3}an/\pi+2L_V\right)\times 3a$ with a=0.145 nm and L_V=0.5 nm, where L_V is the length of vacuum region in the unit cell. Although a three-dimensional periodic boundary condition is used, the cell sizes in the x and y directions are sufficiently large to avoid any interaction with neighboring image cells. It is confirmed that they have little effect (less than 1%) on the total energy, charge distribution, and energy-band structure of a deformed BNNT when they are larger than the tube diameter by 1.0 nm.

Atomic positions and the cell size in the z direction are first relaxed using the conjugate gradient method until atomic forces and the stress component, σ_{zz}, become less than 0.01 eV/Å and 0.01 GPa, respectively. After obtaining the equilibrium structure, tension, torsion, or flattening deformation is applied where atomic configurations are relaxed until their forces become less than 0.01 eV/Å.

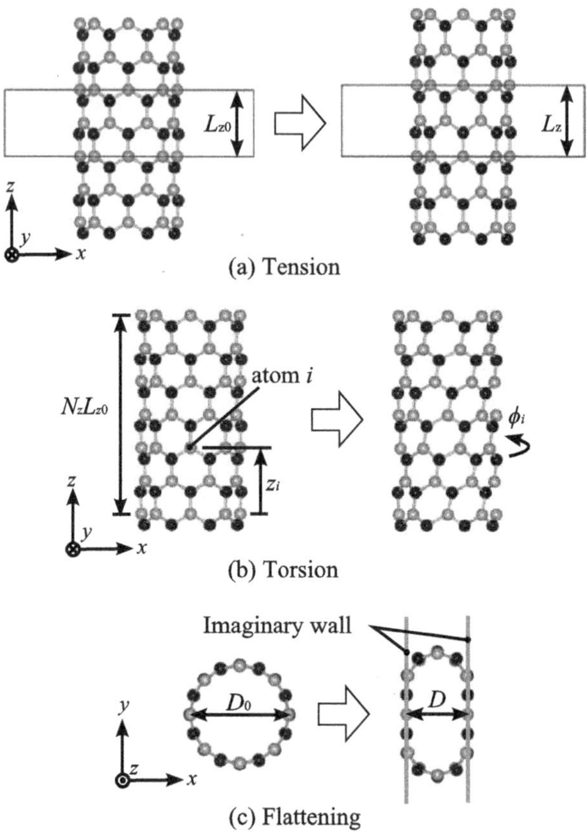

Figure 2. Schematics of tension, torsion, and flattening of SWBNNT.

In tension (Figure 2(a)), the axial strain, ε_{zz}, is defined as

$$\varepsilon_{zz} = \frac{L_z - L_{Z0}}{L_{Z0}} \tag{1}$$

where L_{z0} and L_z are the cell sizes in the z direction of unstretched and stretched BNNTs, respectively. In this study, ε_{zz} is in the range 0.00-0.10 with an increment of 0.02.

In torsion (Figure 2(b)), atom i is rotated φ_i degrees about the z-axis. φ_i is defined as

$$\varphi_i = \theta z_i \tag{2}$$

$$\theta = \frac{360N_\theta}{nN_zL_{z0}} \tag{3}$$

where θ is the specific angle of twist, z_i is the z-coordinate of atom i, N_θ is an integer, and N_z is the number of primitive unit cells in the z direction. The torsion angle must be an integral multiple (N_θ) of 360/n to fulfill the periodic boundary condition. In this study, the value of N_θ is 1 and that of N_z is in the range 3-5.

In flattening (Figure 2(c)), compression in the x direction is applied by reducing the distance between imaginary walls. Once an atom contacts a wall, the atom is allowed to move only on the wall. The flattening ratio, η, is defined as

$$\eta = \frac{D_0 - D}{D_0} \tag{4}$$

where D_0 is the tube diameter at equilibrium and D is the distance between the imaginary walls. In this study, η is in the range 0.00-0.50 with an increment of 0.05.

First-principles DFT calculations are conducted using the Vienna *ab initio* Simulation Package (VASP) [32, 33]. The wave functions are expanded in a plane-wave basis set with a cut-off energy of 350 eV. The ultrasoft pseudopotential proposed by Vanderbilt [34] is used and the exchange-correlation energy is evaluated by the generalized gradient approximation of Perdew and Wang [35]. The Brillouin zone integration is performed by the Monkhorst-Pack scheme [36] using a 1 × 1 × 4 k-point mesh for atomic and electronic relaxations. After the relaxation, the energy-band structure is obtained by calculating energy eigenvalues of 30 points on Γ-X line in the Brillouin zone.

2.2. Results and discussion

2.2.1. Energy-band structures

It is well known that the DFT underestimates the energy gap. For a quantitative discussion of the energy gap, a modified theory such as the GW approximation (GWA) is necessary [37-40]. Nonetheless, previous studies on bulk hexagonal BN and an isolated BN sheet showed that the shape of the energy bands by the DFT is quite similar to that by the GWA except for the magnitude of the energy gap [37, 38]. Thus, the DFT can qualitatively predict energy-band structures of BNNTs.

Figure 3 shows the change in the energy-band structures of the (8,0) SWBNNT under tension, torsion, and flattening. The (6,0) and (10,0) show a changing trend similar to the (8,0) band structure. A common feature among the three deformation modes is that both the valence-band maximum (VBM) and conduction-band minimum (CBM) are located at the Γ point (k=0) during the deformations. Another common feature is that the change in the energy of the VBM, E_{VBM}, is almost zero. Note that while the tension and flattening obviously decrease the energy of the CBM, E_{CBM}, the torsion hardly decreases E_{CBM}. The results suggest that all the three deformation modes decrease energy gaps, $E_g = E_{CBM} - E_{VBM}$, of SWBNNTs, but torsion has less of an effect on the energy gap than tension and flattening.

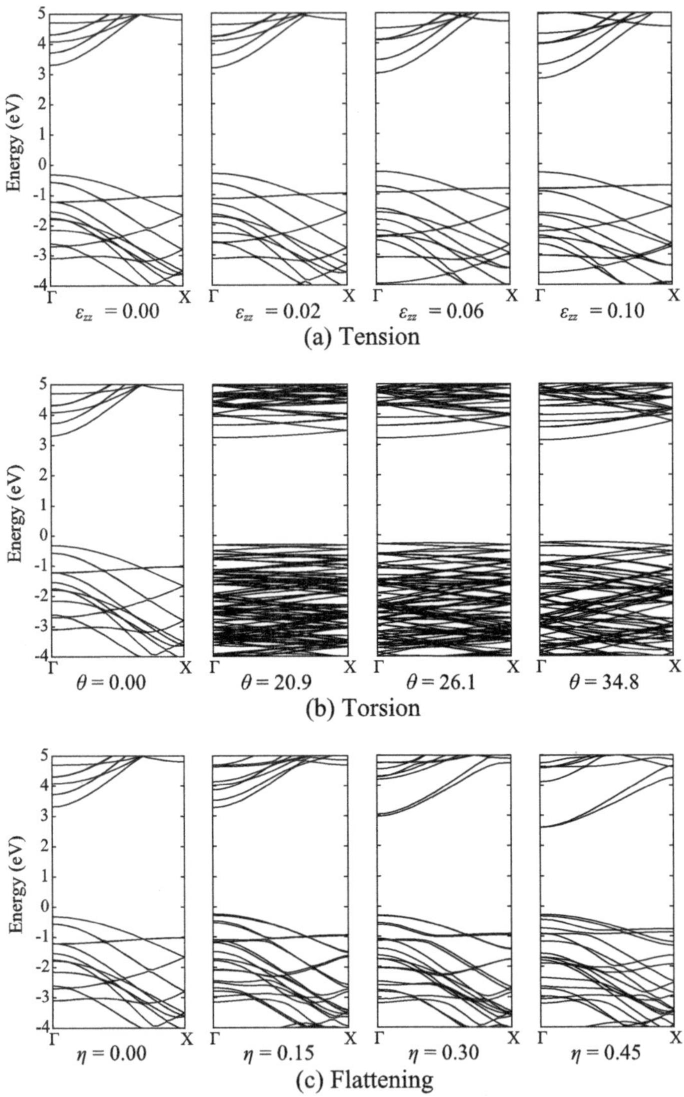

Figure 3. Change in the band structure of an (8,0) SWBNNT. The origin of the energy scale is set at the Fermi level.

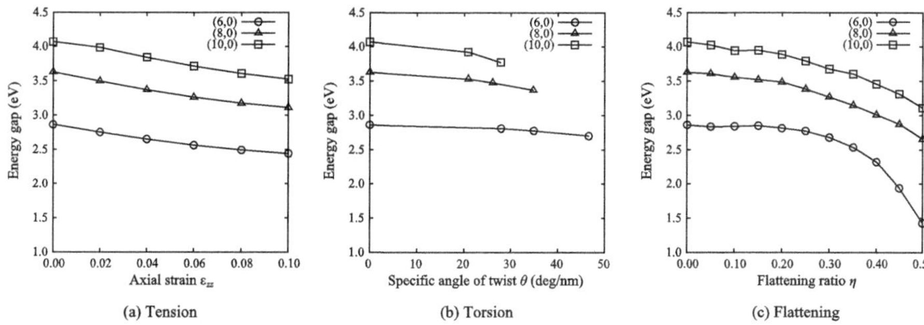

Figure 4. Energy gaps of the SWBNNTs as a function of (a) axial strain, (b) specific angle of twist, and (c) flattening ratio.

Figure 4 shows the energy gaps of the (6,0), (8,0), and (10,0) SWBNNTs as a function of the axial strain, specific angle of twist, and flattening ratio. The energy gap of the (10,0) with θ=41.7 deg/nm (N_z=5) is not shown in the figure because it collapsed. Under tension and torsion except for θ=20.8-27.8 deg/nm in the (10,0), the energy gap decreases almost linearly and the rate of decrease hardly depends on the diameter. In contrast, under flattening, the energy gap decreases quadratically or exponentially and the amount of decrease significantly depends on the diameter; a SWBNNT with the smaller diameter shows a larger decrease in the energy gap. It is also shown that flattening results in a few times larger decrease in the energy gap than tension and torsion.

Although the discussion so far in this section has dealt with the SWBNNTs under the three simple deformation modes, BNNTs would be subjected to combined deformation in their practical use. Therefore, the energy gap of the SWBNNTs subjected to flattening following axial tension is further analyzed (Figure 5). It is found that preceding tension shifts an E_g-η curve downward without dramatic changes in its shape, and that the extent of the shift almost corresponds to the energy gap decrease induced by simple tension (Figure 4(a)). This result suggests that the energy gap of the SWBNNTs under a combination of the three deformation modes can be deduced from Figure 4. In the rest of this section, therefore, only the simple deformation modes will be discussed.

2.2.2. Charge densities at the CBM

Here the mechanism of deformation-induced electronic changes in the SWBNNTs is discussed in terms of charge densities at the CBM (Figure 6). The CBM is composed of boron-derived states. In fact, CBM charge densities are high around boron atoms, while they are low around nitrogen atoms. It is found that the π^* state (p_z orbitals of boron atoms) hybridizes with the σ^*

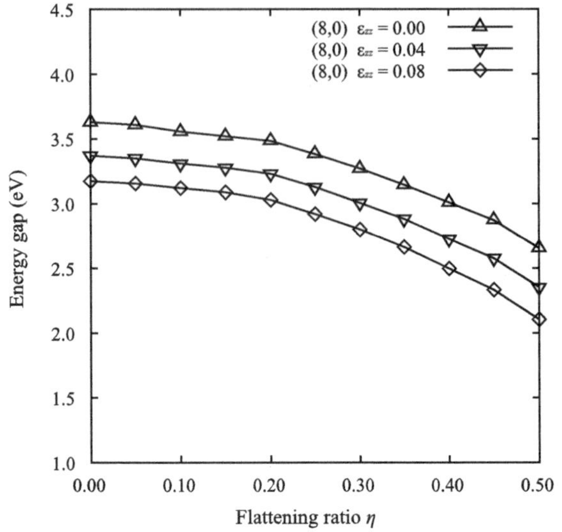

Figure 5. Change in the energy gap of (8,0) SWBNNT under flattening following axial tension.}

state along a circumference passing through boron atoms under no deformation (Figure 6(a), ε_{zz}=0.00), and that the tension, torsion, and flattening induce the change in the CBM state.

With increasing flattening deformation (Figure 6(d)), charges are transferred from the flattened to the curved regions, resulting in an overlap of the charge densities and formations of electronic bonds between neighboring boron atoms in the curved regions. It is this mechanism that results in the decrease in E_{CBM} in the flattened SWBNNTs. Comparing the three SWBNNTs with η=0.45 (Figures 6(d)-(f)), the electronic bonds become stronger as the diameter becomes smaller. Therefore, a flattened SWBNNT with a smaller diameter shows a larger decrease in the energy gap.

Under tension (Figure 6(a)), the tube curvature increases because of Poisson contraction, leading to the enhancement of π^*-σ^* hybridizations and the decrease in E_{CBM}. Figure 6(a) shows the narrowing white center area of zero-charge densities and the spreading gray area of π^*-σ^* hybridizations. The same is true for the torsion (Figure 6(b)), but it induces less change in charge densities than tension (the size of the white center area changes little in Figure 6(b)), resulting in a smaller decrease in the energy gap under torsion than under tension (Figures 4(a),(b)). It should be noted that elastic buckling occurred at a θ between 20.8 and 27.8 deg/nm in the (10,0), leading to local flattening (Figure 6(c)). Therefore, the relation of E_g versus θ deviates from the linear decrease at θ of 20.8-27.8 deg/nm in the (10,0) (Figure 4(b)). It is obvious that the overlap of charge densities is much stronger under flattening than under tension or torsion. Therefore, the decrease in the energy gap in the former is much larger than in the latter.

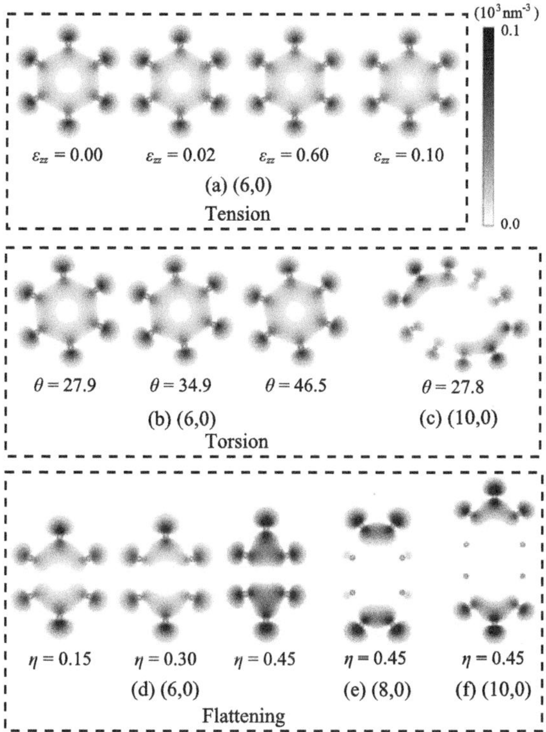

Figure 6. Change in the CBM charge density. Cross sections passing through boron atoms are shown.

2.2.3. Deformation forces

Figure 7 shows the deformation energy as a function of axial strain, specific angle of twist, and flattening ratio. The curves in tension, torsion, and flattening are fitted by cubic, quadratic, and quartic polynomials, respectively. The first and second derivatives of each curve provide the deformation force (Figure 8) and the elastic modulus, respectively. Young's moduli of the (6,0), (8,0), and (10,0) are thus calculated to be 0.759, 0.794, and 0.811 TPa, respectively. They are in good agreement with those measured in experiments (1.22±0.24 TPa [10] and 0.722 TPa [16]) and other first-principles calculations (0.762, 0.785, and 0.803 TPa for (6,0), (8,0), and (10,0), respectively [18]). It is found in Figure 8 that forces under flattening are smaller than under tension and torsion, because strong in-plane B-N covalent bonds prevent in-plane tension and torsion. It is also found that forces rapidly increase later under flattening. The rapid increase starts from around η=0.3 and 0.4 in the (6,0) and (8,0), respectively, where the imaginary wall distances are 0.35 and 0.38 nm, respectively. Because the interlayer distance of hexagonal BNs and MWBNNTs is around 0.34 nm, the rapid increase would be attributed to the repulsive force between the two flattened regions.

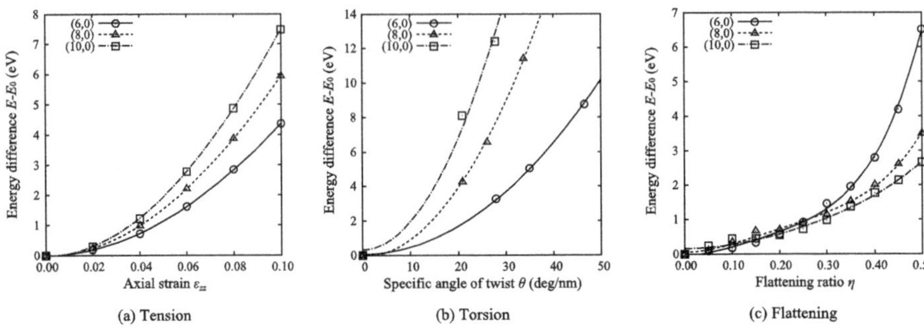

Figure 7. Deformation energy of the SWBNNTs as a function of (a) axial strain, (b) specific angle of twist, (c) flattening ratio.

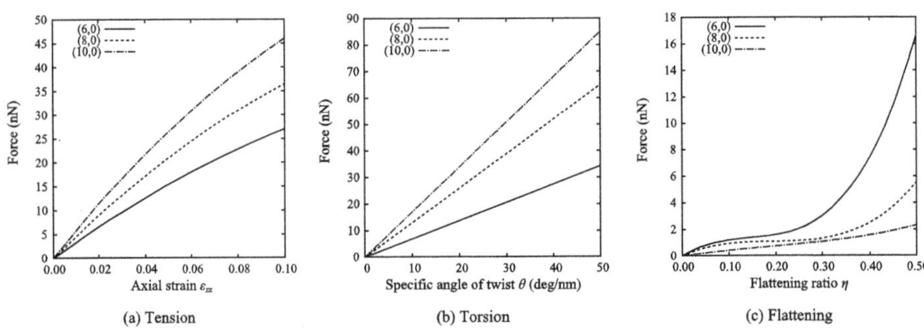

Figure 8. Forces required to deform (6,0), (8,0) and (10,0) SWBNNTs.

Figure 9 shows the relationship between energy gap and deformation force. The three bands to the right illustrate the obtainable range of the energy gap by introducing tension, torsion, and flattening. In tension and torsion, a larger force is required for a larger tube to induce the same amount of energy gap decrease. The opposite is true in flattening, i.e., a larger force for a smaller tube. The key findings from Figure 9 are that (i) the flattening with a force smaller than that applied for tension or torsion leads to the larger decrease in the energy gap, and (ii) flattening offers a larger obtainable range of the energy gap than tension and torsion: 1.4-4.0 eV under flattening, 2.5-2.8 eV and 3.1-4.0 eV under tension, 2.7-2.8 eV, 3.4-3.6 eV, and 3.8-4.0 eV under torsion. These findings indicate that flattening has the potential to enable BNNTs to be used as nanoelectronic devices. However, a valid question is whether flattening BNNTs is experimentally feasible.

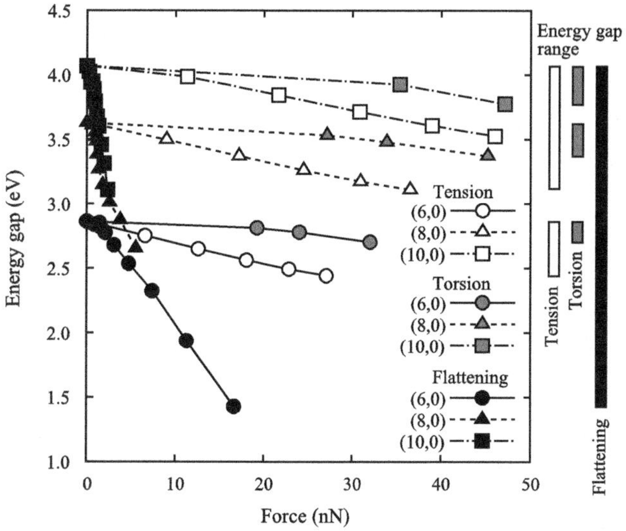

Figure 9. Relationship between energy gap and force of (6,0), (8,0) and (10,0) SWBNNTs.

In order to answer this question, the estimated flattening forces are compared with those of SWCNTs that Barboza et al. have already experimentally succeeded in flattening by means of an atomic force microscopy (AFM) tip [41]. Although they did not actually measure flattening forces of $(n,0)$ SWCNTs with $n \le 10$, they proposed and validated a universal relationship among the applied force, SWCNT diameter, AFM tip radius, and flattening ratio:

$$\frac{F D_0^{3/2}}{(2R)^{1/2}} = \frac{\alpha}{(1-\eta)^{3/2}}\left[\sqrt{2\eta+\eta^2} + tg^{-1}\left(\sqrt{\frac{\eta}{1-\eta}}\right)\right] \tag{5}$$

where R is the AFM tip radius and α is a constant (=1.2×10^{-18} J). Equation (5) indicates that the quantity $FD_0^{3/2}(2R)^{-1/2}$ should be universal to any SWCNT. They showed that all experimental data fall on a single curve obtained by Equation (5) up to $\eta \cong 0.4$. From Equation (5) and the geometric contact conditions between a tube and an AFM tip, the flattening force per unit length of a (6,0) SWCNT (D_0=0.470 nm) is calculated to be 15.4 N/m when η=0.4 and R=30 nm. In contrast, from Figure 8, that of the (6,0) SWBNNT (=F/L_{z0}) is estimated to be 16.8 N/m at η=0.4. The results demonstrate that the flattening force is almost equal in SWCNTs and SWBNNTs, indicating that the same experiments as Barboza et al. would be feasible for SWBNNTs. The fact that CNTs and BNNTs almost have the same tube shape and size when their chiral indexes are the same ($a \cong$ 0.142 nm in CNTs and $a \cong$ 0.145 nm in BNNTs) also encourages the feasibility of flattening BNNTs. It is therefore concluded that the flattening forces estimated are not unrealistic and strongly expected that the same or similar experimental technique also applies to BNNTs.

3. MWBNNTs under flattening

3.1. Simulation procedure

This section focuses on (5,0), (13,0), and (21,0) SW, (5,0)@(13,0) and (13,0)@(21,0) DW, and (5,0)@(13,0)@(21,0) TWBNNTs. Figure 10 shows the simulation model of the (13,0)@(21,0) DWBNNT. The initial nearest interatomic distance between boron and nitrogen atoms is set as 0.145 nm. Boron (nitrogen) atoms in the outer tube are stacked above nitrogen (boron) atoms in the inner tube [42]. The axial direction of the BNNT is parallel to the z-direction. The BNNT is located at the center of the unit cell with a size of 3.637 nm × 3.637 nm × 0.435 nm. Even though a three-dimensional periodic boundary condition is employed, the cell sizes in the x- and y-directions are large enough to avoid interaction with neighboring image cells, because they have little effect (less than 1%) on the energy, charge distribution, and energy band structure of a flattened BNNT, when they are greater than the diameter of the BNNT plus 1.0 nm.

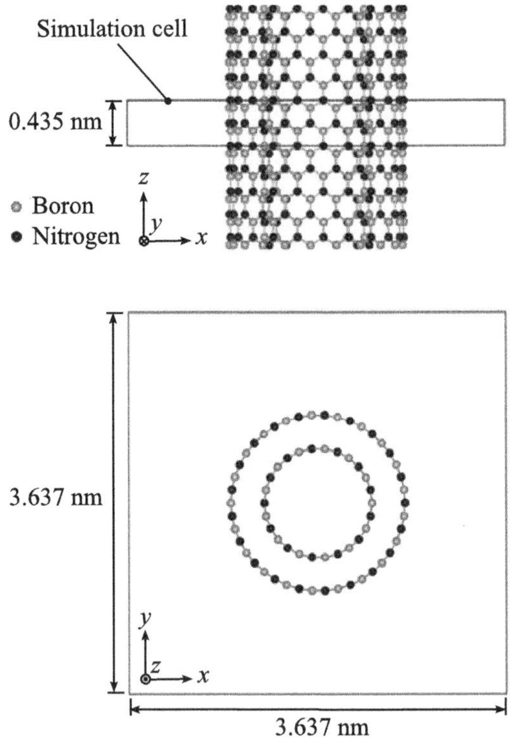

Figure 10. Simulation model of (13,0)@(21,0) DWBNNT.

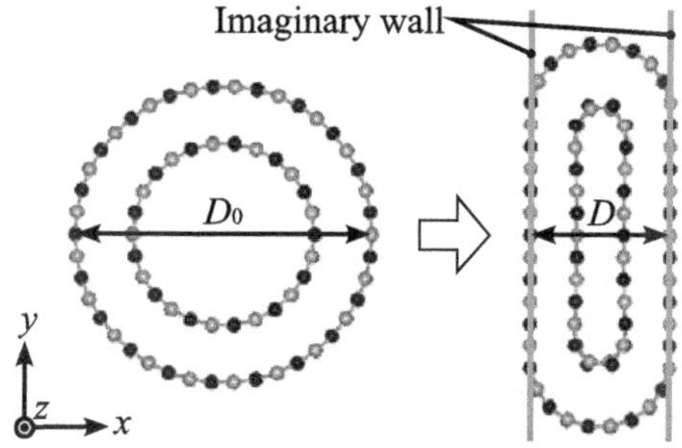

Figure 11. Schematic illustration explaining flattening compression of BNNTs.

Atomic positions and the cell size in the *z*-direction are first relaxed using the conjugate gradient method until atomic forces and the stress component, σ_{zz}, become less than 0.01 eV/Å and 0.01 GPa, respectively. After obtaining the equilibrium structure, a flattening compression in the *x*-direction is applied by reducing the distance between imaginary walls until the BNNT collapses (Figure 11). Once an atom contacts a wall, the atom is allowed to move only on the wall. During compression, the cell sizes are fixed and atomic configurations are relaxed until their forces become less than 0.01 eV/Å. To investigate the degree of deformation, the flattening ratio, η, is defined as

$$\eta = \frac{D_0 - D}{D_0} \tag{6}$$

where D_0 is the diameter of the outermost tube at equilibrium, and D is the distance between the imaginary walls.

First-principles DFT calculations are conducted using the Vienna Ab Initio Simulation Package (VASP) [32, 33]. The wave functions are expanded in a plane-wave basis set with a cut-off energy of 350 eV. The ultrasoft pseudopotential proposed by Vanderbilt [34] is used, and the exchange-correlation energy is evaluated by the generalized gradient approximation of Perdew and Wang [35]. The Brillouin zone integration is performed by the Monkhorst-Pack scheme [36] using a 1 × 1× 4 *k*-point mesh.

3.2. Results and discussion

3.2.1. Energy-band structures

Figure 12 shows the change in the energy band structures of the (13,0) SWBNNT and (13,0)@(21,0) and (5,0)@(13,0) DWBNNTs during flattening deformation. The other SWBNNTs and the (5,0)@(13,0)@(21,0) TWBNNT show a similar changing trend of the band structure to the (13,0) SWBNNT and the (5,0)@(13,0) DWBNNT, respectively. Both the valence band maximum (VBM) and the conduction band minimum (CBM) of the (13,0) SWBNNT and

(13,0)@(21,0) DWBNNT are located at the Γ point (k=0) during the deformation, but those of the (5,0)@(13,0) DWBNNT move to k≠0 midway during the deformation and then return to the Γ point. In each BNNT, the energy of the VBM, E_{VBM}, hardly changes, while that of the CBM, E_{CBM}, changes, indicating that the change in the energy gap, E_g, is mainly caused by a change in E_{CBM}. In the (13,0) SW and (13,0)@(21,0) DWBNNTs (Figure 12(a), (b)), E_{CBM} decreases monotonically. In contrast, in the (5,0)@(13,0) DWBNNT (Figure 12(c)), E_{CBM} first increases and then decreases.

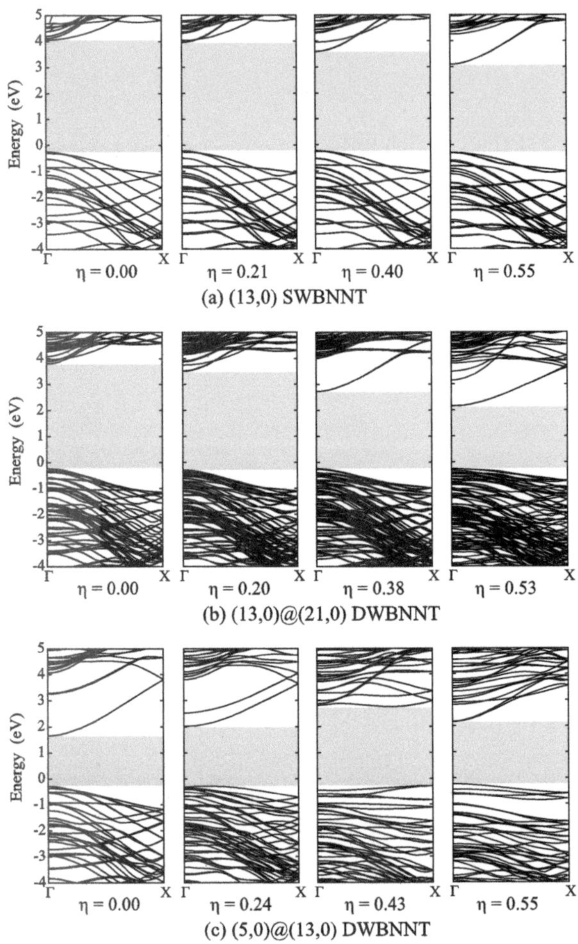

Figure 12. Change in the band structure of (13,0) SWBNNT and (13,0)@(21,0) and (5,0)@(13,0) DWBNNTs in flattening deformation. The origin of the energy scale is set at the Fermi level.

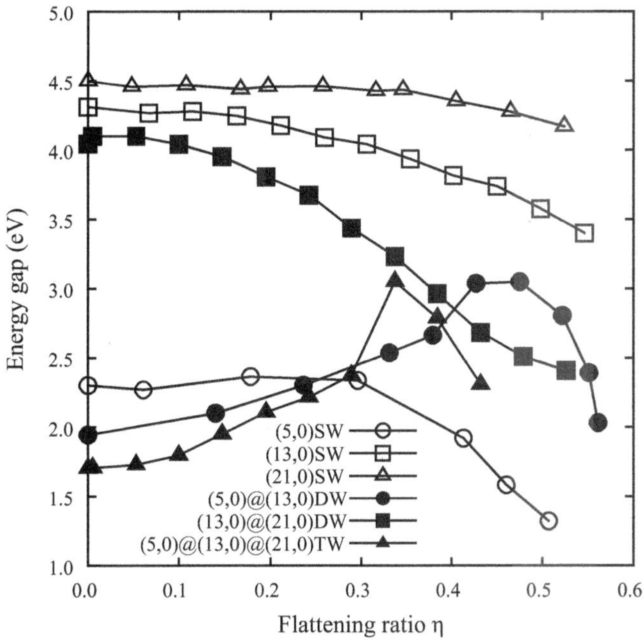

Figure 13. Relationship between energy gap, E_g, and flattening ratio, η, of (5,0), (13,0), and (21,0) SWBNNTs, (5,0)@(13,0) and (13,0)@(21,0) DWBNNTs, and (5,0)@(13,0)@(21,0) TWBNNT.

Figure 13 shows the energy gaps of the (5,0), (13,0), and (21,0) SWBNNTs, (5,0)@(13,0) and (13,0)@(21,0) DWBNNTs, and (5,0)@(13,0)@(21,0) TWBNNT as a function of the flattening ratio. The energy gap of the three SWBNNTs decreases almost monotonically, and the amount of decrease becomes smaller with increasing tube diameter. The energy gap of the (13,0)@(21,0) DWBNNT also decreases monotonically, but it exhibits a more rapid decrease than the (13,0) and (21,0) SWBNNTs. It should be noted that the energy gaps of the (5,0)@(13,0) DWBNNT and (5,0)@(13,0)@(21,0) TWBNNT increase during the early stage and then decrease. This shift occurs earlier in the latter (η=0.34) than in the former (η=0.48). The fact that the (5,0)@(13,0) DWBNNT and (5,0)@(13,0)@(21,0) TWBNNT show different changing trends of E_g from the SWBNNTs proves that interwall interactions significantly affect the electronic structures of the flattened (5,0)@(13,0) DWBNNT and (5,0)@(13,0)@(21,0) TWBNNT.

3.2.2. Charge densities at the CBM

Figure 14 shows charge densities at the CBM of the flattened (5,0), (13,0), and (21,0) SWBNNTs at a cross section passing through boron atoms. The characteristics of the nearly free electron (NFE) state are observed in the BNNTs with a small curvature (Figure 14(b): η=0.00, (c): η=0.00, 0.20), while π^*-σ^* hybridizations appear in the others. The reason the CBM of the (13,0) and

(21,0) SWBNNTs changes from a NFE-like state to a π^*-σ^* hybridized state is that the tube curvature increases locally as the flattening deformation increases (in (n,0) SWBNNTs under no deformation, the CBM is a NFE-like state when $n \geq 13$ and a π^*-σ^* hybridized state when $n < 13$, and the hybridization becomes stronger with increasing tube curvature) [8]. The energy gap of the (5,0) SWBNNT is much smaller than those of the (13,0) and (21,0) SWBNNTs because of its strong π^*-σ^* hybridization. With increasing flattening deformation, charge is transferred from flattened regions to curved ones, leading to an overlap of the charge densities. The E_{CBM} of the SWBNNTs decreases under flattening because of the formation of electronic bonds between neighboring boron atoms in the curved regions. The charge density distribution in curved regions of the (13,0) SWBNNT at η=0.21 is similar to that of the (21,0) SWBNNT at η=0.52, which results in them having almost the same energy gap of 4.2 eV. This is because they have almost the same value of D, namely the same curvature of the curved region. Figure 15 shows the relationship between the energy gap and imaginary wall distance of the (13,0) and (21,0) SWBNNTs. Their energy gaps are almost equal under a same wall distance.

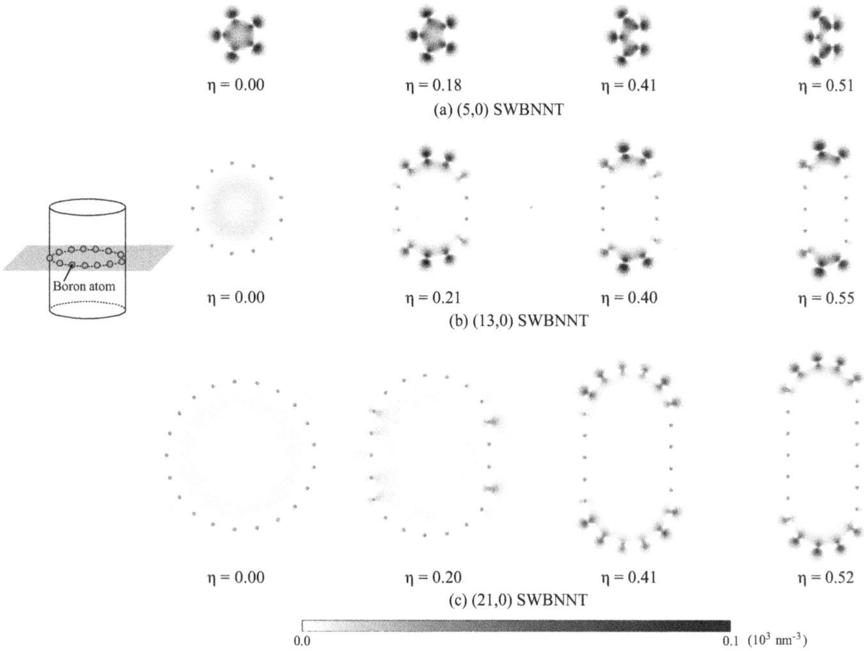

Figure 14. Change in the CBM charge density of (5,0), (13,0), and (21,0) SWBNNTs in flattening deformation.

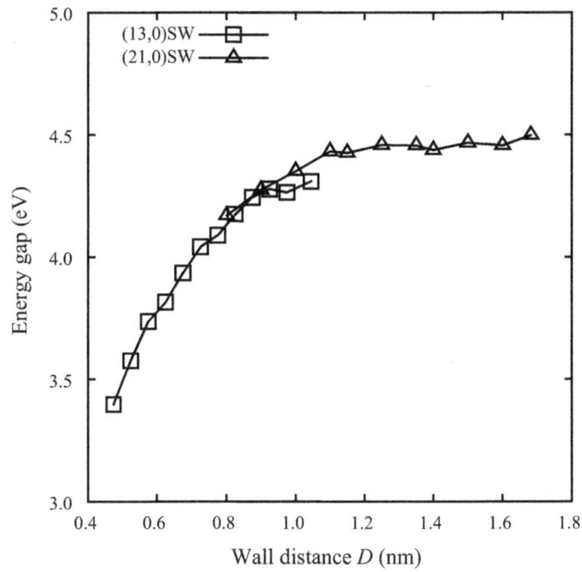

Figure 15. Energy gap of (13,0) and (21,0) SWBNNTs as a function of imaginary wall distance.

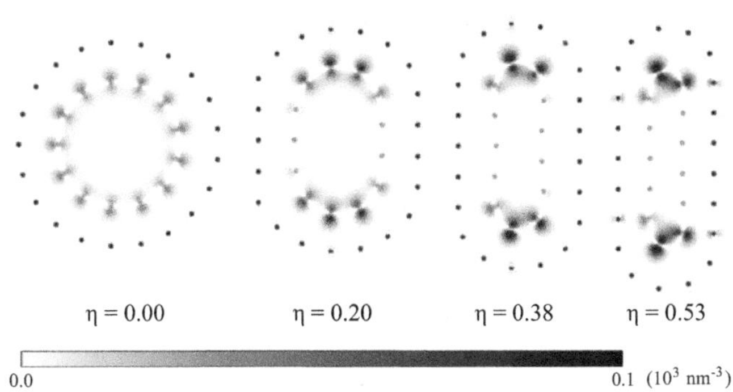

Figure 16. Change in the CBM charge density of (13,0)@(21,0) DWBNNT during flattening deformation.

The CBM charge density distribution of the (13,0)@(21,0) DWBNNT is similar to that of the SWBNNTs (Figure 16). In the inner tube, charge transfer from flattened to curved regions is observed and an overlap of the charge densities is induced in the curved regions. The decrease in E_{CBM} of the (13,0)@(21,0) DWBNNT is caused by the same mechanism as in the SWBNNTs mentioned above. Because the charge densities are distributed almost entirely in the inner tube during deformation, one might expect that the E_g-η curve of the (13,0)@(21,0) DWBNNT coincides with that of the (13,0) SWBNNT. However, E_g of the former is in fact smaller than that of the latter under the same η. As shown in Figure 16, the flattening ratio of the innermost tube, η_{in}, must be larger than η to maintain the interwall spacing constant. This means that the B-B bonds in the flattened (13,0)@(21,0) DWBNNT are stronger than those in the flattened (13,0) SWBNNT under the same η, resulting in a larger decrease in E_{CBM} in the former than in the latter.

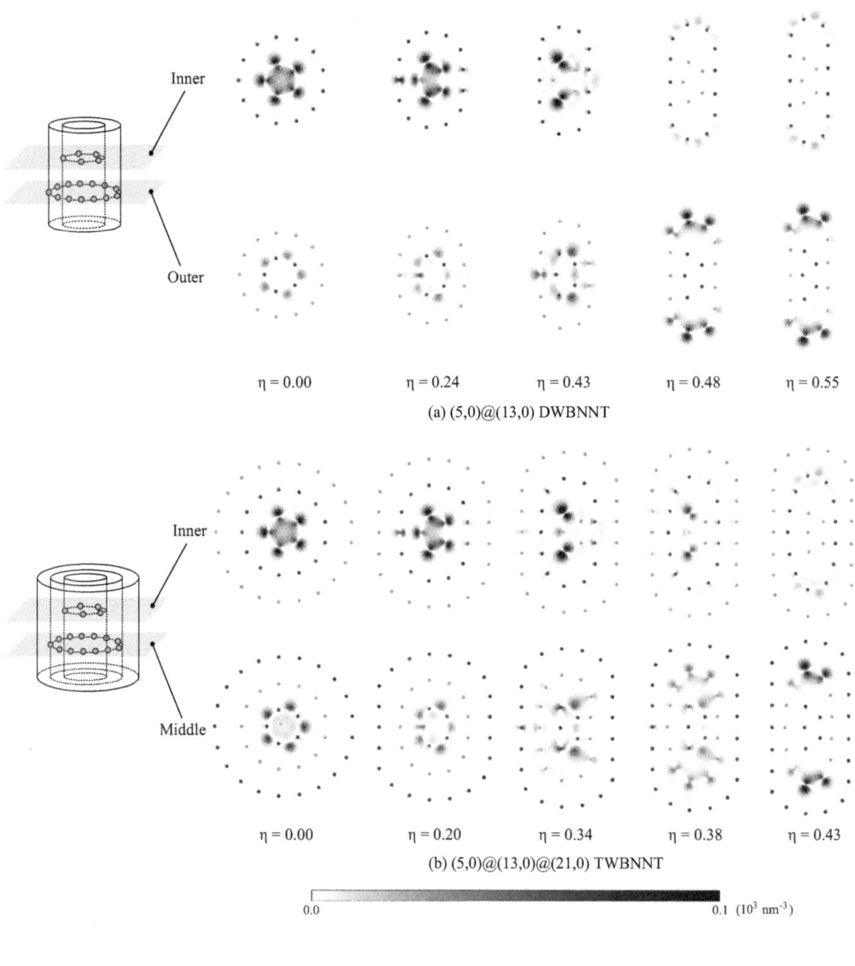

Figure 17. Change in the CBM charge density of (5,0)@(13,0) DWBNNT and (5,0)@(13,0)@(21,0) TWBNNT during flattening deformation.

It should be noted that the (5,0)@(13,0) DWBNNT and (5,0)@(13,0)@(21,0) TWBNNT show different changes in CBM charge densities from those of the SWBNNTs and (13,0)@(21,0) DWBNNT (Figure17). First, the CBM charge densities gradually transfer from boron atoms in the innermost tube to boron atoms in the second innermost tube. This charge delocalization and spreading account for the increase in E_{CBM} in the (5,0)@(13,0) DWBNNT and (5,0)@(13,0)@(21,0) TWBNNT during the early stage of the deformation. Then, overlap of the charge densities in curved regions in the second innermost tube is induced with increasing deformation. Consequently, E_{CBM} of the (5,0)@(13,0) DWBNNT and (5,0)@(13,0)@(21,0) TWBNNT decreases later in the deformation. It is this mechanism that results in the initial increase and subsequent decrease in E_g in the (5,0)@(13,0) DWBNNT and (5,0)@(13,0)@(21,0) TWBNNT. Because the latter has a larger η_{in} than the former under the same η (see the atomic positions in the innermost tube at η of 0.43 in Figure 17), charge spreading from the first to the second innermost tube is completed earlier in the latter. Therefore, the latter shows an earlier shift from increase to decrease in E_g.

3.2.3. Critical diameter of the innermost tube

It is evident that a critical chiral index, (n_c, 0), exists for flattened zigzag MWBNNTs. Consider the innermost tube of a zigzag MWBNNT, denoted as (n_{in}, 0). In the case of $n_{in} > n_c$, E_g decreases monotonically as the flattening deformation increases. In the reverse case ($n_{in} < n_c$), E_g first increases and then decreases. From the results obtained in this study, n_c is proven to be an integer between 5 and 13. Furthermore, if zigzag BNNTs have the same n_{in}, a zigzag BNNT containing more walls shows a more rapid change in E_g (compared with the change in E_g between (13,0) SW and (13,0)@(21,0) DWBNNTs and between (5,0)@(13,0) DW and (5,0)@(13,0)@(21,0) TWBNNTs in Figure 13).

In the aforementioned experimental study on a bent MWBNNT, a notable tendency for a zigzag atomic arrangement and local flattening have been observed [30]. Judging from the HRTEM images, n_{in} of the MWBNNT is larger than n_c. Therefore, it can be said that a possible reason for the change from insulating to semiconducting in the bent MWBNNT is electronic changes, as shown in Figure 16. To the best of our knowledge, there has been no experimental study on deformed BNNTs with $n_{in} < n_c$, but the results for the (5,0)@(13,0) DWBNNT and (5,0)@(13,0)@(21,0) TWBNNT obtained in this study are expected to be good predictions. Bending experiments on BNNTs with $n_{in} < n_c$ are greatly anticipated.

4. Summary

In Section 2, the electronic structures of (n,0) zigzag SWBNNTs subjected to tension, torsion, and flattening were investigated by first-principles DFT calculations. The results revealed that the three deformation modes decrease the energy gaps of the SWBNNTs because of the decrease in the CBM energy caused by an overlap of CBM charge densities between circumferentially neighboring boron atoms. The key findings of Section 2 are that flattening with a force smaller than that applied for tension or torsion causes a larger decrease in the energy gap and that the force required for flattening SWBNNTs is not unrealistic.

In Section 3, electronic structures of flattened MWBNNTs with the zigzag chiral index (n,0) were investigated by first-principles DFT calculations. The key findings obtained in Section 3 are summarized as follows:

1. When the chiral index of the innermost tube, n_{in}, of a zigzag MWBNNT is larger than the critical one, n_c, the energy gap decreases monotonically with increasing flattening compression.

2. When $n_{in} < n_c$, the energy gap first increases and then decreases with increasing flattening compression. This initial increase and subsequent decrease are caused by a charge spreading from the first to the second innermost tube and by bond formation in curved regions in the second innermost tube, respectively.

3. The n_c is found to be an integer between 5 and 13.

Author details

Nobutada Ohno, Dai Okumura and Yusuke Kinoshita

Department of Computational Science and Engineering, Graduate School of Engineering, Nagoya University, Nagoya, Japan

References

[1] Hohenberg P, Kohn W. Inhomogeneous Electron Gas. Physical Review 1964;136(3B) B864-B871.

[2] Kohn W, Sham LJ. Self-Consistent Equations Including Exchange and Correlation Effects. Physical Review 1965;140(4A) A1133-A1138.

[3] Jones RO, Gunnarsson O. The density functional formalism, its applications and prospects. Reviews of Modern Physics 1989;61(3) 689-746.

[4] Lundqvist S, March N. Theory of the Inhomogeneous Electron Gas. New York: Plenum Press; 1983.

[5] Martin R. M. Electronic Structure. Cambridge: Cambridge University Press; 2004.

[6] Kinoshita Y, Hase N, Ohno N. Flattening-induced electronic changes in zigzag single- and multi-walled boron nitride nanotubes: A first-principles DFT study. Physical Review B 2009;80(12) 125114.

[7] Kinoshita Y, Ohno N. Electronic structures of boron nitride nanotubes subjected to tension, torsion, and flattening: A first-principles DFT study. Physical Review B 2010;82(8) 085433.

[8] Blase X, Rubio A, Louie S. G, Cohen M. L. Stability and Band Gap Constancy of Boron Nitride Nanotubes. Europhysics Letters 1994;28(5) 335-340.

[9] Rubio A, Corkill J. L, Cohen M. L. Quasiparticle band structures of short-period superlattices and ordered alloys of AlN and GaN, Physical Review B 1994;49(3) 1952-1956.

[10] Chopra N. G, Luyken R. J, Cherrey K, Crespi V. H, Cohen M. L, Louie S. G, Zettl A. Boron nitride nanotubes. Science 1995;269(5226):966-967.

[11] Golberg D, Bando Y, Tang C, Zhi C. Boron Nitride Nanotubes. Advanced Materials 2007;19(18) 2413-2432.

[12] Chopra N. G, Zettl A. Measurement of the elastic modulus of a multi-wall boron nitride nanotube. Solid State Communications 1998;105(5) 297-300.

[13] Dumitrica T, Bettinger H. F, Scuseria G. E, Yakobson B. I. Thermodynamics of yield in boron nitride nanotubes. Physical Review B 2003; 68 (8), 085412.

[14] Golberg D, Bando Y, Kurashima K, Sato T. Synthesis and characterization of ropes made of BN multiwalled nanotubes. Scripta Materialia 2001;44(8-9) 1561-1565.

[15] Iijima S. Helical microtubules of graphitic carbon Nature 1991;354 56 - 58.

[16] Suryavanshi A. P, Yu M. F, Wen J, Tang C, BandoY, Elastic modulus and resonance behavior of boron nitride nanotubes. Applied Physics Letters 2004;84(14) 2527-2529.

[17] Akdim B, Pachter R, Duan X, Adams W. W. Comparative theoretical study of single-wall carbon and boron-nitride nanotubes. Physical Review B 2003;67(24) 245404.

[18] Baumeier B, Kruger P, Pollman J. Structural, elastic, and electronic properties of SiC, BN, and BeO nanotubes Physical Review B 2007;76(8) 085407.

[19] Kudin K. N, Scuseria G. E, Yakobson B. I. C2F, BN, and C nanoshell elasticity from ab initio computations. Physical Review B 2001;64(23) 235406

[20] Peng Y. J, Zhang L. Y, Jin Q. H, Li B. H, Ding D. T. Ab initio studies of elastic properties and electronic structures of C and BN nanotubes. Physica E: Low-dimensional Systems and Nanostructures 2006;33(1) 155-159

[21] Hernandez E, Goze C, Bernier P, Rubio A. Elastic Properties of C and BxCyNz Composite Nanotubes. A. Physical Review Letters 1998;80(20).

[22] Hernandez E, Goze C, Bernier P, Rubio A. Applied Physics A 1999;68(3) 287-292.

[23] Griebel M, Hamaekers J, Heber F. A molecular dynamics study on the impact of defects and functionalization on the Young modulus of boron-nitride nanotubes. Computational Materials Science 2009;45(4):1097-1103.

[24] Li C, Cho T. W. Journal of Nanoscience and Nanotechnology;2006.

[25] Santosh M, Maiti P. K, Sood A. K. Elastic properties of boron nitride nanotubes and their comparison with carbon nanotubes. Journal of Nanoscience and Nanotechnology 2009;9(9):5425-5430.

[26] Song J, Wu J, Huang Y, Hwang K. C, Jiang H. Stiffness and thickness of boron-nitride nanotubes. Journal of Nanoscience and Nanotechnology 2008;8(7) 3774-3780.

[27] Verma V, Jindal V. K, Dharamvir K. Elastic moduli of a boron nitride nanotube. Nanotechnology 2007;18(43),435711.

[28] Treacy M. M. J, Ebbesen T. W, Gibson J. M. Exceptionally high Young's modulus observed for individual carbon nanotubes. Nature 1996;381(678-680).

[29] Krishnan A, Dujardin E, Ebbesen T. W, Yianilos P. N, Treacy M. M. J. Young's modulus of single-walled nanotubes. Physical Review B 1998;58(20) 14013-14019.

[30] Bai X, Golberg D, Bando Y, Zhi C, Tang C, Mitome M, Kurashima K. Deformation-driven electrical transport of individual boron nitride nanotubes. Nano Letters 2007;7(3)632-637.

[31] Kim Y. H, Chang K. J, Louie S. G. Electronic structure of radially deformed BN and BC3 nanotubes. Physical Review B 2001;63(20)205408.

[32] Kresse G, Furthmuller J. Efficient iterative schemes for ab initio total-energy calculations using a plane-wave basis set. Physical Review B 1996;54(16) 11169-11186.

[33] Kresse G, Furthmuller J. Ab initio molecular dynamics for liquid metals Physical Review B 1993;47(1) 558-561.

[34] Vanderbilt D. Soft self-consistent pseudopotentials in a generalized eigenvalue formalism. Physical Review B 1990;41(11) 7892-7895.

[35] Perdew J. P, Wang Y. Accurate and simple analytic representation of the electron-gas correlation energy. Physical Review B 1992;45(23) 13244-13249.

[36] Monkhorst H. D, Park J. D. On Special Points for Brillouin Zone Integrations. Physical Review B 1976;13(12) 5188-5192.

[37] Arnaud B, Lebegue S, Rabiller P, Alouani, M. Huge excitonic effects in layered hexagonal boron nitride. Physical Review Letters 2006;96(2) 026402.

[38] Blase X, Rubio A, Louie S. G, Cohen M. L. Quasiparticle band structure of bulk hexagonal boron nitride and related systems. Physical Review B 1995;51(11) 6868-6875.

[39] Park C. H, Spataru C. D, Louie S. G. Excitons and Many-Electron Effects in the Optical Response of Single-Walled Boron Nitride Nanotubes. Physical Review Letters 2006;96(12) 126105.

[40] Wirtz L, Marini A, Rubio A. Excitons in boron nitride nanotubes: dimensionality effects. Physical Review Letters 2006;96 126104.

[41] Barboza A. P. M, Chacham H, Neves B. R. A. Universal Response of Single-Wall Carbon Nanotubes to Radial Compression. Physical Review Letters 2009;102(2) 025501.

[42] Jhi S. H, Roundy D. J, Louie S. G, Cohen M. L. Formation and electronic properties of double-walled boron nitride nanotubes. Solid State Communications 2005;134(6) 397-402.

Chapter 8

Tissue Engineering and Regenerative Medicine

Minoru Ueda

1. Introduction

Tissue specific adult stem cells are the most hopeful cell at this moment for clinical use because it may represent Mother Nature's repair cells. Such cells are potentially present within all of the tissue of the body and may remain dormant until they are activated in response to tissue injury. Initially, the chemical environment at the site of any injury is very hostile. These adult stem cells, having a low oxygen requirement, appear to have the ability to survive this environment. When adequate numbers of cells have been achieved by multiplication, they are then programmed to mature and repair tissue damage of a certain magnitude. If this is the case, with the development of appropriate tissue-specific scaffolds and the use of the optimal cell type, I believe that physicians and scientists will ultimately be able to repair or replace any tissue in the human body that is injured or damaged as a result of disease or trauma. Studies involving the use of stem cells and mature cells, in combination with genetic manipulation and determination of the efficacy cellular delivery systems and scaffoldings, should be enable rapid progression to human treatments. It is my belief that exploring the use of appropriate vehicles and cell types will ultimately lead to resolution of stroke symptoms, such as paralysis, and may help reverse symptoms associated with such central nervous system diseases as Parkinson's disease and Alzheimer's disease.

2. Research projects

This chapter was edited by collecting all the achievement performed in the laboratory of oral and maxillofacial surgery and it brings together the specific experiences of the scientific community in these experiences of our scientific community in this field as well as the clinical experiences of the most renowned experts in the fields from all over Nagoya University. The editors are especially proud of bringing together the leading biologists and material scientists together with dentist, plastic surgeons and surgeons of all specialities from all department of the medical school of Nagoya University. Taken together, this unique collection of worldwide expert achievement and experiences represents the current spectrum of possibilities in tissue engineered substitution.

2.1. Bone regeneration with self assembling peptide nanofiber scaffolds in tissue engineering for osseointegration of dental implants

The aim of this study was to evaluate the correlation between the osseointegration of dental implants and tissue-engineered bone using a nanofiber scaffold, *PuraMatrix* (PM). The first molar and all premolars in the mandible region of dogs were extracted, and three bone defects were prepared with a trephine bar on both sides of the mandible after 4 weeks. The experimental groups were as follows: 1) PM, 2) PM and dog mesenchymal stem cells (MSCs), 3) PM, dog MSCs, and platelet-rich plasma (PRP), and 4) a control (defect only). Implants were placed with in the prepared areas 8 weeks later, and assessed by histological and histomorphometric analyses (bone-implant contact (BIC)). The BICs for groups 1, 2, 3, and 4 were 40.77%, 50.35%, 55.64% and 30.57%, respectively. The findings indicate that PM may be useful as a scaffold for bone regeneration around dental implants.

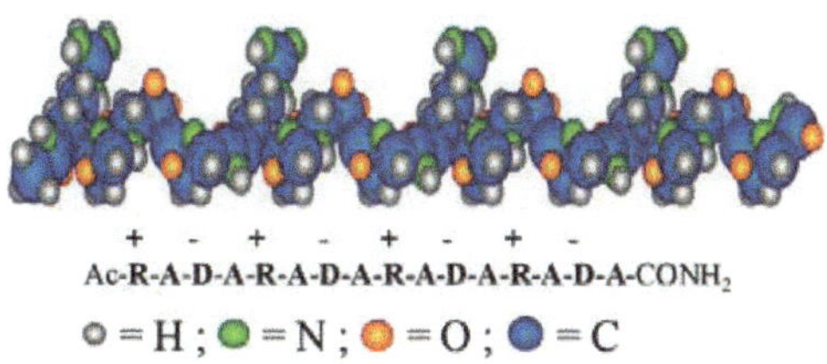

Figure 1. Molecular structure of PM. Molecular models of amiphiphilic self-complementary peptide have 16 amino acids with an alternating polar and nanopolar pattern. A= alanine; R= arginine; D= aspartic acid; + and – refer to the positively and negatively charged residues, respectively (From Kohgo et al. [1]. Reprinted with permission from Quintessence Publishing Co, Inc, Chicago).

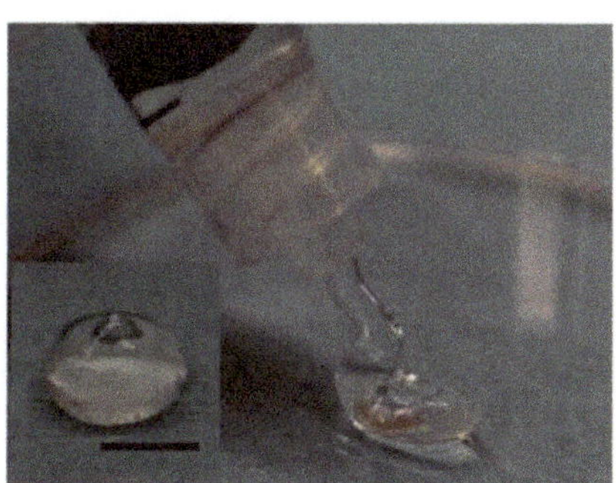

Figure 2. PM is an injectable scaffold and shows good plasticity. Bar = 5 mm. (From Kohgo et al. [1]. Reprinted with permission from Quintessence Publishing Co, Inc, Chicago).

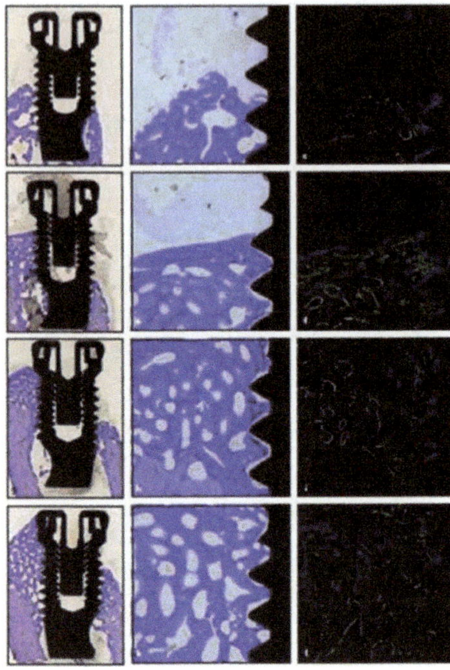

Figure 3. Photographs of histologic sections, as seen on light microscopy, 8 weeks after implant placement. (a to c) In the control group, the buccal and lingual walls were not sufficiently for dental implants. (d to f) In the PM group, slight bone regeneration in the lingual wall was observed. (g to i) However, slightly more could be seen in the PM/dog MSCs group. (j to l) On the other hand, the amount of regenerated bone was greatest in the PM/dog MSCs/PRP group (a, d, g and j, magnification tion rat, e, h, and k, magnification ×200; c, f, i, and l, maginification ×200) (From Kohgo et al. [1]. Reprinted with permission from Quintessence Publishing Co, Inc, Chicago).

Graft materials	BIC
Control	30.57% ± 2.50%
PM	40.77% ± 12.85%
PM/dMSCs	50.35% ± 8.02%
PM/dMSCs/PRP	55.64% ± 4.97%

BIC = bone-to-implant contact; SD = standard deviation; PM = PuraMatrix; dMSCs = dog mesenchymal stem cells; PRP = Platelet-rich plasma * $P < 0.01$ **$P < 0.05$ (From Kohgo et al. 2011. Reprinted with permission from Quintessence Publishing Co, Inc, Chicago).

Table 1. Results of BIC (mean ±SD)[†]

The results suggest that tissue-engineered bone can integrate well around dental implants. PM is a 3D structure may have the potential to be a scaffold applicable in bone tissue engineering.

2.2. Self-assembling peptide nanofiber scaffolds, platelet rich plasma, and mesenchymal stem cells for injectable bone regeneration with tissue engineering

The ideal biomaterial scaffolds, particularly in cranio-maxillofacial, plastic, or orthopedic fields with complicated bone-defect shapes, should have excellent plasticity for fitting into complex defect shapes and the rate of absorption needs to be fast in order to avoid infection. Moreover, these materials and their internal structures may not provide a particularly favorable environment for cell survival and bone regeneration, and internal, microscale environment of these materials needs to serve as an extracellular matrix (ECM). ECM is a dynamic organized nanocomposite that not only provides mechanical support for embedded cells but also interacts with cells and promotes and regulates cellular functions such as adhesion, migration, proliferation, and differentiation and is consequently involved in three-dimensional morphogenesis. This study considered the use of the matrix material PM, which is synthesized by chemical peptide methods and has similarities to the fibers and pore sizes found in the ECM. We investigated a capability of PM as a scaffold for bone regeneration in combination with dog MSCs and/or PRP using tissue engineering and regenerative medicine technology. First, teeth were extracted from an adult hybrid dog's mandible region. After 4 weeks, bone defects were prepared on both sides of the mandible with a trephine bar. The listed graft materials were implanted into these defects: 1) control (defect only), 2) PM, 3) PM/PRP, 4) PM/dog MSCs, and 5) PM/dog MSCs/PRP. At 2, 4, and 8 weeks after implantation, each sample was collected from the graft area with a trephine bar and assessed by histological and histomorphometric analyses.

From histological evaluation, It was observed that the bone regenerated by PM/dog MSCs/PRP was excellent quality, and it was found that mature bone had been formed. Histometrically, at 8 weeks newly formed bone areas comprised 12.39 ± 1.29% (control), 25.28 ± 3.92% (PM), 27.72 ± 3.15% (PM/PRP), 50.07 ± 3.97% (PM/dog MSCs), and 58.43 ± 5.06% (PM/dog MSCs/PRP). The PM/dog MSCs and PM/dog MSCs/PRP groups showed a significant increase at all weeks compared with the control, PM, or PM/PRP. These results showed that MSCs might keep their own potential and promote new bone regeneration in the three-dimensional structure by PM scaffolds. Taken together, it is suggested that PM might be useful as a scaffold of bone regeneration in cell therapy, and these results might lead to an effective treatment method for bone defects. In the future this tissue-engineered bone grafting material could be used as a minimally invasive method instead of traditional grafting procedures.

2.3. Effects of self-assembling peptide hydrogel scaffold on bone regeneration with recombinant human bone morphogenetic protein-2

Various biomaterials have been tested as scaffolds for bone regeneration, such as beta-tricalcium phosphate, hydroxyapatite, and polymers. However, a scaffold has still not been found that has the characteristics of biologic safety, absorbability, cell interaction, and bone inductivity. A self-assembling peptide hydrogel scaffold is made of artificial synthetic

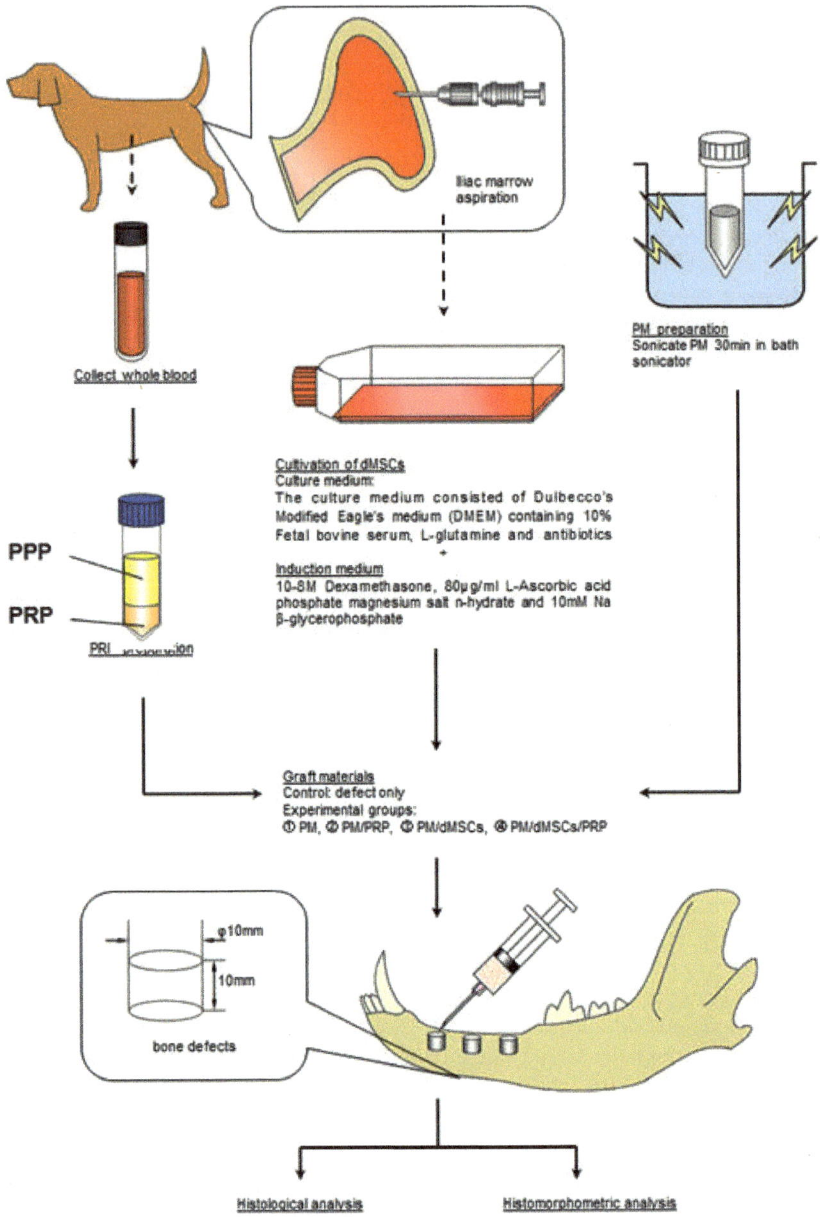

Figure 4. Schema of the experimental protocol (From Yoshimi et al. [2, 3]. Reprinted with permission).

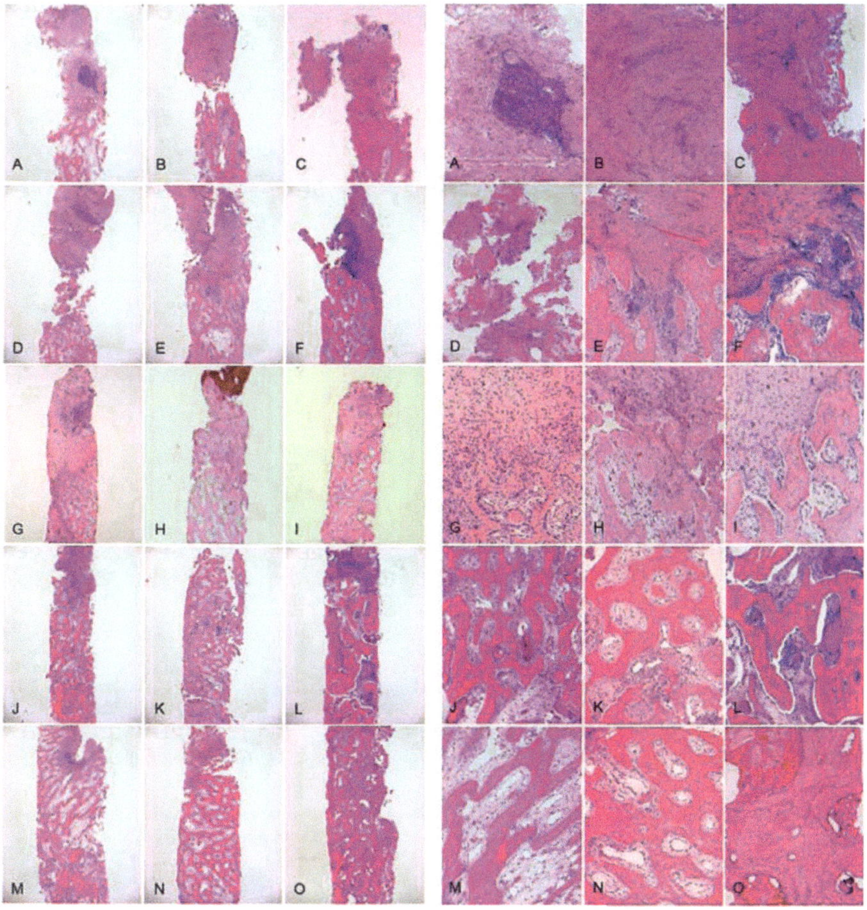

Figure 5. Histological evaluation of control, PM, PM/PRP, PM/dog MSCs, and PM/dog MSCs/PRP implantations at each time point (center figure:lower magnification, right figure: higher magnification). Sections of representative implants are shown from the respective group. The sections were stained with hematoxylin and eosin. Original magnification, ×25 for all photographs. (A) 2 weeks Control group, (B) 4 weeks Control group, (C) 8 weeks Control group, (D) 2 weeks PM group, (E) 4 weeks PM group, (F) 8 weeks PM group, (G) 2 weeks PM/PRP group, (H) 4 weeks PM/PRP group, (I) 8 weeks PM/PRP group, (J) 2 weeks PM/dog MSCs group, (K) 4 weeks PM/dog MSCs group, (L) 8 weeks PM/dog MSCs group, (M) 2 weeks PM/dog MSCs/PRP group, (N) 4 weeks PM/dog MSCs/PRP group, and (O) 8 weeks PM/dog MSCs/PRP group (From Yoshimi et al. [4, 5]. Reprinted with permission).

materials featuring biologic safety and absorbability. PM is expected to be a candidate as a scaffold for bone regeneration. Recombinant human bone morphogenetic protein-2 (rhBMP-2) has exhibited high osteogenic activity in experimental studies. The objective of this pilot study was to histologically evaluate bone regeneration using a self-assembling peptide hydrogel scaffold with rhBMP-2 on the bone augmentation in a rabbit calvaria model (Figure 6).

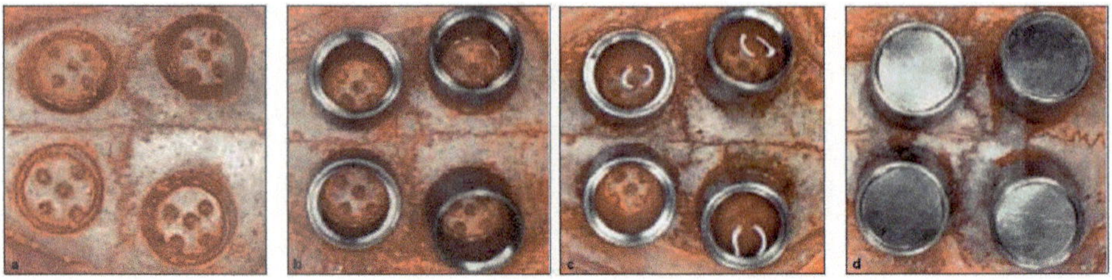

Figure 6. Photographs of surgical procedure. In the left and right parietal and frontal bones, 4 circular slits were prepared. Five holes were prepared in outer cortical bone inside this circle (a). Apical end of the 4 titanium cylinders was pressed into each slit, and primary fixation was obtained (b). Four titanium cylinders were filled with respective materials (c). The top of the cylinders was closed with a titanium lid (d) (From Ikeno et al. [4]. Reprinted with permission from Quintessence Publishing Co, Inc, Chicago).

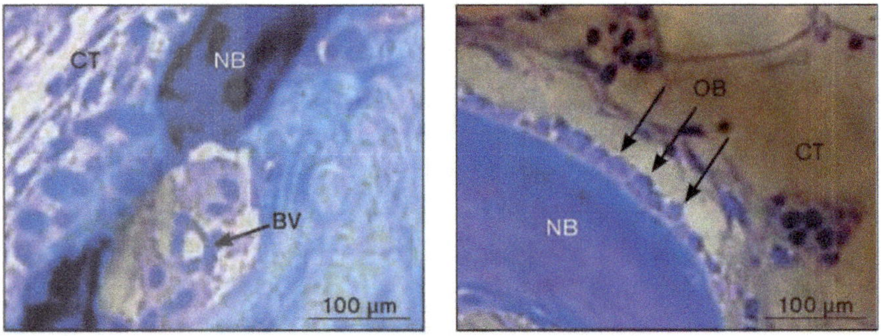

Figure 7. Higher magnification of a histological section Bar = 100μm. NB, newly formed bone; CT, connective tissue; BV, blood vessel; OB, osteoblast-like cells (From Ikeno et al. [4]. Reprinted with permission from Quintessence Publishing Co, Inc, Chicago).

New bone formation seemed to occur from the calvarial bone through the perforations in the outer cortical bone; newly formed bone was also observed in all groups. Under higher magnification, newly formed bone, including cells and some blood vessels, was observed in the connective tissue and the edges of bone were lined with osteoblast-like cells in the cylinder (Figure 8). Regenerated tissue in the sample treated with PM/rhBMP-2 was observed in about two thirds of the cylinders (Figure 9).

Histomorphometric analysis showed that regenerated tissue in the cylinder with PM/rhBMP-2 was significantly increased compared to the empty control (Fig. 8). The mean area values of regenerated tissue in the cylinders were 35.80% ± 10.35% (control), 47.94% ± 5.65% (rhBMP-2), 48.94% ± 11.33% (PM), and 58.06% ± 14.84% (PM/rhBMP-2). The mean area values of newly formed bone in the cylinders were 9.39% ± 4.34% (control), 14.03% ± 2.25% (rhBMP-2), 13.99% ± 2.15% (PM), and 16.61% ± 3.79% (PM/rhBMP-2). Neither rhBMP-2 nor PM alone significantly enhanced bone regeneration compared to the empty control cylinder.

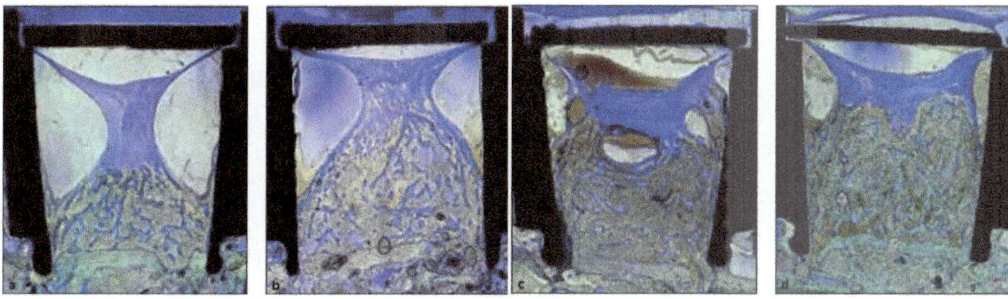

Figure 8. Histological sections of cylinders with empty control (a), rhBMP-2 (b), PM (c), and PM/rhBMP-2 (d) (toluidine blue stain) (From Ikeno et al. [4]. Reprinted with permission from Quintessence Publishing Co, Inc, Chicago).

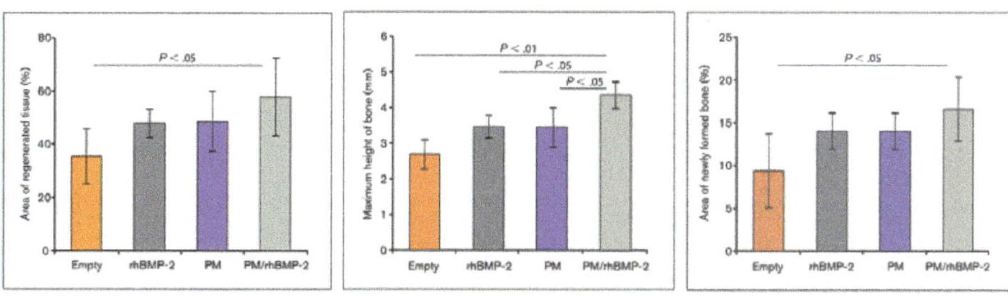

Figure 9. Regenerated tissue area measured by image analysis. The percentage of area of regenerated tissue (left), maximum height of newly-formed bone (middle), and area of newly-formed bone (right) in the cylinder with each group was displayed. Bar = SD. Analysis of variance (ANOVA) (From Ikeno et al. [4]. Reprinted with permission from Quintessence Publishing Co, Inc, Chicago).

PM, a synthetic self-assembling peptide, used in combination with recombinant human bone morphogenetic protein-2 significantly enhanced bone regeneration in a bone augmentation model in rabbits. PM promises to be an alternative synthetic material as a useful carrier for recombinant human bone morphogenetic protein for bone regeneration.

2.4. A study of bone healing around the titanium screw implants in osteoporosis: Can sandblasted surface contribute to the implant stability?

Dental implants are widely performed as a prosthetic treatment in edentulous patients. However, in the aged patients, there are various systemic risk factors such a osteoporosis. The aim of this study was to investigate whether estrogen deficiency interrupts bone healing around titanium implants and to evaluate whether bone healing around implants under the condition of estrogen deficiency is affected by implant surface variance. Forty eight female SD rats were divided into two groups: ovariectomized rats (OVX; n=24) and sham operated rats (SHAM; n=24). Each group was further divided into two groups: a machine polished implants placed group and a sandblasted implants placed group. Both implants were placed in the rat left femur centrifugal site 84 days after OVX or sham surgery. After 28, 56 days, the rats were killed, and non-decalcified section was obtained. BIC and bone area (BA) around the implants were assessed with cortical bone and cancellous bone. Furthermore, bone density (BD) was evaluated in a 500mm wide zone of cancellous bone lateral to the implants (Figure10). At 28 and 56 days after implantation, no significant difference was found between the OVX and SHAM groups for BIC and BA in cortical bone. BIC, BA, and BD with cancellous bone was lower in OVX group than in SHAM group. However, BIC and BA tended to improve by the variance of implant surface (Figure 10.11.12.13).

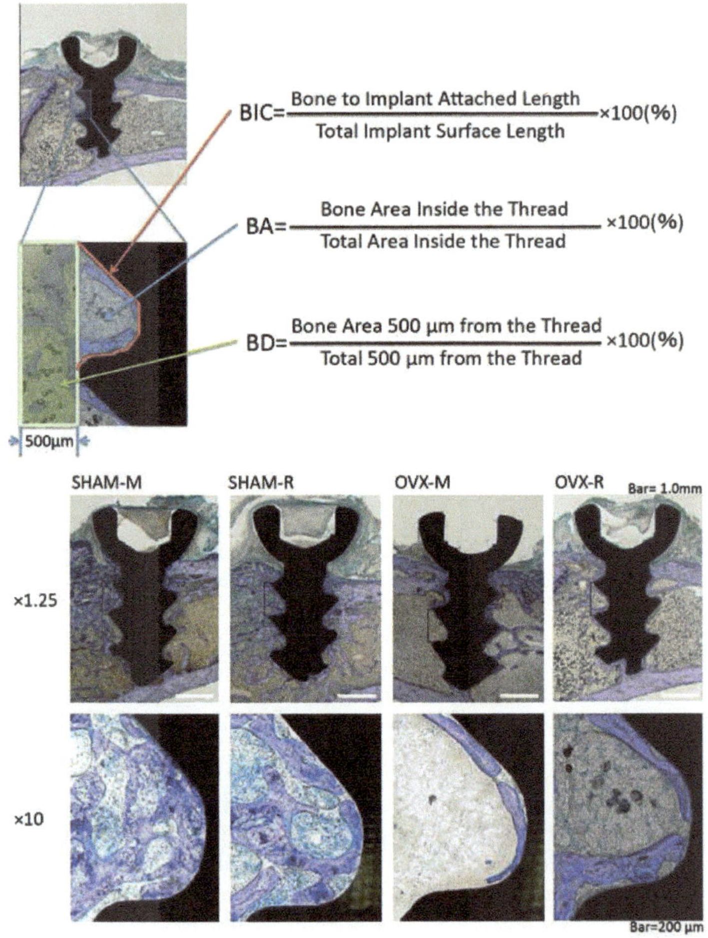

Figure 10. Histomorphological observation and morphological evaluations (From Tateishi et al. [5, 6]. Reprinted with permission from Quintessence Publishing Co, Inc, Chicago).

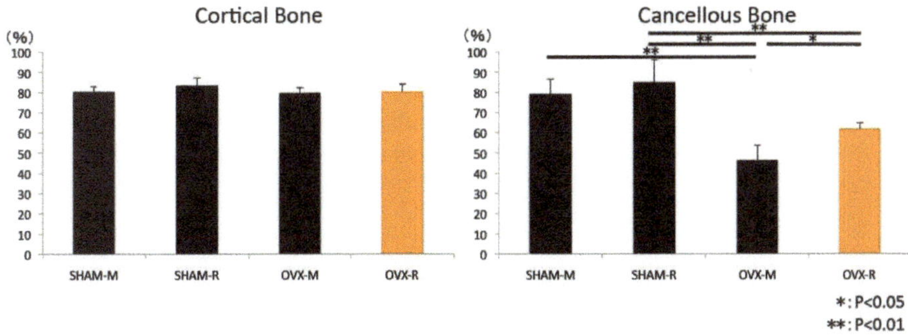

Figure 11. Bone to implant contact (BIC) at 28 days after implantation (From Tateishi et al. [5, 6]. Reprinted with permission from Quintessence Publishing Co, Inc, Chicago).

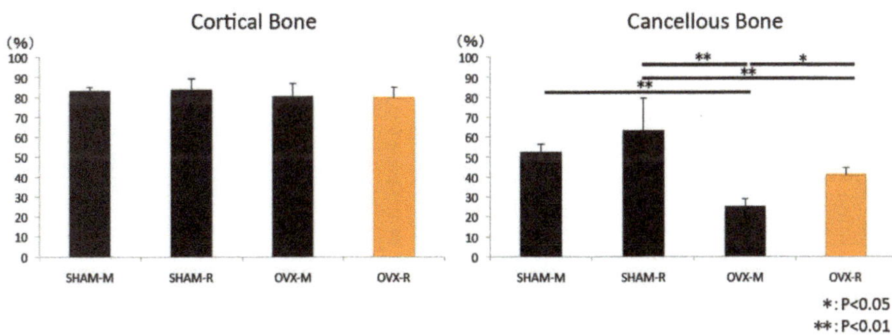

Figure 12. Bone area (BA) inside the implant thread at 28 days after implantation (From Tateishi et al. [5, 6]. Reprinted with permission from Quintessence Publishing Co, Inc, Chicago).

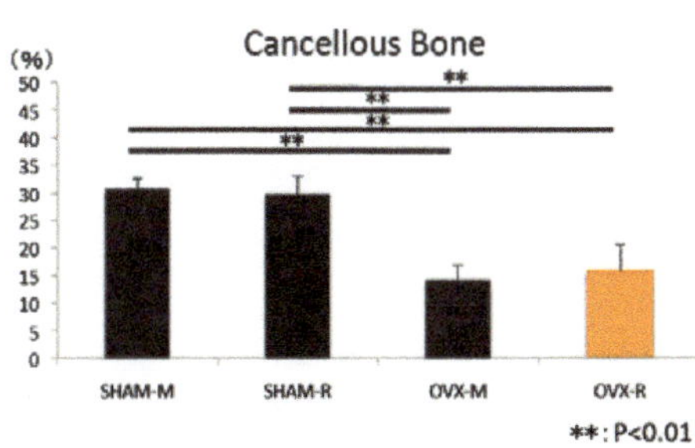

Figure 13. Bone density (BD) around the implants at 28 days after implantation (From Tateishi et al. [5, 6]. Reprinted with permission from Quintessence Publishing Co, Inc, Chicago).

Estrogen deficiency affected bone healing and bone density around the titanium implants, especially in cancellous bone, but sandblasted surface property has the possibility to improve osseointegration. However, the positive effect by rough surface property is limited on the implant surface.

2.5. Osteogenic induction of bone marrow-derived stromal cells on simvastatin-releasing, biodegradable, nano- to microscale fiber scaffolds

Tissue engineering is an effective approach for the treatment of bone defects. Statins have been demonstrated topromote osteoblastic differentiation of bone marrow-derived MSCs. Electrospun biodegradable fibers have also shown applicability to drug delivery in the form of bone tissue engineered scaffolds with nano- to microscale topography and high porosity similar to the natural ECM. The aim of this study was to investigate the feasibility of a simvastatin-releasing, biodegradable, nano- to microscale fiber scaffold (SRBFS) for bone tissue engineering with MSCs. Simvastatin was released from SRBFS slowly (Figure 14). MSCs were observed to spread actively and rigidly adhere to SRBFS. MSCs on SRBFS showed an increase in alkaline phosphatase activity 2 weeks after cell culture (Figure 15). Furthermore, osteoclastogenesis was suppressed by SRBFS *in vitro* (Figure 16a-c). The new bone formation and mineralization in the SRBFS group were significantly better than in the biodegradable fiber scaffold (BFS) without simvastatin 12 weeks after implantation of the cell-scaffold construct into an ectopic site on the murine back (Figure 17a-c). These results suggest that SRBFS promoted osteoblastic differentiation of MSCs *in vitro* and *in vivo*, and demonstrate feasibility as a bone engineering scaffold.

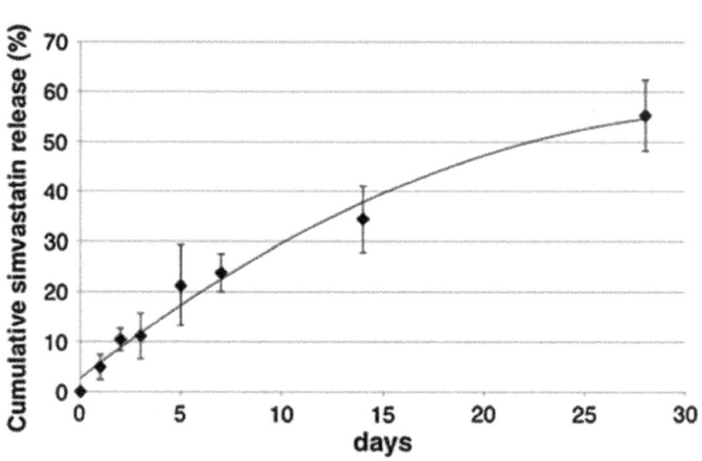

Figure 14. Cumulative release of SRBFS *in vitro* (From Wadagaki et al. [7, 8]. Reprinted with permission).

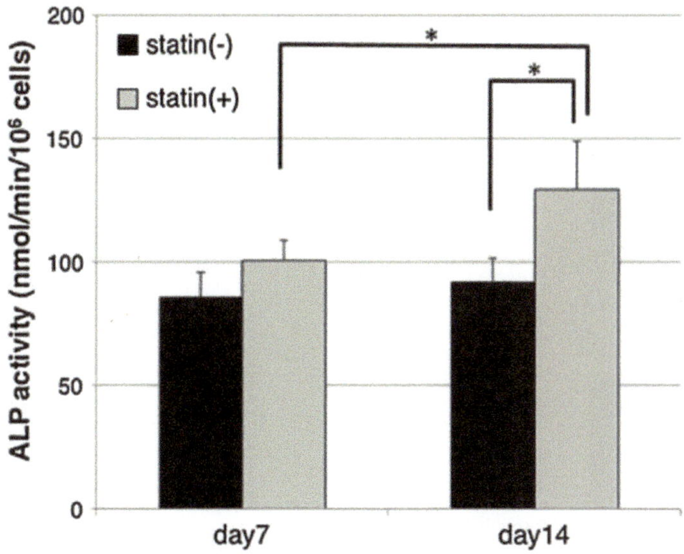

Figure 15. ALP activity of MSCs on SRBFS and BFS measured on days 7 and 14 (From Wadagaki et al. [7, 8]. Reprinted with permission).

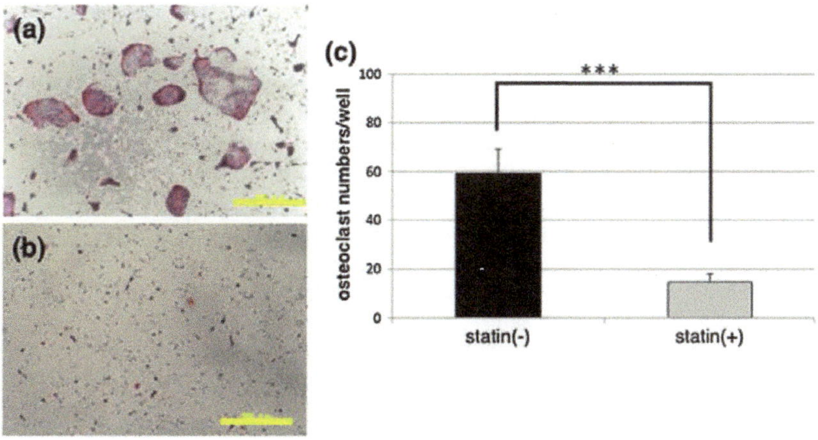

Figure 16. Simvastatin released from SRBFS inhibits RANKL-induced osteoclastogenesis 5 days after stimulation. Cells were cultured for 5 days with (a) SRBFS and (b) BFS after RANKL treatment and stained for TRAP expression. Bar = 500 μm. (c) The total number of TRAP-positive multinucleated osteoclasts (i.e., those containing three nuclei) per well were counted (From Wadagaki et al. [7, 8]. Reprinted with permission).

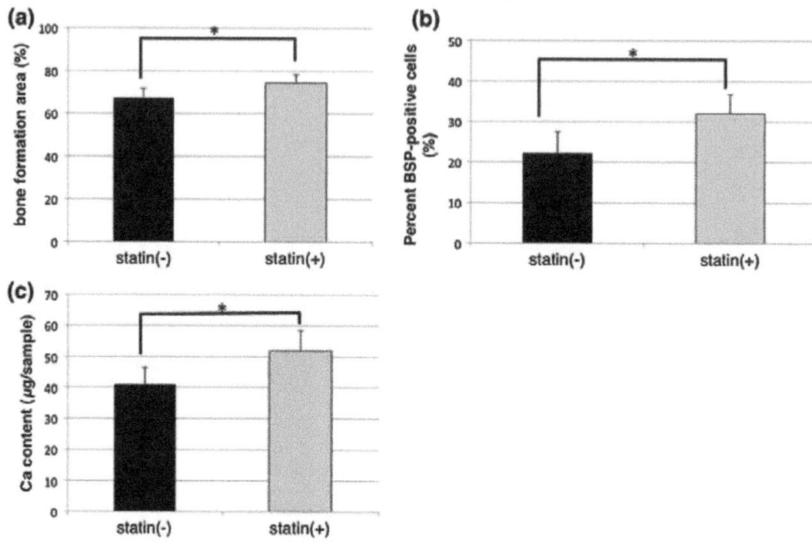

Figure 17. (a) Bone formation area analyzed by H&E staining 12 weeks after implantation. (b) Percent BSP (bone sialo protein) -positive cells as determined by immunohistochemical staining with BSP 12 weeks after implantation. (c) Accumulated calcium content per sample 12 weeks after implantation (From Wadagaki et al. [7, 8]. Reprinted with permission).

2.6. Conditioned media from mesenchymal stem cells enhanced bone regeneration in rat calvarial bone defects

Recently tissue engineering has become available as a treatment procedure for bone augmentation. However, this procedure has several problems such as an expensive cost for capital investment and cell culture, complicated safety and quality management of cell handling and invasiveness of cell collection for patients. On the other hand, it was reported that the stem cells secreted many growth factors and chemokines during their cultivation and that could affect on the cellular characteristics and behavior. This study investigated the effect of stem cell cultured conditioned media on bone regeneration (Fig.18). Cultured conditioned media from human bone-marrow-derived mesenchymal stem cells (MSC-CM) enhanced the migration, proliferation and expression of osteogenetic marker genes, such as *osteocalcin* and *Runx2,* of rat MSCs *in vitro*. MSC-CM included cytokines such as insulin-like-growth factor (IGF)-1 and vascular endothelial growth factor (VEGF).

In vivo, a prepared bone defect of a rat calvarial model was implanted in five different rat groups using one of the following graft materials: human MSCs/agarose (MSCs), MSC-CM/agarose (MSC-CM), and defect only (Defect). After 4 and 8 weeks, implant sections were evaluated using micro-computed tomography (micro-CT) and histological analysis. Micro-CT analysis indicated that the MSC-CM group had a greater area of newly regenerated bone compared with the other groups ($P < 0.05$) (Figure 19) and histological analysis at 8 weeks indicated that the newly regenerated bone bridge almost covered the defect. Interestingly, the effects of MSC-CM were stronger than those of the MSCs group.

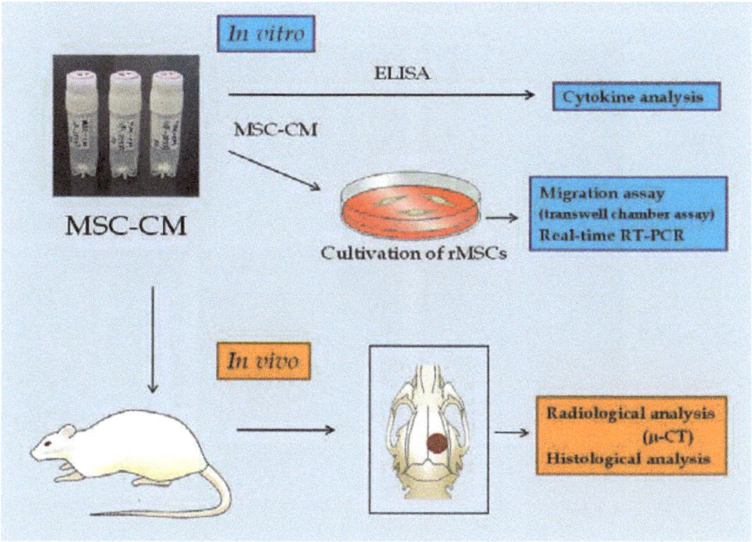

Figure 18. Outline of the experimental protocol.

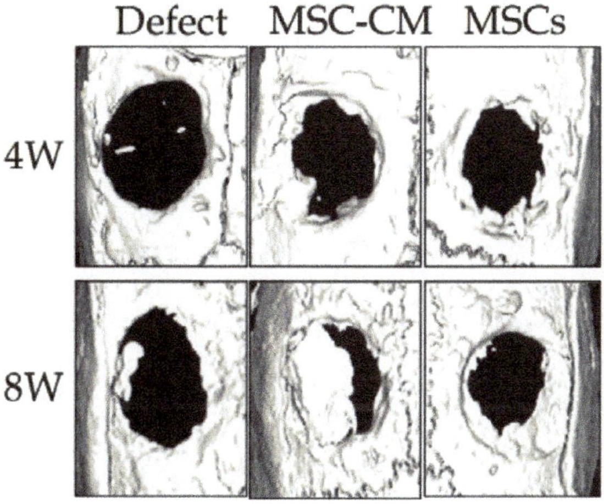

Figure 19. Micro-CT analysis of bone regeneration following implantation of MSC-CM or controls into a bone defect.

In vivo imaging also showed that migration of injected rMSCs to the bone defect in the MSC-CM group was greater than in the other groups (Figure 20).

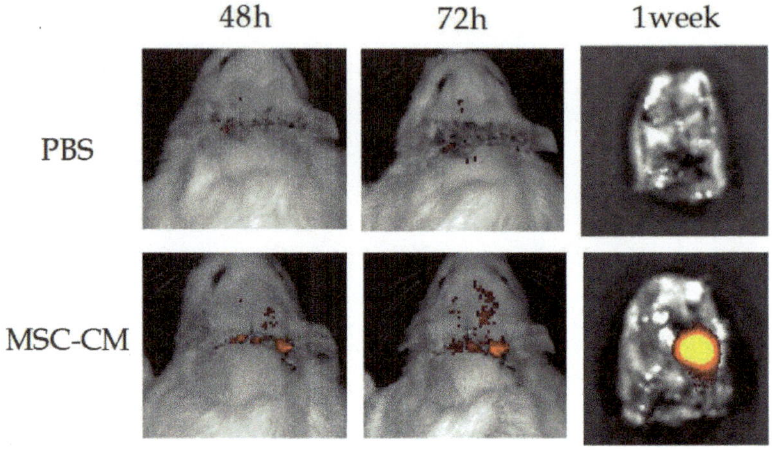

Figure 20. *In vivo* imaging of infected rat MSCs migration to implants.

These results demonstrated that MSC-CM can regenerate bone through mobilization of endogenous stem cells. The use of stem cell cultured conditioned media for bone regeneration will be a unique concept that utilizes paracline factors of stem cells without cell transplantation.

2.7. Recovery of neo-callus formation in the H-DO gap by the angiogenic activities of SDF-1

Distraction osteogenesis (DO) is a unique therapy that induces skeletal tissue regeneration without stem/progenitor cell transplantation. Although the self-regeneration property of DO provides many clinical benefits, the long treatment period required is a major drawback. A high-speed DO mouse model (H-DO), in which the distraction was done two times faster than in control DO (C-DO) mice, failed to generate new bone callus in the DO gap. We found that this was caused by the unsuccessful recruitment of bone marrow endothelial cells (BM-ECs)/endothelial progenitor cells (EPCs) into the gap. We then tested the ability of a local application of stromal cell-derived factor-1 (SDF-1), a major chemo-attractant for BM-ECs/EPCs, to accelerate the bone regeneration in H-DO. Our data showed that, in H-DO, SDF-1 induced callus formation in the gap through the recruitment of BM-ECs/EPCs, the maturation of neo-blood vessels, and increased blood flow. These results indicate that the active recruitment of endogenous BM-ECs/EPCs may provide a substantial clinical benefit for shortening the treatment period of DO.

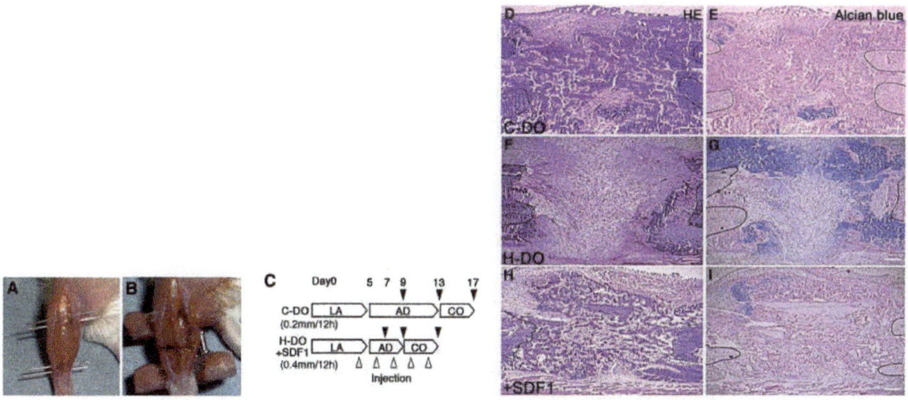

Figure 21. Traction speed affects new bone callus formation in mouse DO models.(A, B) Surgical procedure for the mouse DO model. An anterior longitudinal incision was made on the right leg of the animal. Needles were inserted through the skin into the proximal and distal metaphysis of the tibia (A). Subsequently, sets of needles were fixed to the custom-made fixator with acrylic resin (B). After polymerization of the resin, osteotomy was carried out at the middle of the diaphysis (arrow). (C) Distraction protocols and experimental design. After a 5-day latency period, distraction was started at a rate of 0.2 mm/12 h (C-DO) or 0.4 mm/12 h (H-DO). The lengthening was continued for 8 days in the C-DO and 4 days in the H-DO model, resulting in a length increase of 3.2 mm. Black arrowheads indicate the time points of sacrifice. White arrowheads indicate the time points for injecting 200 ng SDF-1 protein (+SDF-1). LA, latency period; AD, active distraction period; CO, consolidation period. (D–I) Representative micrographs of sections displaying the DO gap stained with Hematoxylin–Eosin (HE) (D, F and H) and Alcian Blue–Fast red (E, G and I) (n = 8). The left and right of each figure correspond to the end of the distal and proximal bone fragment, respectively. Neo-callus formation was evident within the C-DO gap at the end of the consolidation period (D). A little cartilage was observed in the periosteal but not in the endosteal region (E). The H-DO gap was filled with fibrotic tissues and periosteum-derived cartilages (F, G). The local administration SDF-1 rescued the callus formation in the H-DO gap (H, I). Bar = 300 µm (D–I). (From Fujio et al. [9]. Reprinted with permission).

We tested whether the local administration of SDF-1, a chemo- attractant for BM-ECs/EPCs, would rescue the disrupted callus formation and integration of BM-ECs/EPCs in the H-DO gap. A collagen gel matrix containing 200 ng of SDF-1 protein was injected into the H-DO gap every other day (Figure 21C). We found that the high levels of SDF-1 rescued the callus formation and increased the number of BM-ECs/EPCs in the H-DO gap (Figures 21H, I and 22C). Although about 80% of the CD31+ cells co-expressed Sca-1 in the C-DO gap, in the H-DO gap treated with SDF-1, 50% of the CD31+ cells were negative for Sca-1, suggesting that some of the BM-ECs/EPCs recruited by SDF-1 had already differentiated into mature endothelial cells (Figure 22E).

In summary, our study demonstrated that locally administered SDF-1 promotes the recruitment of endogenous BM-ECs/EPCs and neo callus formation in the DO gap. We propose that the regulation of endogenous stem/progenitor cell trafficking is a powerful therapeutic strategy in skeletal regeneration.

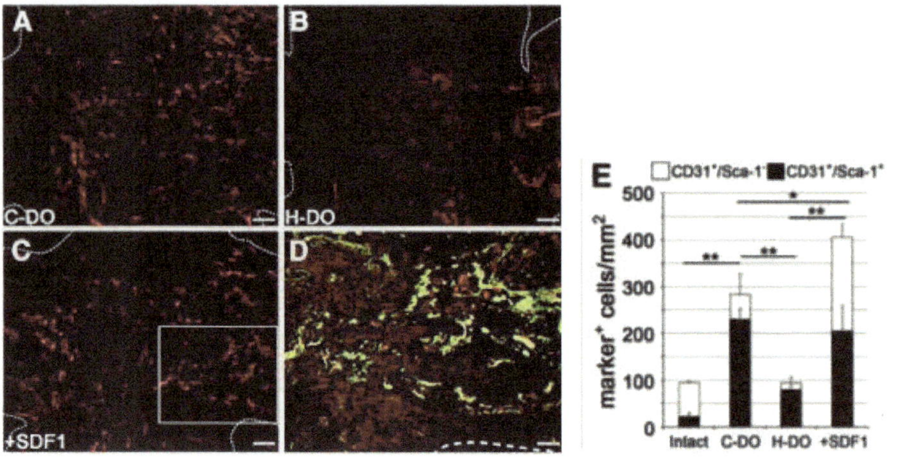

Figure 22. Contribution of BM-ECs/EPCs to DO healing. Mice were sacrificed at the middle of the active distraction period in each group: day 9 and day 7, respectively, for the C-DO and H-DO group. (A, B, and C) The recruitment of BM-ECs/EPCs was evaluated by immunofluorescence staining for CD31 (red). CD31+ BM-ECs/EPCs accumulated in the C-DO (A), but not in the H-DO gap (B). Local administration of SDF- 1 rescued the recruitment of BM-ECs/EPCs to the H-DO gap (C) (n = 8). (D) Boxed area in (C) is shown in higher-magnification micrographs, in which the CD31 signals are seen together with Sca-1 (green). Note that SDF-1 treatment increased the number of CD31+Sca-1 cells in the gap. (E) CD31-single-positive and CD31+Sca-1+ cells were counted by Image J software. The number of CD31+Sca-1+ BM-ECs/EPCs in the C-DO gap was significantly higher than that in intact bone marrow, whereas that in the H-DO gap was significantly lower. SDF-1 treatment rescued the number of BM-ECs/EPCs in the H-DO gap. The dotted line represents native bone. Data represent the mean ± SD. $^{**}P > 0.01$ and $^{*}P > 0.05$. Intact: intact bone marrow. Bar = 100 μm (A, B, C) and 50 μm (D) (From Fujio et al. [9]. Reprinted with permission).

2.8. Effect of GDF-5 and BMP-2 on the expression of tendo/ligamentogenesis-related markers in human PDL-derived cells

The effect of growth differentiation factor 5 and bone morphogenetic protein 2 on human periodontal ligament-derived cells was investigated with specialreference to tendo/ ligamentogenesis-related markers.The results from this study showed that both GDF-5 and BMP-2 affect the differentiation of PDL-derived cells in vitro. However, the effect on those differential ligament-makers was not identical and the underlying mechanisms might be complex.This study focused on the tendo-/ligamentogenesis related markers and confirmed the effect of those factors not only on crude PDL-derived cells but also on STRO-1+ and STRO-1- PDL-derived cells. The results from our study showed some potential beneficial effect of GDF-5 on periodontal tissue regeneration. However, the underlying mechanisms appear to be complicated, and the overall benefit of the clinical application of the factor requires further analyses.

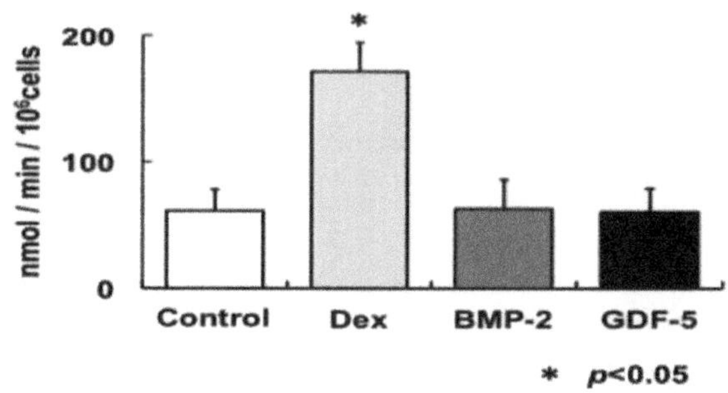

Figure 23. ALP activity of crude PDL-derived cells in passage 2. ALP activity of the Dex group was significantly greater compared with control, BMP-2 and GDF-5 groups. *$P < 0.05$. Values are the mean ± standard deviation of five experiments (From Inoue, M. et al. [10]. Reprinted with permission).

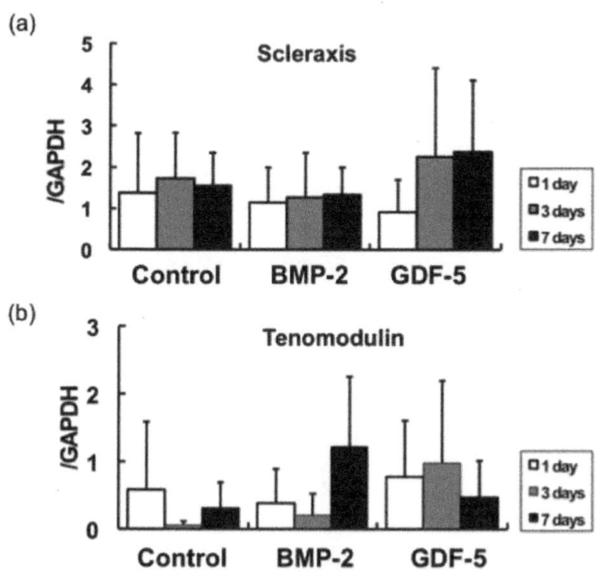

Figure 24. Quantitative RT-PCR analysis for scleraxis (a) and tenomodulin (b) gene expression of crude PDL-derived cells in passage 2. Cells were cultured with culture medium with or without BMP-2, GDF-5 for 1, 3 and 7 days. There were no significant differences among control, BMP-2 and GDF-5 groups on any time points. Values are the mean ± standard deviation of five experiments (From Inoue, M. et al. [10]. Reprinted with permission).

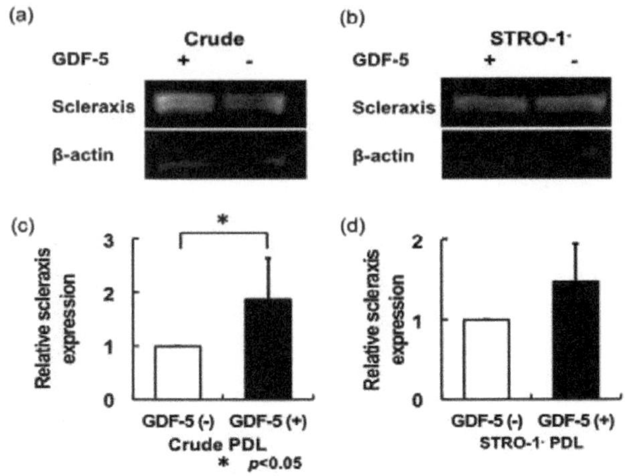

Figure 25. Western blot analyses of scleraxis crude PDL-derived cells (a) and STRO-1-PDL-derived cells (b) at passage 2. Expression of scleraxis was detected in all samples. An experiment representative of five similar studies is shown. (c) GDF-5 treated crude PDL-derived cells had significantly higher scleraxis expression than the other groups. (d) There was a similar tendency to crude PDL-derived cells, but no significant difference in STRO-1- PDL-derived cells among all groups. Values are the mean ± standard deviation of five experiments (From Inoue, M. et al. [10]. Reprinted with permission).

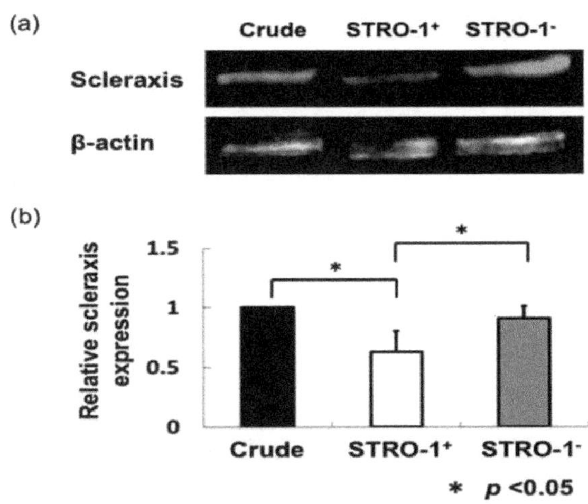

Figure 26. The results from western blot analyses of scleraxis in crude PDL-derived cells,STRO-1+ and STRO-1- PDL-derived cells at passage 2. All samples were treated with rmGDF-5. (a) The expression of scleraxis proteins was detected in all groups for 7 days. An experiment representative of six similar studies is shown. (b) Reduced scleraxis expression in STRO-1+PDL-derived cells was statistically significant compared to crude P2 and STRO-1- PDL-derived cells. Values are the mean ± standard deviation of six experiments (From Inoue, M. et al. [10]. Reprinted with permission).

2.9. Transient TWEAK overexpression leads to a general salivary epithelial cell proliferation

Tumor necrosis factor-like weak inducer of apoptosis (TWEAK) is a multifunctional cytokine that has pro-apoptotic, pro-angiogenic and pro-inflammatory effects. In liver, TWEAK leads to proliferation of progenitor oval cells, but not of mature hepatocytes. This study evaluated the hypothesis that TWEAK overexpression in salivary glands would lead to the proliferation of a salivary progenitor cell. A recombinant, serotype 5 adenoviral vector encoding human TWEAK, AdhTWEAK, was constructed, initially tested *in vitro*, and then administered to male Balb/c mice via cannulation of Wharton's duct. TWEAK expression *in vivo* was monitored as protein secreted into saliva and serum by enzyme-linked immunosorbent assays. Salivary cell proliferation was monitored by proliferating cell nuclear antigen staining and apoptosis was monitored using TUNEL staining. AdhTWEAK administration led to a dose-dependent, transient TWEAK protein expression (Figure 27), detected primarily in saliva. Salivary epithelial cell proliferation was generalized, peaking on approximately days 2 and 3 (Figure 28, 29). TWEAK expression had no detectable effect on apoptosis of salivary epithelial cells. Transient overexpression of TWEAK in murine salivary glands leads to a general proliferation of epithelial cells vs a selective stimulation of a salivary progenitor cell.

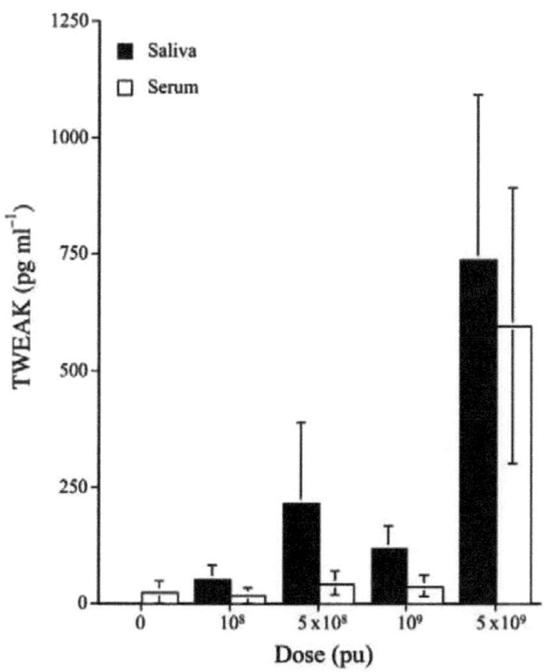

Figure 27. Effect of the AdhTWEAK dose administered on the detection of hTWEAK in murine saliva and serum (From Sugito et al. [11]. Reprinted with permission).

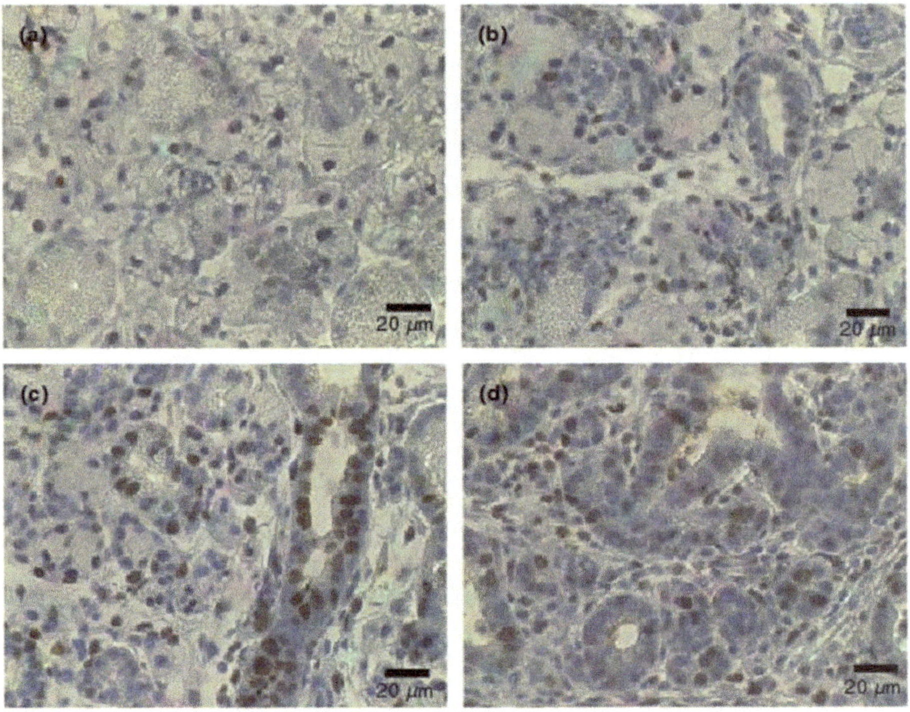

Figure 28. Detection of proliferating cell nuclear antigen (PCNA) staining in submandibular glands of mice following AdhTWEAK administration. AdhTWEAK (10^9 particles per gland), or saline, was delivered to both submandibular glands (n = 4 mice per group) and PCNA staining performed on gland sections as described in Materials and methods to evaluate cell proliferation. Brown staining represents PCNA-positive nuclei. Sections are counterstained with hematoxylin. (a) Day 0 after saline administration; (b) Day 1 after AdhTWEAK administration; (c) Day 2 after-AdhTWEAK administration; (d) Day 3 after AdhTWEAK administration. Bar = 20 μm (From Sugito et al. [11]. Reprinted with permission).

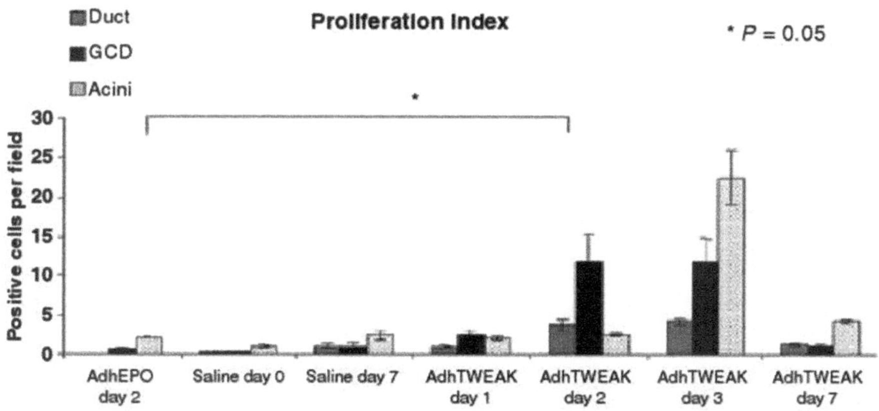

Figure 29. Quantification of AdTWEAK-induced salivary epithelial cell proliferation (From Sugito et al. [11]. Reprinted with permission).

2.10. Injectable soft-tissue augmentation by tissue engineering and regenerative medicine with human mesenchymal stem cells, platelet-rich plasma, and hyaluronic acid scaffolds

The restoration of soft tissue by adequate implant material is needed in case of functional and aesthetic impairments from loss of soft connective tissue. The implants vary, and repeated injections are generally required for varying degrees of material resorption. Recently it was reported that autologous cell injection was useful to improve soft tissue. The aim of this study was to evaluate the possibility of soft-tissue augmentation adopting tissue engineering and regenerative medicine (TERM) technology for a longer duration of injected implants. TERM is the combination and reorganization of three types of injection materials to regenerate organs or tissues: 1) living cells, including cultured human MSCs or human fibroblasts (Fibro); 2) scaffolds of hyaluronic acid (HA); and 3) growth factors of PRP. The experimental combinations were as follows: HA, HA/Fibro, HA/MSCs, HA/PRP, HA/PRP/Fibro and HA/PRP/MSCs. These were intradermally injected into immunodeficient rats and evaluated by histological analysis, the percentage of original volume and the maintenance volume (Figure 30).

The percentage of original volume values at 14 days showed significant differences between groups with and without PRP upon comparison (Table 2). As for the maintenance volume values, HA/PRP/Fibro, and HA/PRP/MSCs from 7 to 14 days were higher than others (Figure 31). HA/PRP/MSCs groups maintained the shape and dimensions of the injected implant, indicating that the injected cells produced type I collagen (Figure 32). The findings suggest that a soft tissue-engineered procedure with MSCs may be useful for longer-lasting soft-tissue augmentation.

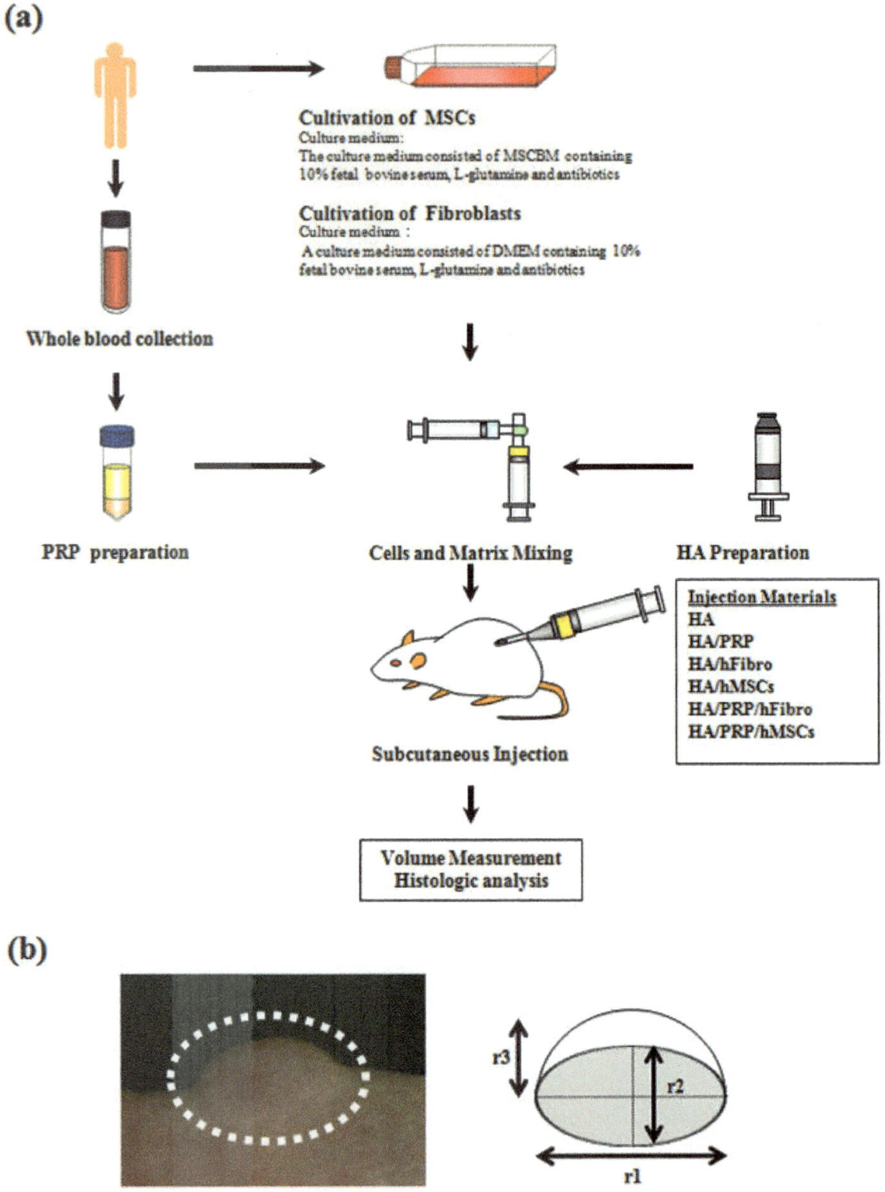

Figure 30. Schema of experimental protocol (a) and dermal mound formed after sample injection and the measurement diagram (b) (From Okabe et al. [12]. Reprinted with permission).

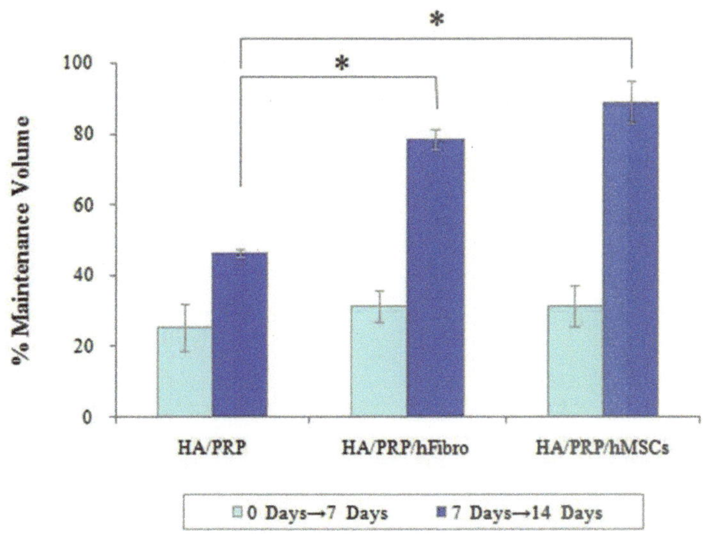

Figure 31. The percentage of maintenance volume values (%) in HA/PRP, HA/PRP/Fibro and HA/PRP/MSCs. Bar = SD. *P < 0.05 (From Okabe et al. [12]. Reprinted with permission).

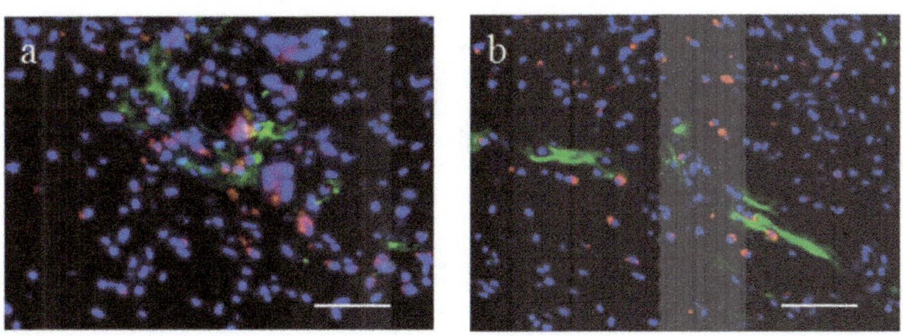

Figure 32. Distribution of type I collagen produced by MSCs (a) or Fibro (b) in the HA/PRP gels. Bar = 50 μm (From Okabe et al. [12]. Reprinted with permission).

2.11. Potentiality of new cell therapy for skin regeneration in wound healing

The process of wound healing depends upon a variety of interactions between cells and the extracellular matrix. There is hyaluronic acid in one of the extracellular matrix. It is well known that hyaluronic acid not only supports tissue architecture as a passive structural component of the matrix in various connective tissues but is also involved in dynamic cellular processes during wound healing. Recently, cell therapies which is a low aggression is paid attention in place of surgical operation and medical treatment of wound healing. It is well known that human dental pulp cell shows the property that resembled MSCs. In this research, we examined the benefit of stem cells from human exfoliated deciduous teeth (SHEDs) in wound healing (Figure 33).

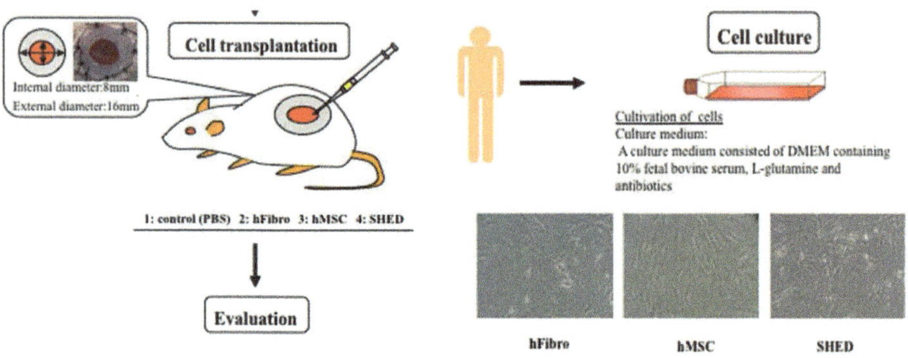

Figure 33. Schema of experimental protocol (From Nishino et al. [13]. Reprinted with permission).

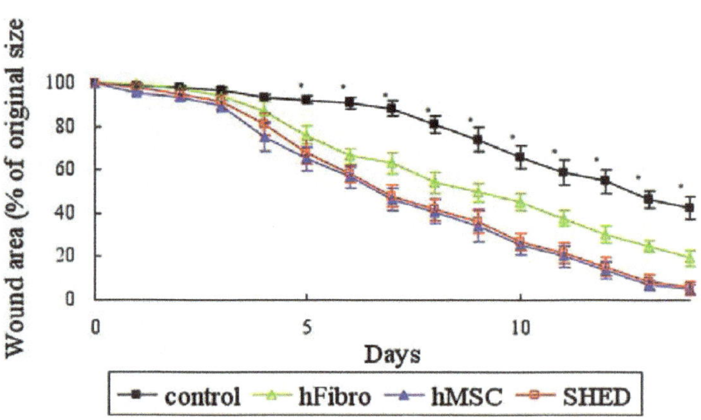

Figure 34. Wound measurement (From Nishino et al. [13]. Reprinted with permission).

SHEDs and MSCs significantly accelerated wound closure compared with Fibro and control treatment (Figure 34). At day 7 and 14, the evaluation by fluorescence microscope showed that PKH 26 positive cells (Fibro, MSCs, SHEDs) were surrounded by hyaluronic acid binding protein production of hyaluronic acid. The plasma membrane of transplanted cells showed red fluorescence by PKH 26. Hyaluronic acid was visualized with green fluorescence. DAPI was used to visualize the nuclei (blue fluorescence) (Figure 35).

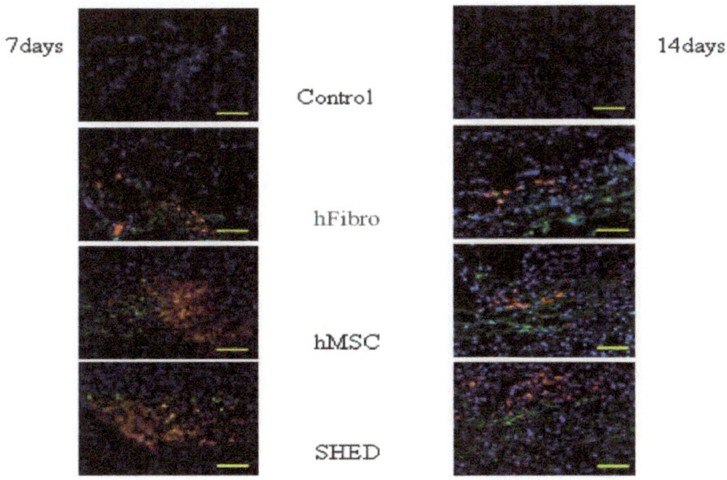

Figure 35. Histological evaluation (From Nishino et al. [13]. Reprinted with permission).

	Day 7 (ng/mg)	Day 14 (ng/mg)
control	1558.41 ± 60.33	704.17 ± 40.26
hFibro	2092.75 ± 42.56	1425.79 ± 56.31
hMSC	2342.07 ± 188.10	1653.10 ± 120.84
SHED	2314.85 ± 164.91	1644.98 ± 120.70

Table 2. Hyaluronic Acid Contents at Day 7 and 14 (From Nishino et al. [13]. Reprinted with permission).

Quantitate of hyaluronic acid was determined by measuring with ELISA. Significantly increased amounts of hyaluronic acid in wounded tissues were observed at day 7 and 14 in MSCs, SHEDs, and Fibro as compared with control ($P < 0.05$). This study demonstrated that deciduous teeth, considered as medical waste, would be novel therapeutic approaches in the treatment of wounds and new stem cell source for wound healing.

2.12. Umbilical cord Wharton's Jelly: A new potential cell source of mesenchymal stem cells for wound healing

Neonatal congenital disease, such as cleft and lip palate, involves soft tissue defect as well as skeletal abnormality. Thus the development of the therapeutic approaches accelerating both skeletal regeneration and wound healing is valuable for the treatment of neonatal congenital abnormality (Figure 36). Umbilical cords are routinely discarded as medical wastes in clinic. We have succeeded in isolating stem cells from the Wharton's Jelly in umbilical cords (so called umbilical cord matrix stem cells: UCMSCs). UCMSCs exhibited multipotential differentiation activities. In this research, we focused on the isolation and identification of MSCs from the Wharton's jelly of umbilaical cord. The effect of local injection of UCMSCs on cutenaous wound healing was examined using excisional wound model in mice.

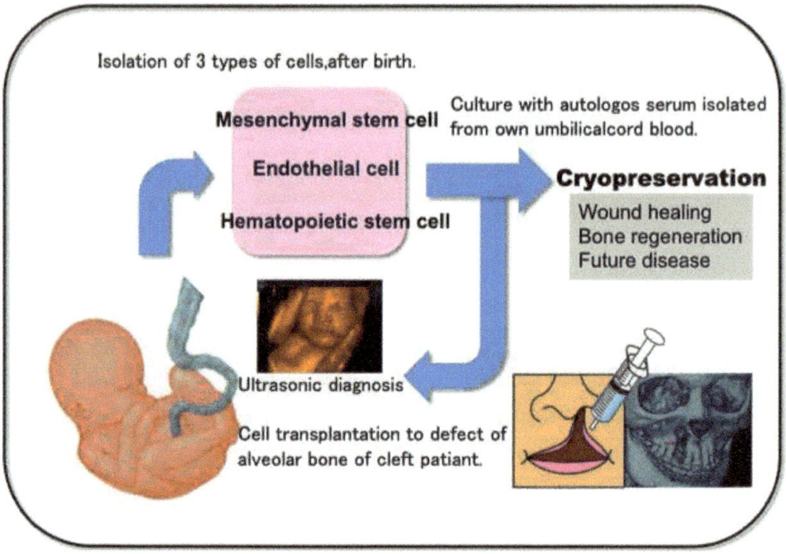

Figure 36. New concept of treatment of cleft lip & plate (From Shohara et al. [14, 15]. Reprinted with permission).

We have succeeded in isolating stem cells from umbilical cords (Figure 37A, B). Proliferation of UCMSCs was significantly greater than that of bone marrow stromal cells. UCMSCs proliferated much faster than did bone marrow stromal cells. UCMSCs exhibited multipotential differentiation activities toward osteogenic, adipogenic. We found that UCMSCs shared most of their immunophenotype with bone marrow stromal cells, including positivity for CD90, CD73, CD105, but negativity for CD11b, CD34(endothelial progenitor cell marker), CD45(hematopoietic markers), and HLA-DR. It shows that UCMSCs expressed high levels of mesenchymal stem cell markers. To investigate the wound repair activity of UCMSCs, we transplanted them in mouse excisional wound splinting model and the acceleration of the wound closure was evaluated (Figure 38). We found that the wound receiving the UCMSCs exhibit significantly faster healing compared with PBS-injected control (Figure 39A). After 14 days from operation, the closed wound area was 99.72 ± 0.17 % in the UCMSCs-group, while that in the control was 82.13 ± 5.85 % (Figure 39B). Histological analysis of wounds on day 14 indicated that granulation tissue of UCMSCs-group appeared to be thicker and larger as compared with the untreated group. Thus, these results demonstrated that the engrafted UCMSCs accelerated wound healing process.

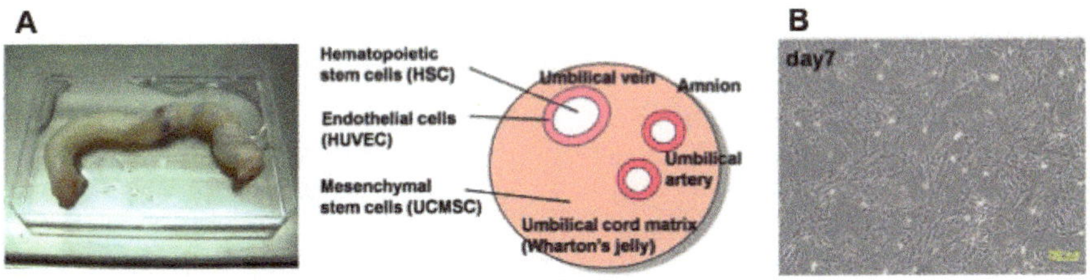

Figure 37. (A) Morphology and cross sectional image of umbilical cord. (B) UCMSCs exhibited fibroblastic morphology with bipolar spindles shape (From Shohara et al. [14, 15]. Reprinted with permission).

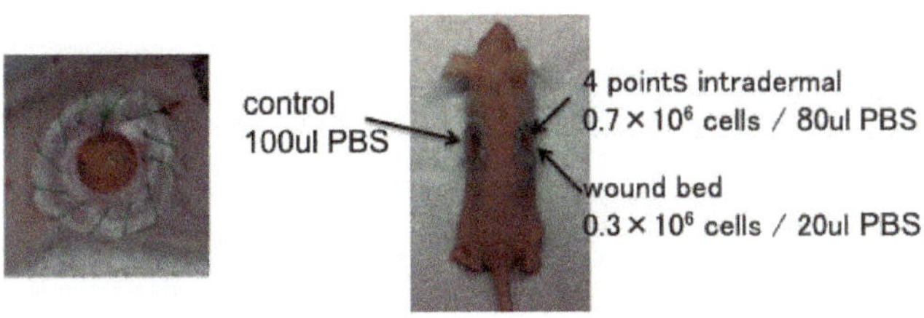

Figure 38. 6 mm full-thickness excicional wound splinted with silicone plate (From Shohara et al. [14, 15]. Reprinted with permission).

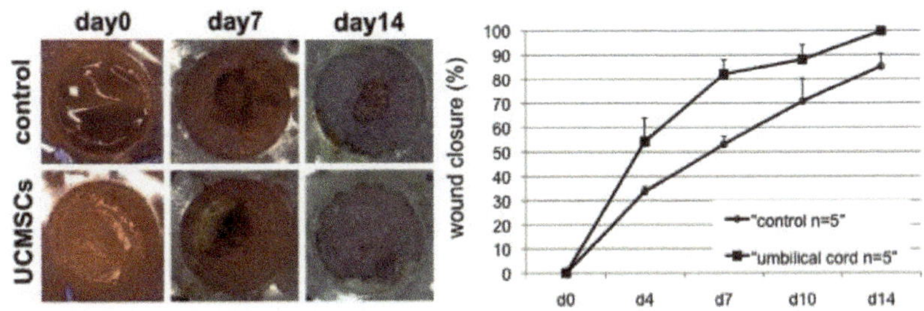

Figure 39. (A) Representative photograph of the wound at day 0, 7, 14 after UCMSCs transplantation. (B) Measurement of wound closure at different time points (From Shohara et al. [14, 15]. Reprinted with permission).

In summary, the present study describes the isolation and primary characterization of stem cells from medical wastes, such as umbilical cord. Furthermore, local injection of UCMSCs accelerated cutenaous wound healing process. This stem cells can be isolated without invasive surgical procedures, providing unique cell resources for regenerative medicine. Together with the distinct advantages of UCMSCs, such as accessibility, painless procedures to donors, possible source for autologous cell therapy and lower risk of viral contamination, we suggest that UCMSCs should be considered a promising cell resource for cell therapy.

2.13. Acceleration of wound healing with stem cell–derived growth factors

Recently, it has been revealed that bone marrow-derived MSCs accelerate skin wound healing, and it is attracting attention as a new cell therapy. However, MSCs are isolated from bone marrow from patient, and collection of the bone marrow is considerably invasive. In addition, there is a problem that proliferative capacity of MSCs decreases with aging. Thus, application of MSCs for elderly patients and fresh cases was difficult because it takes a long time to cultivate the cells. On the other hand, it has been known that MSCs secrete many growth factors. We presumed that growth factors secreted by MSCs play a main role in wound healing effect, and examined effect of MSC-CM on wound healing.

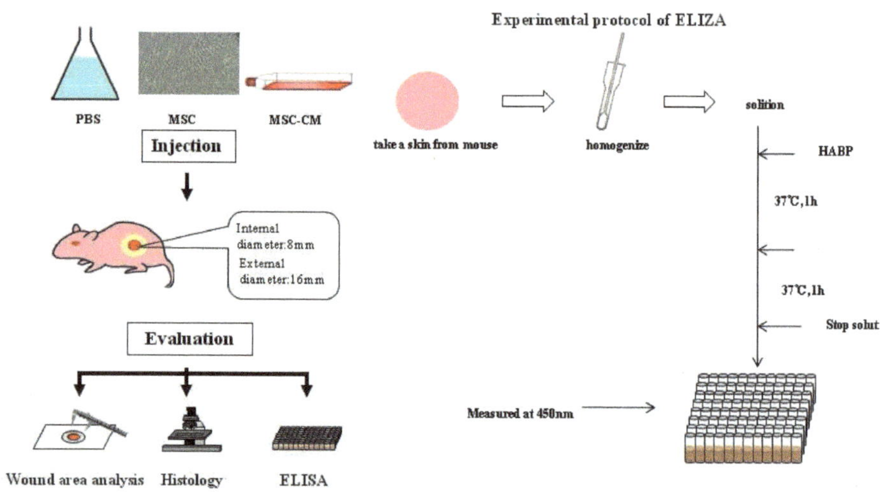

Figure 40. Scheme of the experiment(From Tamari et al. [16-18]. Reprinted with permission from Quintessence Publishing Co, Inc, Chicago).

We observed wound healing process macroscopically and histologically using an excisional wound splinting mouse model, and examined expression level of hyaluronic acid related to the wound healing process to evaluate wound-healing effect of MSCs, MSC-CM, and control (PBS) (Figure 39). The MSCs and MSC-CM groups accelerated wound healing as compared with the control group. The area of wound in the MSCs and MSC-CM groups at 5 days and later indicated statistically significant difference as compared with the control group.(Figure

40, 41) At 7 days after administration of MSCs or MSC-CM, epithelialization was accelerated, thick connective tissue was formed in the skin defective area, and the area of wound was reduced in the MSCs and MSC-CM groups as compared with the control group. Hyaluronic acid was expressed in the marginal part of the wound significantly. At 14 days after operation, infiltration of inflammatory cells was decreased as compared with 7 days, and the wound was closed in the MSCs and MSC-CM groups, while defective part of epithelium was observed in the control group. Expression of hyaluronic acid was decreased from 7 days. At 7 and 14 days after operation, expression levels of hyaluronic acid were 1512.0 ± 84.10 ng/mg and 683.7±56.4 ng/mg in control group, 2450.3 ± 225.7 ng/mg and 1690.2 ± 170.2 ng/mg in MSCs group, and 2360.6 ± 230.0 ng/mg and 1570.1 ± 142.5 ng/mg in MSC-CM group. At 7 and 14 days, the MSCs and MSC-CM groups expressed significantly high level of hyaluronic acid as compared with the control group ($P < 0.05$). The expression level of hyaluronic acid was lower at 14 days than that at 7 days in all three groups.

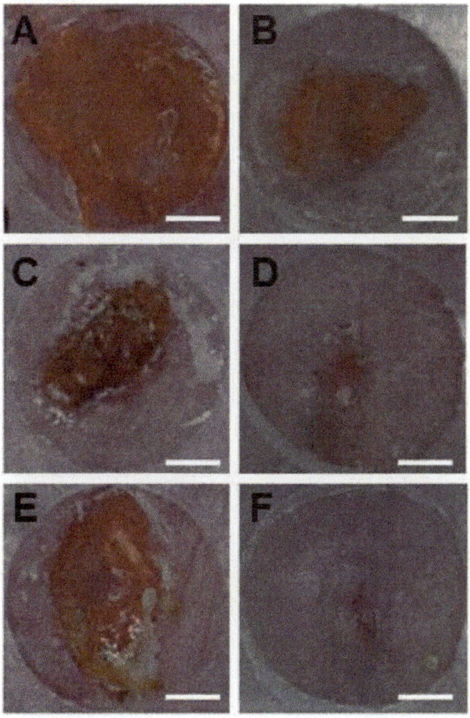

Figure 41. Macroscopic findings at 7 and 14 days after injection. Left panels (A, C, and E) at 7 days after administration; Right panels (B, D, and F) at 14 days after administration. (A and B) control group; (C and D) MSCs group; (E and F) MSC-CM group. Bar = 3 mm (From Tamari et al. [16-18]. Reprinted with permission from Quintessence Publishing Co, Inc, Chicago).

These experimental results indicated that both MSCs and MSC-CM groups have wound healing acceleration effect as compared with the control group. The wound healing acceleration effects of the MSC-CM group and the MSCs group were equivalent. Accordingly, it is suggested that the MSC-CM contains growth factor derived from stem cells and is able to provide wound healing acceleration effect equivalent to stem cell transplantation, and may become new therapeutic method for wound healing in the future.

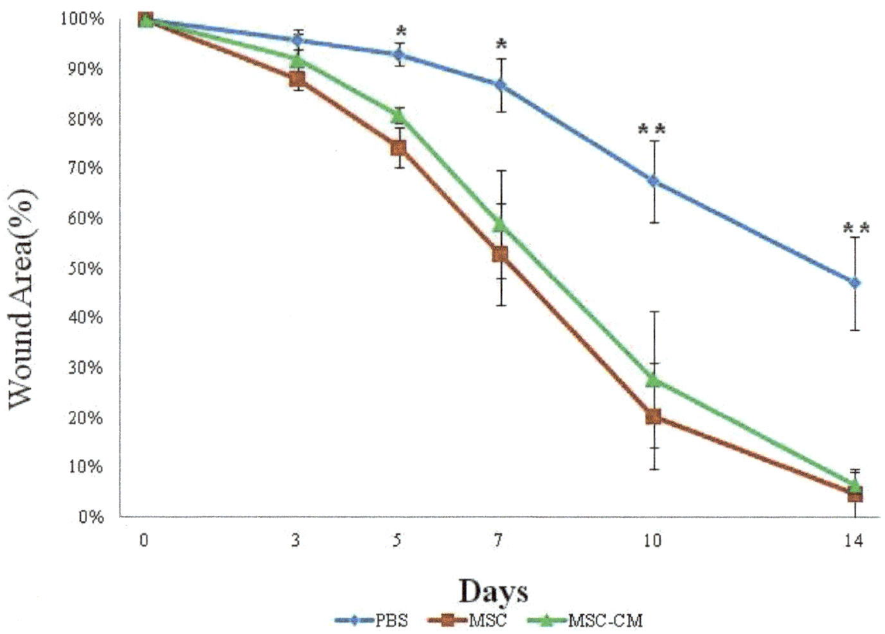

Figure 42. Time-dependent changes in the area of wound area determined with image analysis software. area of actual wound/area of original wound ×100. Test for significant difference (ANOVA). MSC-CM group vs. control group, $*P < 0.05$. $**P < 0.01$ (From Tamari et al. [16-18]. Reprinted with permission from Quintessence Publishing Co, Inc, Chicago).

2.14. Human dental pulp-derived stem cells promote locomotor recovery after complete transection of the rat spinal cord by multiple neuro-regenerative mechanisms

Spinal cord injury resulted in severe functional disability. In worldwide, 2.5 million people live with spinal cord injury More than 200,000 new injuries reported each year. Currently no curative therapy is available. Recent study has demonstrated that transplantation of stem cells in injured spinal cord would support functional recovery. The cellular resources of stem cell are Human embryonic stem cell, Induced pluripotent stem cells, Human embryonic neural stem cell, and adult mesenchymal stem cells from bone marrow. These stem cells have been shown to be valuable cellular resources for treatment of animal model of SCI. However, to use them for patient, there are significant ethnical, safety and invasive problems. Thus, an ideal stem cell resource for SCI treatment is still an elusive subject. Here we have examined a neuro-regenerative activity of tooth derived stem cells in a rat model of SCI. Merits of teeth derived stem cells for SCI treatment are 1, They can be easily isolated from medical wast such as wisdom teeth and milk teeth. 2, AS they are originated from Neural crest, they can be a valuable cellular resources for treatment of neuro-degenerative disease. 3, as they are autologous cellular resources, must be safe for cell therapy.

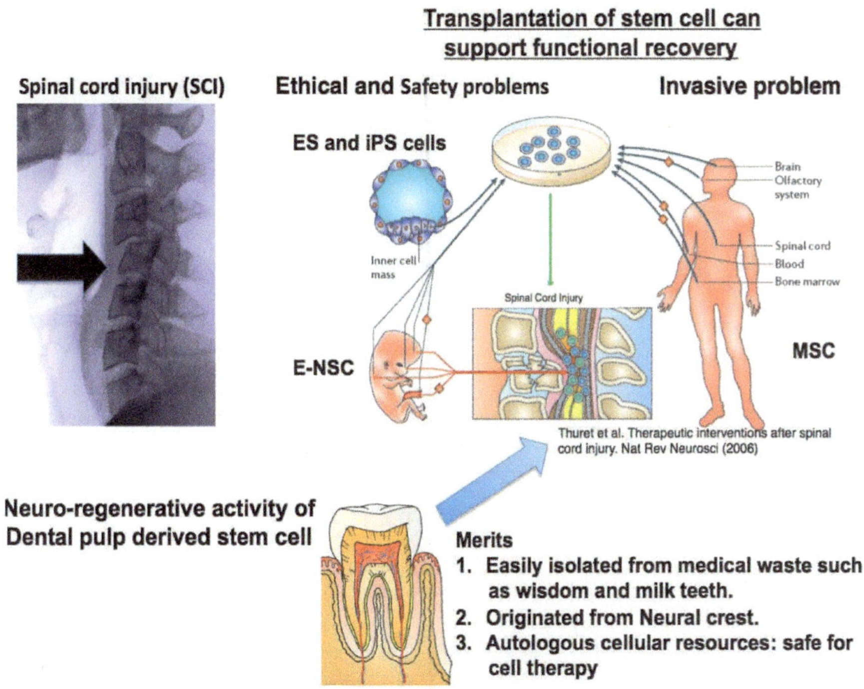

Figure 43. Merits of teeth derived stem cells for SCI treatment.

Flow cytometry analysis showed that SHEDs and DPSCs expressed a set of mesechymal stem cell marker, CD90, CD73, CD105 and CD90, but not endothelial/hematopoietic marker, CD54, CD34, CD45, CD11b/c or HLA-DR, as describe previously. Majority of these stem cells uniquely co-expressed several neural linage markers, including Nestin and Doublecortin (neural stem cell marker), GFAP (astrocyte marker), CNPase (immature oligodendrocyte maeker) and A2B5 (oligodendrocyte precursor marker) altogether. Next we have examined expression of neurotropic factors by real-time PCR. Comparing with skin-derived fibroblasts SHEDs expressed NT-3 (Neurotrophin-3) and BDNF (Brain Derived Neurotrophic Factor), more than 3 and 5 times, respectively. These results demonstrate that SHED and DPSC, stem cells derived dental pulps, are valuable cellular resources for neuro-regeneration therapy. SHEDs or DPSCs were transplanted into the transected spinal cord as described in the Materials and Methods. At 8 weeks post-engraftment, BBB scoring suggested a recovery of hindlimb locomotor function both in SHEDs or DPSCs-transplanted rats (n = 11 for SHEDs and n=10 for DPSCs) in comparison with vehicle controls (n = 10) analysis revealed a higher frequency of locomotion recovery (BBB score > 7 for SHEDs and > 6 for DPSCs) in SHEDs or DPSCs vs. vehicle controls. Thus, both SHED and DPSC promote functional recovery of completely transected spinal cord. As axonal myelination promotes functional recovery of injured spinal cord, we studied whether transplanted SHEDs preserved myelinated area in injured spinal cord. In the PBS-treated control group, the transverse area of the lesion site exhibited no or little staining of Fluoro-myelin. In SHEDs-transplanted, however, myelin positive area covered 15% and 80% of entire spinal cord, at epicenter and 5mm caudal lesion site, respectively, showing that transplanted SHEDs play significant role in preservation of axonal myelination in injured spinal cord.

Our study revealed that engrafted SHEDs exhibited three major therapeutic benefits for recovery after SCI, including inhibition of the SCI-induced apoptosis of neurons, astrocytes, and oligodendrocytes, which promoted the preservation of neural fibers and myelin sheaths, regeneration of the transected axon through the direct inhibition of multiple AGI signals, such as chondroitin sulfate proteoglycans and MAG, by paracrine mechanisms, and replacement of lost or damaged oligodendrocytes after SCI through specific differentiation into mature oligodendrocytes under the extreme conditions of SCI. To our knowledge, the neuro-regenerative activities and are unique to tooth-derived stem cells, and are not exhibited by any other previously described stem cells. Thus, our data demonstrate that tooth-derived stem cells may provide significant therapeutic benefits for treating the acute phase of SCI through both cell-autonomous and paracrine/trophic regenerative activities. We demonstrated multifaceted neuro-regenerative activities of tooth-derived stem cells that fulfill many requirements for functional recovery after SCI. In addition to their remarkable neuro-regenerative activities, we did not observe the malignant transformation of engrafted SHEDs 8 weeks after their implantation (data not shown). Furthermore, SHEDs and DPSCs can be obtained from exfoliated deciduous and impacted adult wisdom teeth without adverse health effects. Thus, there are few ethical concerns regarding their clinical use. We propose that tooth-derived stem cells may be an excellent and practical cellular resource for the treatment of SCI.

2.15. Growth factors derived from dental pulp stem cells: A new potential clinical benefits for CNS regeneration therapy

There is no effective treatment for SCI. In the acute phase of SCI, damaged neurons are going to die within a day after the injury by inflammatory reaction. Until a weak later, multiple axon growth inhibitors (AGIs) are produced by the astroglial scar and degenerated myelin surrounding the injured CNS (Figure 44). We have been reported the clinical benefits of the engrafted human DPSCs (hDPSCs) and stem cells from human SHEDs in the treatment of acute phase rat SCI model. Recently, growth factors secreted from stem cells were important to stem cell therapy. So we focused on the paracrine effect of growth factors derived from SHEDs. Here, we show that local administration of a serum-free conditioned media of SHEDs (SHED-CM) into the rat SCI resulted in remarkable recovery of hindlimb locomotor functions. We found that this effect is associated with several paracrine effect of SHED-CM.

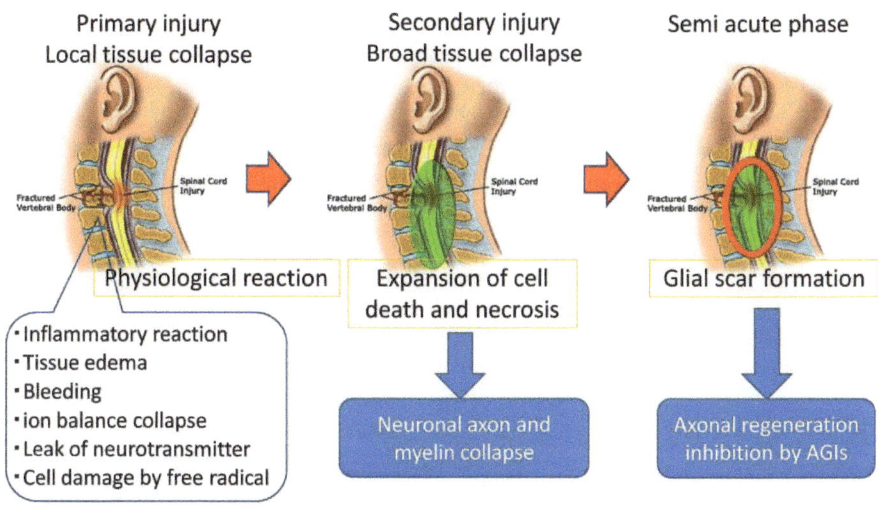

Figure 44. Acute phase of SCI.

We examined the effects of the treatment with SHED-CM, MSC-CM or Fibro-CM on the functional recovery after contusion SCI. CMs were continuously delivered intrathecally by infusion pump into the SCI epicenter. The level of recovery of hindlimb locomotion was evaluated using the Basso, Beattie, Bresnahan locomotor rating scale (BBB scale). After the recovery period, the rats that had received SHED-CM were able to support their weight through the plantar surface of the paw and step with fore-hindlimb coordinated manner In contrast, although the MSC-CM or Fibro-CM treated rats move 3 joints of hindlimb, they were not able to walk with weight support. These results demonstrate that SHED-CM, but not MSC-CM and Fibro-CM, provides significant therapeutic benefit for the treatment of the acute phase of SCI. In the acute phase of SCI, much inflammatory cytokines are released. These inflammatory cytokines activate microglia and leads to expansion of inflammatory reaction. We got rat spinal cord mRNA after SCI, and analyzed expression of inflammatory and anti-inflammatory cytokines. We found that inflammatory cytokines expression down regulated MSC-CM, Fibro-CM, and SHED-CM groups rather than control groups at three time points. On the other hand,importantly, anti-inflammatory cytokines expression up regulated only SHED-CM groups. We got mouse primary microglia mRNA after LPS stimulation and analyzed the expression. Similarly *in vivo* results, inflammatory cytokines expression down regulated MSC-CM, Fibro-CM, and SHED-CM culture groups rather than control groups. But,uniquely, glutamate and NO level in microglia culture medium reduced only SHED-CM groups. Damaged neurons and glial cells are going to die within a day after injury, and expand cell death in one week. These cell death leads to degeneration of myelin and chronic atrophy of spinal cord. So we tried TUNEL staining. Compared to the control groups, SHED-CM groups were down regulated TUNEL positive cells significantly at 24hour and 1week after SCI. And we stained myelin by fluoromyelin at eight weeks after SCI. At control groups, spinal cord was going to atrophy from epicenter, but on the other hand SHED-CM groups were maintained myelin and kept form of spinal cord from epicenter. Until a week after SCI, multiple AGIs are produced by astroglial scar and degenerated myelin surrounding injury site. These AGIs accelerate the neuronal apoptosis and inhibit axonal regrowth. We got primary cerebellum granule neurons (CGNs) and tried neurite outgrowth assay on AGIs coating dish. And we found that CGNs only SHEDs and DPSC-CM culture groups inhibited AGIs activity and strong promoted their neurite. So we stained by NFM and 5-HT eight weeks after SCI. And we found that neuronal axon extended beyond the injury epicenter to caudal. Our studies demonstrate potential clinical benefits of SHED-CM for the treatment of the acute phase of SCI, providing a novel, safe and effective neuro-regenerative therapy that protects patient's CNS from the traumatic and ischemic CNS injury.

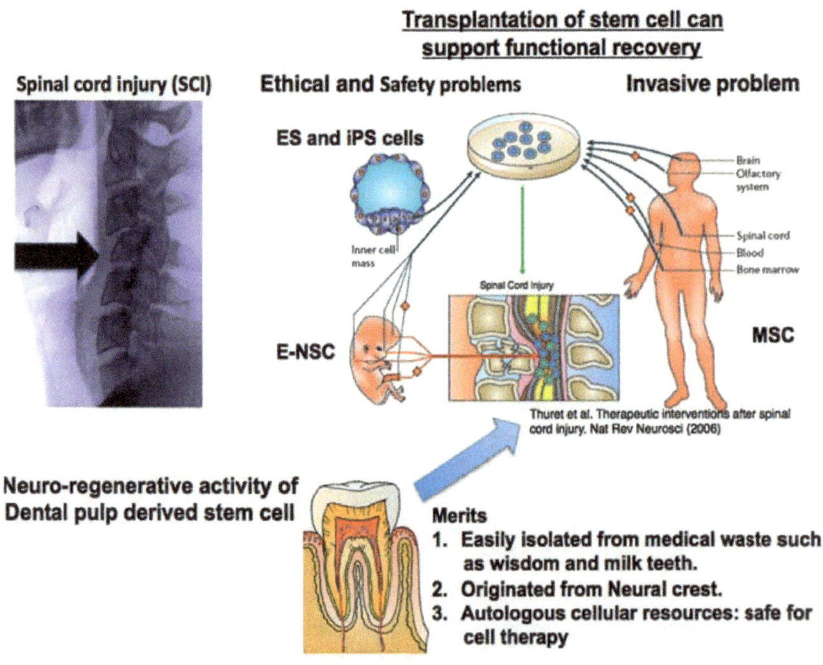

Figure 45. Multifaceted treatment strategy.

2.16. SHED-CM rnhances recovery of focal cerebral ischemia in rats

Regenerative therapy using stem cells is a promising approach for treatment of stroke. Recently, we reported that DPSCs ameliorated ischemic tissue injury in rat brain and accelerated functional recovery after middle cerebral artery occlusion (MCAO). In this study, we investigated the effects of SHED-CM after pMCAO (permanent middle cerebral artery occlusion). Adult male Sprague-Dawley rats were subjected to pMCAO. SHED-CM was then administered intranasally and motor function and infarct volume evaluated. SHEDs were cultured in DMEM serum-free medium. Conditioned medium of SHEDs was collected after 48 h of culture and centrifuged at 1500 rpm for 5 min. The supernatant was recentrifuged at 3000 rpm for 3 min followed by collection of the second supernatant, named SHED-CM. Seventy-two hours after pMCAO, the rats were anesthetized again with 1.5% isoflurane in a mixture of 70% N_2O and 30% O_2. A total of 100 μL of SHED-CM was administered to rat via the olfactory pathway using a Hamilton microsyringe (Figure 46). The SHED-CM preparation was administered in 10 μL at a time, with an interval of 2 min between each administration. Intranasal administration was performed everyday from days 3 to 15.

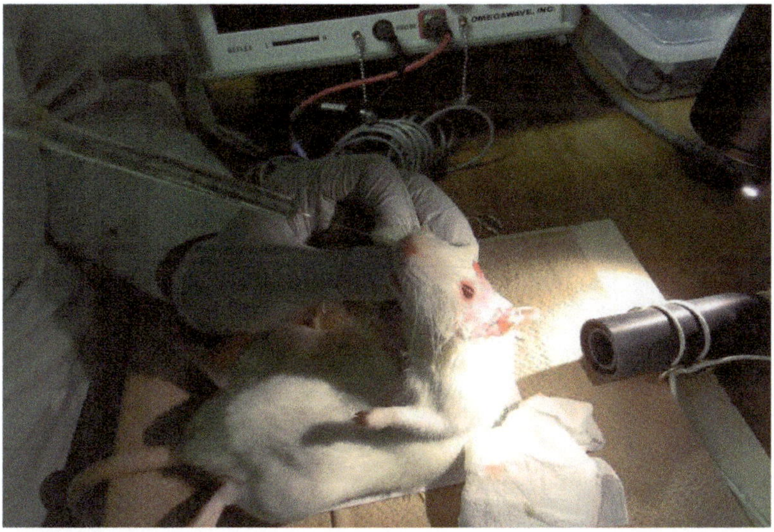

Figure 46. Intranasal administration.

The rats were blindly examined on days 1, 3, 6, 9, 12, and 15 using a standardized motor disability scale with slight modifications. The cryosections obtained from samples on day 16 were stained with hematoxylin and eosin. Image J was used to determine each infarct area in 12 coronal sections at 1.00-mm intervals. The entire infarction area was covered by these 12 coronal sections. Regional infarct volumes were calculated by summing the infarct areas and multiplying these areas by the distance between sections (1.00 mm). The two groups displayed almost the same high score for motor function in the early stages as shown in Figure 47 right. Differences in the score appeared gradually between the two groups during the middle stage. On day 15, progressive improvement in motor disability in the SHED-CM group became significant compared to the PBS group. As shown in Figure 47 left, there was a significant decrease in infarct volume on day 16 in the SHED-CM group compared to the PBS group. These results suggest that SHED-CM promoted regeneration. Recently, we reported the characteristics of SHEDs compared with DPSCs and MSCs. The results indicated that SHEDs possessed high proliferation ability and were enriched with extracellular matrix, suggesting it may be a useful source for stem cell-based therapy. In addition, using micro array analysis, we showed that SHEDs had higher expression levels of several growth factors, such as fibroblast growth factor, transforming growth factor, connective tissue growth factor, nerve growth factor, and bone morphogenetic protein. Taken together, these findings indicate SHEDs are a more potentially useful source of stem cells for cell therapy than DPSCs and MSCs.

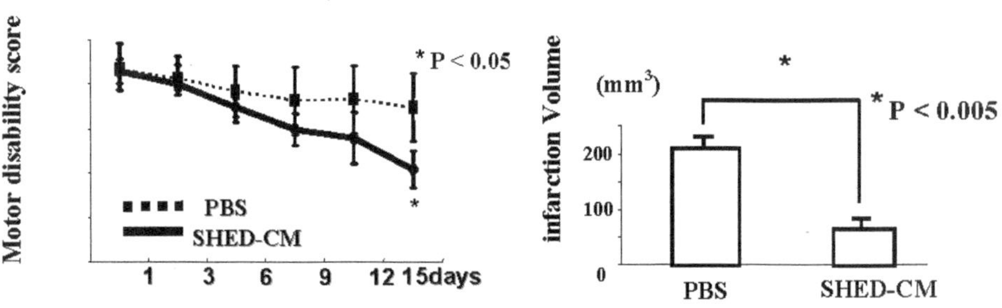

Figure 47. Evaluation of motor function and Reduction in infarct volume.

However, cell therapy is always associated with problems, such as canceration, immune rejection, and ethical issues. Therefore, it is necessary to find alternative treatments to cell therapy. Studies in recent years have resulted in the recognition of a paracrine function in factors, and have suggested that stem cell transplantation may also be regarded as "cell-based" cytokine therapy. Accordingly, we investigated two steps for developing a new treatment for cerebral ischemia. In the first step, we used SHED-CM as a new source for treatment of cerebral ischemia. We have reported previously in a rat model of stroke that needle administration of DPSCs induced recovery of motor disability and reduction in infarct volume, proliferation of their presumptive progeny in SVZ, migration to the infarct, and differentiation into the appropriate neurons. As several growth factors involved in neural regeneration are secreted from DPSCs, we hypothesized in the current study that SHED-CM may improve recovery of motor disability and reduce infarct volume. In the second step, we investigated intranasal administration of SHED-CM. Using this administration, therapeutic molecules traverse the BBB through the olfactory pathway and the less-studied trigeminal neural pathway. An important advantage of intranasal administration is that it is less invasive with the factors being delivered directly to the brain.

Our results suggested that SHED-CM including some growth factors may produce effects in stroke model. The success of the two steps investigated suggest it may be possible to use a shortcut for clinical application. Administration of SHED-CM resolves the ethical issues involved with cell therapies, as SHED-CM is not a cell but rather a conjugate of many growth factors. As SHED-CM can be stocked, it is possible to use it for acute stages of stroke, either alone or with ready-made treatment, such as recombinant tissue plasminogen activator, anticoagulation, and antiplatelet therapy. This study suggested that intranasal administration of SHED-CM may help recovery in acute stroke patients in the future. In conclusion, regeneration therapy using SHED-CM is a very safe method with no associated problems and is therefore a potential candidate for innovative treatment of cerebral ischemia.

3. Conclusion

The endpoint of all our studies is to establish successful novel therapies for patients suffering from diseases or disorders. These studies suggest that engineered tissues may have an expanded clinical applicability in the future and may represent a viable therapeutic option for those who require tissue replacement or repair. The future outlook is positive with the recent discovery of iPSCs and stem cells from human SHEDs. We especially focus on the potential of SHEDs, which were identified as a population of highly proliferative, clonogenic cells capable of differentiating into a variety of cell types including neural cells, adipocytes, and odontoblasts. The stem cell field is also advancing rapidly, providing new therapeutic options. In the future, banking these stem cells may provide a convenient source for autologous therapy and for matching recipients with histocompatible donors. Tissue engineering holds great promise for the future of medicine, as experimental efforts are currently under way for virtually every type of tissue and organ within the human body.

Author details

Minoru Ueda

Department of Oral and Maxillofacial Surgery, Graduate School of Medicine, Nagoya University, Nagoya, Japan

References

[1] Kohgo T, Yamada Y, Ito K, Okabe K, Yoshimi R, Yajima A, Baba S, Ueda M. Bone regeneration for dental implants using tissue engineered bone with a novel scaffold. In: 25th Annual Meeting of Academy of Osseointegration. 4-6 March 2010, Orland, FL, USA; 2010.

[2] Yoshimi R, Katagiri W, Osugi M, Inukai T, Yamamoto A, Hibi H, Ueda M. Novel xenograft technique using the stem cells cultured conditioned media for bone regeneration. In: 20th Annual Scientific Meeting of the European Association of Osseointegration, 13-15 October 2011, Athens, Greece; 2011.

[3] Yoshimi R, Katagiri W, Osugi M, Inukai T, Hibi H, Ueda M. Stem cell cultured conditioned media of stem cells from exfoliated deciduous teeth enhanced cell proliferation: association with Wnt/beta-catenin pathway. In: The 56th Congress of the Japanese Society of Oral and Maxillofacial Surgeons. 21-23 October 2011, Osaka, Japan; 2011.

[4] Ikeno M, Hibi H, Kinoshita K, Hattori H, Ueda M. Effects of the permeability of shields with autologous bone grafts on bone augmentation. Oral & Craniofacial Tissue Engineering 2011;1(3) 198-204.

[5] Tateishi H, Okamoto Y, Kinoshita K, Hibi H, Ueda M. Effects of Implant Surface on Bone Healing AroundTitanium Screw Implants in Ovariectomized Rats. Oral and Craniofacial Tissue Engineering 2011;1(1) 42-49.

[6] Tateishi H, Okamoto Y, Kinoshita K, Hibi H, Ueda M. A Study of Bone Healing Around the Titanium Screw Implants in Ovariectomized Rats.: Can Rough Surface Contribute to The Implant Stability? In: 26th Annual Meeting of Academy of Osseointegration, 3-5 March 2011, Washington D.C., USA; 2011.

[7] Wadagaki R, Mizuno D, Yamawaki-Ogata A, Satake M, Kaneko H, Hagiwara S, Yamamoto N, Narita Y, Hibi H, Ueda M. Osteogenic Induction of Bone Marrow-Derived Stromal Cells on Simvastatin-Releasing, Biodegradable, Nano- to Microscale Fiber Scaffolds. Annals of Biomedical Engineering 2011;39(7) 1872-1881.

[8] Wadagaki R, Mizuno D, Yamawaki-Ogata A, Satake M, Kaneko H, Yamamoto N, Narita Y, Hibi H, Ueda M. Osteoinduction of bone marrow-derived stromal cells on simvastatin-releasing fiber scaffolds. In: The 10th Congress of Japanese Society for Regenerative Medicine. 1-2 Mar 2011, Tokyo, Japan; 2011.

[9] Fujio M, Yamamoto A, Ando Y, Shohara R, Kinoshita K, Kaneko K, Hibi H, Ueda M. Stromal cell-derived factor-1 enhances distraction osteogenesis-mediated skeletal tissue regeneration through the recruitment of endothelial precursors. Bone 2011;49(4) 693-700.

[10] Inoue M, Ebisawa K, Itaya T, Sugito T, Yamawaki-Ogata A, Sumita Y, Wadagaki R, Narita Y, Agata H, Kagami H, Ueda M. Effect of GDF-5 and BMP-2 on the expression of tendo/ligamentogenesis-related markers in human PDL-derived cells. Oral Diseases 2011;

[11] Sugito T, Mineshiba F, Zheng C, Cotrim A. P, Goldsmith C. M, Baum B. J. Transient TWEAK overexpression leads to a general salivary epithelial cell proliferation. Oral Diseases 2008;15(1) 76-81.

[12] Okabe K, Yamada Y, Ito K, Kohgo T, Yoshimi R, Ueda M. Injectable soft-tissue augmentation by tissue engineering and rebeneratibe medicine with human mesenchy-

mal stromal cells, platelet-rich plasma and hyaluronic acid scaffolds. Cytotherapy 2009;11(3) 307-316.

[13] Nishino Y, Ebisawa K, Yamada Y, Okabe K, Kamei Y, Ueda M. Human Deciduous Teeth Dental Pulp Cells (hDPC) with Basic Fibroblast Growth Factor(b-FGF) Enhance Wound Healing of Skin Defect. Journal of Craniofacial Surgery 2011;22(2) 438-442.

[14] Shohara R, Yamamoto A, Takikawa S, Iwase A, Hibi H, Kikkawa F, Ueda M. Umbilical cord Wharton's jelly: A new potential cell source of mesenchymal stem cells for wound healing. In: Proceedings of 9th International Society Stem Cell Research. 15-18 June 2011, Toronto, Canada; 2011.

[15] Shohara R, Yamamoto A, Takikawa S, Iwase A, Hibi H, Kikkawa F, Ueda M. Umbilical cord Wharton's jelly: A new potential cell source of mesenchymal stem cells for wound healing. In: Proceedings of International Symposium on Micro-NanoMechatronics and Human Science. 7 November 2011, Nagoya, Japan; 2011.

[16] Tamari M, Nishino Y, Yamamoto N, Ueda M. Acceleration of Wound Healing with Stem Cell–Derived Growth Factors. Oral & Craniofacial Tissue Engineering 2011;1(3) 181–187.

[17] Tamari M, Nishino Y, Yamamoto N, Furue H, Shigetomi T, Ueda M. Effect of mesenchymal stem cell derived paracrine factor on wound healing in mice skin. In: The 56th congress of the Japanese Society of Oral and Maxillofacial Surgeons. 21-23 October 2011, Osaka, Japan; 2011.

[18] Tamari M, Nishino Y, Yamamoto N, Ueda M. Wound Healing Acceleration by Stem Cell–Derived Growth Factors. In: Proceedings of International Symposium on Micro-NanoMechatronics and Human Science, 7 November 2011, Nagoya, Japan; 2011.

[19] Ikeno M, Hibi H, Kinoshita K, Hattori H, Ueda M. Effects of self-assembling peptide hydrogel scaffold on bone regeneration with recombinant human bone morphogenetic protein-2. Oral & Craniofacial Tissue Engineering 2011;1(2) 91-97.

[20] Ikeno M, Hibi H, Kinoshita K, Yajima A, Hattori H, Ueda M. The effect of shields on the bone augmentation model in rabbit calvariae. In: The 63th Annual Meeting of The Japanese Stomatological Society. 16-17 April 2009, Hamamatsu, Japan; 2009.

[21] Ikeno M, Hibi H, Kinoshita K, Yajima A, Hattori H, Ueda M. The effect of shields on the bone augmentation model in rabbit calvariae. In: XIII Biennial International Congress of the International Society of Craniofacial Surgery. 26-30 Sepember 2009, Oxford, UK; 2009.

[22] Ikeno M, Hibi H, Kinoshita K, Yajima A, Hattori H, Ueda M. The effect of shields on the bone augmentation model in rabbit calvariae. In: The 9th Congress of the The Japanese Society of Regenerative Medicine. 18-19 March 2010, Hiroshima, Japan; 2010.

[23] Ikeno M, Hibi H, Kinoshita K, Hattori H, Shibuya H, Yajima A, Ueda M. Effects of self-assembling peptide hydrogel scaffold on the bone augmentation in rabbit calvar-

iae. In: The 55th Annual Meeting of The Japanese Society of Oral and Maxillofacial Surgeons, 16-18 October 2010, Chiba, Japan; 2010.

[24] Ikeno M. The effect of self-assembling peptide hydrogel scaffold on the bone augmentation model in rabbit calvariae. In: Proceedings of International Symposium on Micro-NanoMechatronics and Human Science, 7-10 November 2010, Nagoya, Japan; 2010.

[25] Inoue T, Sugiyama M, Hattori H, Yamamoto A, Hibi H, Ueda M. SH-CM Enhances Recovery of Focal Cerebral Ischemia in Rats. In: 32nd Annual Meeting of the Japanese Society of Inflammation and Regeneration, 2-3 June 2011, Kyoto, Japan; 2011.

[26] Inoue T, Sugiyama M, Hattori H, Hibi H, Ueda M. SH-CM Enhances Recovery of Focal Cerebral Ischemia in Rats. In: The 56th Congress of the Japanese Society of Oral and Maxillofacial Surgeons, 21-23 October 2011, Osaka, Japan; 2011.

[27] Inoue T, Sugiyama M, Hattori H, Hibi H, Ueda M. SH-CM Enhances Recovery of Focal Cerebral Ischemia in Rats. In: 22th IEEE International Symposium on Micro-NanoMehatronics and Human Science, 7 November 2011, Nagoya, Japan; 2011.

[28] Kohgo T, Yamada Y, Ito K, Yajima A, Yoshimi R, Okabe K, Baba S, Ueda M. Bone Regeneration with Self-assembling Peptide Nanofiber Scaffolds in Tissue Engineering for Osseointegration of Dental Implants. International Journal of Periodontics & Restorative Dentistry 2011;31(4) e9-e16.

[29] Matsubara K, Yamamoto A, Sakai K, Ueda M. The potential clinical benefits of the serum-free conditioned media derived from dental pulp stem cells in CNS regeneration therapy. In: 3th Nagoya global retreat Nagoya University Global COE Program "Integrated Functional Molecular Medicine for Neuronal and Neoplastic Disorders", 25-26 February 2011, Nagoya, Japan; 2011.

[30] Matsubara K, Yamamoto A, Sakai K, Ueda M. The potential clinical benefits of the serum-free conditioned media derived from dental pulp stem cells in CNS regeneration therapy. In: The 10th Congress of the Japanese Society for Regenerative Medicine, 1-2 March 2011, Tokyo, Japan; 2011.

[31] Matsubara K, Yamamoto A, Sakai K, Matsushita Y, Ueda M. The potential clinical benefits of the serum-free conditioned media derived from dental pulp stem cells in CNS regeneration therapy. In: 2th Symposium of Nagoya University and National Institute for Physiological Sciences, 20 August 2011, Nagoya, Japan; 2011.

[32] Matsubara K, Yamamoto A, Sakai K, Matsushita Y, Ueda M. The potential clinical benefits of the serum-free conditioned media derived from dental pulp stem cells in CNS regeneration therapy, In: Global COE program The 3rd International Symposium, 8-9 December 2011, Nagoya, Japan; 2011.

[33] Nishino Y, Yamada Y, Ebisawa K, Nakamura S, Okabe K, Umemura E, Hara K, Ueda M. Stem cells from human exfoliated deciduous teeth (SHED) enhance wound Healing and possibility of novel cell therapy. Cytotherapy 2011;13(5) 598-605.

[34] Nishino Y, Ebisawa K, Yamada Y, Okabe K, Kamei Y, Ueda M. Human Deciduous Teeth Dental Pulp Cells (hDPC) with Basic Fibroblast Growth Factor (b-FGF) Enhance Wound Healing of Skin Defect. In: 8th Annual Meeting of the International Society for Stem Cell Research, 16-19 June 2010, San Francisco, CA, USA; 2010.

[35] Nishino Y, Yamada Y, Ebisawa K, Nakamura S, Okabe K, Umemura E, Hara K, Ueda M. Stem cells from human exfoliated deciduous teeth (SHED) enhance wound healing and possibility of novel cell therapy. In: 54th congress of the Japanese Society of Oral and Maxillofacial Surgeons, 9-11 October 2010, Sapporo, Japan; 2010.

[36] Okabe K, Yamada Y, Ito K, Yoshimi R, Kohgo T, Yajima A, Hibi H, Ueda M. Experimental study for soft-tissue regeneration with tissue engineering and regenerative medicine. In: The 52nd Congress of the Japanese Society of Oral and Maxillofacial Surgeons, 29-30 September 2007, Nagoya, Japan; 2007.

[37] Osugi M, Katagiri W, Yoshimi R, Inoue M, Hara K, Inukai T, Hibi H, Ueda M. The stem cells cultured conditioned media enhanced bone regeneration. In: 20th Annual Scientific Meeting of the European Association of Osseointegration, 13-15 October 2011, Athens, Greece; 2011.

[38] Osugi M, Katagiri W, Yoshimi R, Inoue M, Hara K, Inukai T, Hibi H, Ueda M. The stem cells cultured conditioned media enhanced bone regeneration. In: The 56th Congress of the Japanese Society of Oral and Maxillofacial Surgeons, 21-23 October 2011, Osaka, Japan; 2011.

[39] Osugi M, Katagiri W, Yoshimi R, Inoue M, Hara K, Inukai T, Hibi H, Ueda M. The stem cells cultured conditioned media enhanced bone regeneration. In: The 32nd Congress of the Japanese Society of Inflammation and Regeneration, 2-3 June 2011, Kyoto, Japan; 2011.

[40] Sakai K, Yamamoto A, Matusbara K, Nakamura S, Tauchi R, Wakao N, Imagama S, Hibi H, Ueda M. Multifaceted neuro-regenerative activities of tooth-derived stem cells promote locomotor recovery after complete transection of the spinal cord. The Journal of Clinical Investigation 2012;122(1) 80-90.

[41] Sakai K, Yamamoto A, Hibi H, Yamada Y, Fujio F, Yamagata M, Ueda M. Transplantation of dental pulp stem cells in spinal cord injury. In: 88th International Association for Dental Research, 4-17 July 2010, Barcelona, Spain; 2010.

[42] Sakai K, Yamamoto A, Hibi H, Yamada Y, Fujio F, Ueda M. Transplantation of human dental pulp stem cells in complete transaction of the rat spinal cord. In: 2nd Nagoya University Global COE Program, Integrated Functional Molecular Medicine for Neuronal and Neoplastic Disorders, 26-27 February 2010, Nagoya, Japan; 2010.

[43] Sakai K, Matsubara K, Yamamoto A, Ueda M. Transplantation of dental pulp stem cells in spinal cord injury. In: The 3rd Symposium of Young Researchers-Innovative MEMS design and the biomedical application- Micro/Nano Global COE, 6 December 2010, Nagoya, Japan; 2010.

[44] Sakai K, Matsubara K, Yamamoto A, Hibi H, Ueda M. Engrafted dental pulp stem cells promoted functional recovery of completely transected rat spinal cord. In: 9th International Society Stem Cell Research, 15-18 June 2011, Toronto, Canada; 2011.

[45] Shohara R. Is it possible? Body repairing with stem cells from medical wastes. In: Proceedings of International cross-disciplinary symposium on Micro-Nano Systems, 12 November 2009, Nagoya, Japan; 2009.

[46] Shohara R, Takikawa S, Yamamoto A, Hibi H, Fujio M, Sakai K, Yamagata M, Goto M, Iwase A, Kikkawa F, Ueda M. Development of bone tissue engineering using human umbilical cord derived tissue regenerative cells. In: Proceedings of International Symposium on Micro-NanoMechatronics and Human Science, 9 November 2010, Nagoya, Japan; 2010.

[47] Tateishi H. An Experimental Study of Bone Healing Around the Titanium Screw Implants in Ovariectomized Rats: Enhancrement of Bone Healing by Bone Marrow Stromal Cells. In: Proceedings of International Symposium on Micro-NanoMechatronics and Human Science, 9 November 2010, Nagoya, Japan; 2010.

[48] Tateishi H, Okamoto Y, Kinoshita K, Hibi H, Ueda M. A study of bone healing around the titanium screw implants in ovariectomized rats. Effects of bone healing by variance of implant surface property. In: 30th Annual Meeting of Japanese Society of Oral Implantology Kinki Hokuriku Section, 19-20 November 2010, Fukui, Japan; 2010.

[49] Ueda M, Nishino Y. Cell based cytokine therapy for skin rejuvenation. Journal of Craniofacial Surgery 2010;21(6) 1861-1866.

[50] Wadagaki R. Osteogenic Induction of Bone Marrow-Derived Stromal Cells on Simvastatin-Releasing, Biodegradable, Nano-Micro Fiber Scaffolds. In: The 3rd Symposium of Young Researchers -Innovative MEMS design and the biomedical application, 6 December 2010, Nagoya, Japan; 2010.

[51] Yoshimi R, Yamada Y, Ito K, Nakamura S, Abe A, Nagasaka T, Okabe K, Kohgo T, Baba S, Ueda M. Self-assembling peptide nanofiber scaffolds, platelet-rich plasma, and mesenchymal stem cells for injectable bone regeneration with tissue engineering. Journal of Craniofacial Surgery 2009; 20(5) 1523-1530.

[52] Yoshimi R, Yamada Y, Ito K, Kohgo T, Okabe K, Yajima A, Hibi H, Ueda M. Application for bone regeneration by regenerative medicine with new three dimension matrix using nanotechnology. In: 52nd Annual Meeting of Japanese Society of Oral and Maxillofacial Surgeons, 29 September 2007, Nagoya, Japan; 2007.

Chapter 9

Measurement of Frictional Properties on the Micro/Nanometer Scale

Kenji Fukuzawa

1. Introduction

Scanning probe microscopes (SPMs) were invented by Binnig and Rohrer in 1980s and have been one of the key methods for measurements and manipulations in nanotechnologies since then [1, 2]. The first SPMs were a scanning tunnelling microscope (STM), which uses a tunnelling current. The vertical resolution of sub-nm comes from the extremely short length of interaction between the probe and sample surface. If a sharp tip can be prepared, the lateral resolution of sub-nm is possible. STM can see sample surfaces at an atomic resolution and manipulate a single atom. The concept of STM can be extended to various fields and wide varieties of properties can be measured on the nano scale by changing the interaction between the probe and sample. It should be noted that micro fabrication techniques based on micro-machining techniques has been indispensable to the development of SPMs. Atomic force microscopes (AFM) are one of the most successful microscopes in SPMs [3]. As a probe, they use micro cantilevers with a tiny tip, whose apex radius is around several nm. Typical dimensions of the cantilever are 100-μm long and a few-μm thick. Micro structures at these dimensions are not easy to fabricate for conventional machining. Therefore, micro fabrication methods have been developed. Micromaching not only improved the performance of SPMs but also made them a popular technology because the fabrication methods were suitable for mass production.

SPM have been extending its application to mechanical fields. One of the successful applications is a friction force microscope (FFM), which has been a key technology of micro/nano tribology [4]. FFM can measure the local frictional or rheological properties at a micro/nano scale although AFM is limited to measurement of the topography. The performance of micro/nano mechanical devices, such as micro/nano electro mechanical systems (MEMS/NEMS), computer hard disk drives (HDDs), is dominated by surface forces such as friction and viscous forces rather than volume forces such as gravitational forces. Therefore, the measurement of the local frictional properties is very important for the performance analysis and design of micro/nano mechanical devices. FFM is also applied to local characterization of the sample surface by using that different materials usually have different frictional properties. If the sample consists of composite materials such as laminated materials or carbon-fiber-reinforced materials, the local composition can be evaluated at a nanometer scale by measuring the frictional property distribution. Recently, the local chemical composition and properties such as hydrophobicity can be mapped by measuring friction force distribution with a chemically functionalised probe. In this chapter, the basis and drawbacks of FFM are briefly reviewed and recent trials to overcome the drawbacks are described.

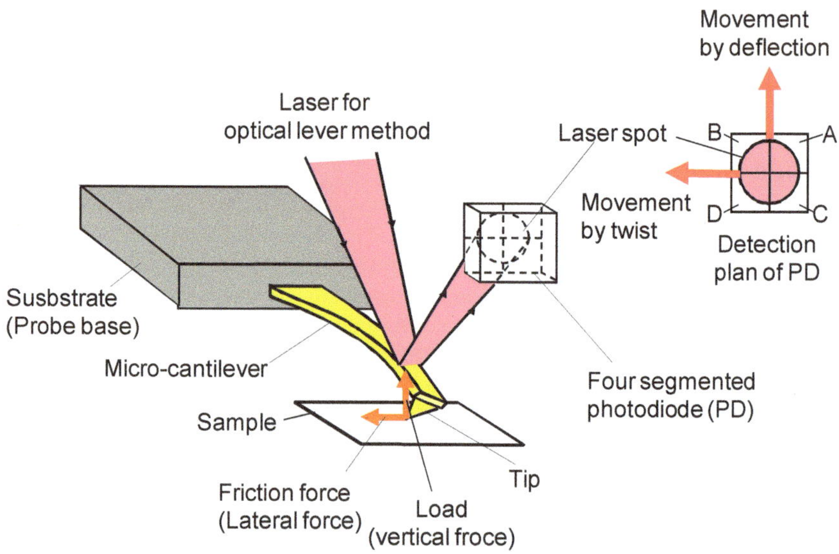

Figure 1. Schematic of conventional micro-cantilever-based FFM.

2. Conventional FFM

Conventional FFMs use a micro cantilever as a probe, which is used in AFM. By slightly modifying an AFM setup, replacing a two-segmented photodiode (PD) as a position sensitive detector (PSD) with a four-segmented one, FFM measurement become possible. Because of this convenience, many of the commercial AFMs have a four-segmented photodiode and can make a FFM measurement with a micro cantilever probe for AFM. However, probes are designed for AFM, not for FFM. This limits the performance of FFM. In SPMs, the development of the probe is significant as described above. Although many trials have been done to overcome the drawbacks, a standard probe for FFM has not been established yet. Here, measurement principle of conventional FFMs and then their drawbacks are explained.

2.1. Measurement principle of conventional FFM

When the probe is scanned under a constant load condition as shown in Figure 1, the micro cantilever twists along its long axis due to the friction force applied to the probe tip. FFM measures the friction forces by detecting this torsion of the cantilever. In addition to the friction force, the load, the force in the vertical direction, is simultaneously applied to the tip. This causes the vertical deflection of the micro cantilever. Lettering θ and Δz be torsion angle and vertical deflection, respectively. The friction force in lateral direction, F_L and the load in the vertical direction, F_V are respectively given by

$$F_L = \frac{k_l}{l}\theta \quad (1)$$

$$F_V = k_v \Delta z \quad (2)$$

where k_l and k_v are to the tensional rigidity and spring constant of the cantilever. In addition, l is the length of the cantilever. The friction force and load can be simultaneously obtained from measured torsion angle θ and vertical deflection Δz by using Eqs. (1) and (2). If the sharp tip whose apex radius is around one nanometer is prepared and the probe is scanned on the sample surface with a piezo scanner, the friction coefficient, which is the most fundamental frictional properties, can be mapped on a nanometer scale. Next, a method for measuring the torsion angle θ and vertical deflection Δz is explained. This is based on the method for measuring the vertical deflection in AFM, which is called optical lever method. A schematic setup is shown in Figure 1. The light from the laser diode is focused on to the micro cantilever surface and the reflected light forms the light spot on the four-segmented PD. The four PDs detects the light intensities separately and the difference in PD signals, (I_A+I_B)-(I_C+I_D) and (I_A+I_C)-(I_B+I_D) are obtained. Here, I_A, I_B, I_C, and I_D are the signals from PD of A, B, C, and D, respectively. When the load is applied to the probe and micro cantilever deflects vertically, the light spot moves in the vertical direction. This causes the change in the differential signal (I_A+I_B)-(I_C+I_D). In a similar fashion, the friction force is applied, the differential signal (I_A+I_C)-(I_B+I_D) changes due to the lateral displacement of the light spot, which is generated by the torsion of the micro cantilever. The mode of probe scanning in FFM is the same as the contact AFM mode. In this mode, the probe is scanned under a constant load condition, which is achieved by controlling the probe height with a piezo actuator so that the vertical deflection signal (I_A+I_B)-(I_C+I_D) keeps a set value. Typical probes for FFM are I- and V-shaped microcantilevers as shown in Figure 2. Usually, I-shaped cantilever is made of single crystal silicon and V-shaped one is made of silicon nitride. Typical spring constant in vertical direction, K_V is around 1 N/m. Both probes are basically designed for contact mode AFM. Thus, only the introduction of the four-segmented PD can provide FFM measurement with usual AFM setups.

2.2. Drawbacks of conventional FFM probes

Conventional FFM probes have been very useful for mapping friction properties and producing fruitful results in various fields, however, it has a fundamental problem to be solved. It uses the torsion angle θ and vertical deflection Δz for detecting the friction force F_L and load F_V. Two deformations have to be independent of each other for accurate measurement of the friction force and load. Considering that both deformations occur at the end of the micro cantilever, they are possible to interfere with each other. The independent deformations are valid only when the deformations are small enough, which means that the friction force and load are small enough. This mechanical interference cause significant drawbacks in FFM.

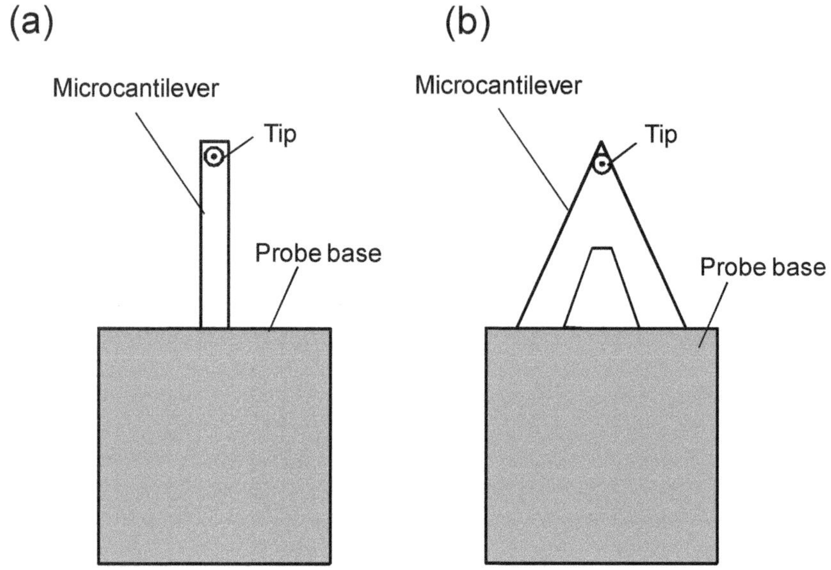

Figure 2. Typical types of micro cantilever for FFM. (a) I-shaped and (b) V-shaped cantilevers. The figures are viewed from the tip side.

The first drawback is the decrease of the measurement accuracy [5, 6]. As described in Section 2.1, in Eqs. (1) and (2), the tensional rigidity k_l and spring constant k_v are assumed to be constant. However, when the friction force is large, the torsional rigidity k_l and the spring constant k_v depends on the load and friction force. These limit FFM applications for conventional probe. Therefore, conventional FFMs remain at the level of qualitative evaluation means in many applications.

The second drawback is the decrease of the measurement performance. The fundamental equations, Eqs. (1) and (2) indicate that the improvement of the force sensitivity requires the decrease of the rigidity, k_l or k_v. In contrast, the reduction of the mechanical interference requires the high rigidity for the counterpart deformation. For example, the torsional rigidity has to be set high in order to measure the load accurately. Since the V-shaped cantilevers are designed for a contact AFM mode, crossed cantilever structure is adopted in order to increase

k_l and reduce the torsion of the cantilever. This leads to the decrease of the sensitivity of the friction force. In order to measure the friction force accurately, the vertical spring constant k_v has to be higher, which decreases the load sensitivity. Thus, the design for the high force sensitivities of both the friction force and load is difficult for conventional cantilever probe due to the mechanical interference. In conventional FFMs, since probes are designed for AFM, most of the probe is set so as to reduce torsion of the cantilever. Therefore, friction force sensitivity is made victim of that of the load.

In addition to the drawbacks due to the mechanical interference, conventional FFMs have another fundamental drawback. As described in Section 2.1, the torsion of the micro cantilever is detected to obtain the friction force. This is implemented by the optical lever method. The optical lever method can provide the differential signal $(I_A+I_C)-(I_B+I_D)$ corresponding to the torsion angle θ in Eq. (1). For quantitative measurement, the relationship between the signal $(I_A+I_C)-(I_B+I_D)$ and angle θ has to be known. For this purpose, the torsion angle θ of the micro cantilever has to be measured. In addition, this measurement has to be done for the micro cantilever installed in AFM setup because the signal $(I_A+I_C)-(I_B+I_D)$ may change for the setup to setup. This torsion angle measurement is not easy in usual setups. Therefore, the standard calibration method for the friction force has not been established yet. This is one of the reasons why FFM remains at the level of qualitative evaluation means [7].

3. Dual-axis FFM probes

Many trials have been done in order to overcome the above drawbacks of the conventional FFM. One promising candidate is a dual-axis probe. The mechanical interference problem in the conventional probe is caused by the fact that the probe uses the two deformations of the same part, the probe end. If the different parts of the probe deform due to the friction force and load, the mechanical interference can be reduced or eliminated. Dual-axis probe is based on this idea. Chui et al. presented the probe that serially connected an array of cantilevers to a V-shaped cantilever with a tip [8]. The schematic structure is shown in Figure 3. The surface of the cantilever array is set vertically to the V-shaped cantilever one. The double cantilever array laterally deflects for the friction force and V-shaped cantilever vertically deforms for the load. The friction force and load is obtained from these lateral and vertical deflections. Ando et al. presented the concept of another type of dual-axis probe, where the two double cantilevers that rotate by 90 degrees to each other are serially connected as shown in Figure 4 [9]. The double cantilever with its surface normal to the sample surface deflects for the friction force and another double cantilever does for the load.

The detection method for vertical deflection of the probe for conventional FFM can be applied to the dual-axis probe whereas that for lateral one cannot because the probe does not twist and deflects laterally. In the optical lever method, which is widely used in conventional FFMs, the light does not move on the PD even if the probe deflects laterally. Therefore, the optical lever method cannot be applied to dual-axis probes directly. Moreover, in the both dual-axis probes, the deformation parts are serially connected as shown in Figures 3 and 4. The displacement of the probe end is the sum of the deformations of two deformation parts such as the cantilever array and V-shaped cantilever. This means that the deformations of the two parts cannot be determined by measuring the displacement of the probe end as can be done in conventional FFMs. In the probe that presented by Chui et al., in order to solve this problem, they fabricated a very special type of the probe that has piezoresistive sensors both on the lateral and vertical cantilevers to detect lateral and vertical deflections. In general, the piezo resistive sensor is not easy to fabricate on the double cantilever array because their surface is normal to the base substrate. In the probe that presented by Ando et al., the solution for this problem was on a conceptual level and not demonstrated experimentally. Thus, the standard probe for the dual-axis FFM has not been established yet. Recently, a new type of dual-axis probe called micro mechanical probe (MMP) has been presented [10]. This probe can overcome the significant problem of the dual-axis probe as mentioned above. The details of the MMP are described below.

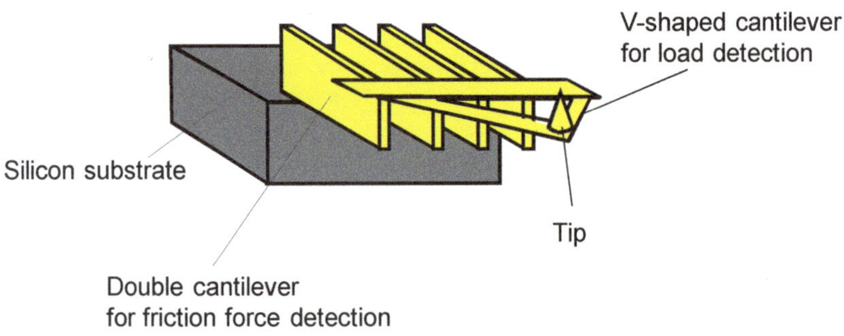

Figure 3. Schematic of dual-axis FFM probe with an array of double cantilevers for the friction force detection and V-shaped cantilever for the load detection.

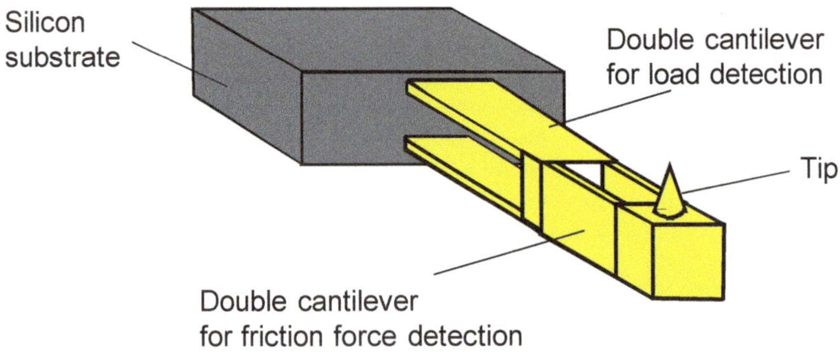

Figure 4. Schematic of a dual-axis FFM probe with serially connected double cantilevers for the friction and load detections

3.1. Concept of Micro Mechanical Probe (MMP)

The structure and set-up of MMP are schematically shown in Figure 5. This probe consists of the double cantilever for the friction force detection and the torsion beam for the load detection. The tip is fabricated at the end of the double cantilever. The double cantilever is supported by the torsion beam and rotates along the torsion beam when the load is applied to the tip. In this probe, the deformation parts are not connected serially as done in the dual-axis probes that above mentioned. This can provide that both the deformations of the double cantilever and torsion beams can be measured by detecting the displacement of the probe end. By detecting the lateral deflection of the double cantilever Δx and the torsion angle of the torsion beam, θ, the friction force F_L and load F_V can be obtained by

$$F_L = k_{dl} \Delta x \tag{3}$$

$$F_V = \frac{k_{tv}}{l_t} \theta_t \tag{4}$$

where k_{Ld} and k_{Vt} are the spring constant of the double cantilever and torsional rigidity of the torsion beam, respectively. In the MMP, the lateral deflection Δx and the torsion angle can be measured by detecting the lateral and vertical displacements of the probe end. As described above, the lateral displacement of the dual-axis probe cannot be detected by the optical lever method. In the MMP, the low reflection pattern called optical lever pattern is fabricated on the probe head surface. As one of the optical lever pattern, a micro roof-shaped structure was presented [11]. The light that focused onto the roof-shaped structure scatter whereas the light focused onto the flat surface reflects and goes to the PD for the optical lever method. Therefore, the reflection light forms the light spot with a dark pattern on the PD. When the load is applied, the light spot on the PD moves in the vertical direction as does in conventional FFM. In contrast, when the friction force is applied, the spot does not move but the dark pattern generated by the optical lever pattern moves laterally. Therefore, the differential signal $(I_A+I_B)-(I_C+I_D)$ gives the vertical displacement of the probe head or the torsion angle of the torsion beam θ whereas $(I_A+I_C)-(I_B+I_D)$ gives the lateral deflection of the double cantilever Δx. This method for measuring the vertical and lateral displacements are the same as that in conventional FFMs as explained in Section 2.1. This indicates that the optical lever method, which is widely used in conventional FFMs, can be applied to the MMP-based dual-axis FFM. Therefore, the MMP can be easily installed to conventional FFM or AFM setups with FFM options and can change conventional FFM to dual-axis FFM.

3.2. Mechanical design of MMP

Let us think about the necessary structure for the MMP [12]. In general, the needed force resolution for the friction force and load is of the order of 1 nN. The typical displacement resolution of the optical lever method is of the order of 0.1 nm. Therefore, a required spring constant of the double cantilever and torsion beam is of the order of 1 N/m. In conventional FFM probes, the typical lateral and vertical spring constants of V-shaped cantilever is around

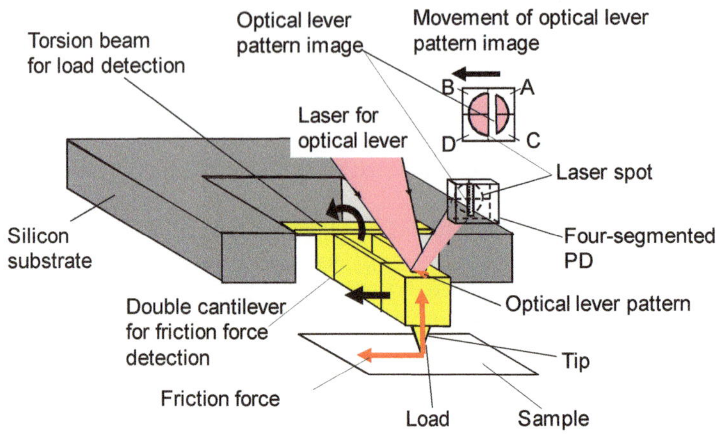

Figure 5. Schematic of the micro mechanical probe (MMP)-based FFM. The MMP consists of a double cantilever for the friction force detection and a torsion beam for the load detection.

100 and 0.1 N/m, respectively, and the typical lateral and vertical spring constants of I-shaped cantilever is around 100 and 1 N/m, respectively. These indicate that the friction force sensitivity is two or three order of magnitude lower than that of the load. This is because these probes are designed for AFM and the mechanical interference is reduced by sacrificing the friction sensitivity as described above.

The spring constant of the double cantilever beam in the lateral direction is given by

$$k_{dl} = 2E\frac{b_d h_d^3}{l_d^3} \tag{5}$$

where E is Young's modulus. the spring constant of the double cantilever beam in the vertical direction is given by

$$k_{dv} = \frac{1}{2}E\frac{b_d^3 h_d}{l_d^3} \tag{6}$$

The effective spring constant of the torsion beam in the vertical direction is given by

$$k_{tv} = 2f\frac{G b_t h_t^3}{l_c^2 l_t} \tag{7}$$

where the torsion angle is converted to the vertical displacement of the probe end. The lateral spring constant of the torsion beam k_{tl} is given by

$$k_{tl} = \frac{2}{3} E \frac{b_t^3 h_t}{l_c^2 l_t} \quad (8)$$

By using Eqs. (5) to (8), the spring constants are obtained as $k_{dl} = 4.3$ N/m, $k_{dv} = 1.5 \times 10^3$ N/m, $k_{tv} = 13.1$ N/m, $k_{tl} = 4.9 \times 10^2$ N/m when $l_d = 1200$ μm, $b_d = 185$ μm, $h_d = 5$ μm, $l_t = 370$ μm, $b_t = 70$ μm, and $h_t = 20$ μm. Here, *l*, *b*, and *h* denote the length, width, and thickness, respectively and the suffixes *d* and *t* mean the double cantilever and torsion beam, respectively. The pitch between the double cantilever beams is set to $w_h = 400$ μm and the tip height *t* is set to 10 μm. The lateral rigidity of the torsion beam k_{tl} is 100 times larger than that of the double cantilever beam k_{dl}, and the vertical rigidity of the double cantilever beam k_{dv} is 100 times larger than that of the torsion beam k_{tv}. The probe with these dimensions can measure the lateral and vertical forces independently with a force resolution of 1 nN without the mechanical interference.

3.3. Fabrication method of MMP

Here, the fabrication method of the MMP is described. As described in the above section, the double cantilever of the MMP requires the structure with a very high aspect ratio of around 40: 1. In the above example, a cantilever with a width of 185 μm and the thickness of 5 μm. The structure at these small dimensions is difficult to fabricate for conventional machining techniques. Moreover, the high aspect structure is not easy for even the micromachining techniques although they are suitable for fabrication of the structure on the micrometer scale. Among the current micromachining techniques, anisotropic chemical etching and deep reactive ion etching (DRIE) are candidates that can fabricate microstructure with a high aspect ratio. Although the DRIE can provide height aspect microstructure, it takes efforts to make the fabricated surface flat. In contrast, anisotropic etching can produce flat surface more easily because a crystal plane appears by etching. In the anisotropic chemical etching the shapes of structures that can be fabricated are limited to the crystal structure of the substrate whereas they are not in the DIE. The case where anisotropic etching is selected is explained below.

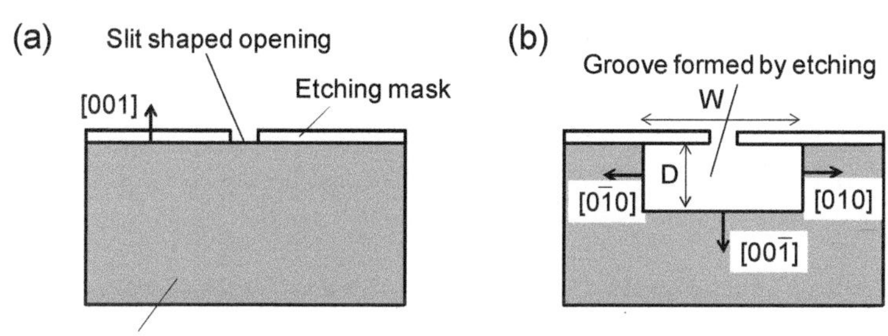

Figure 6. Anisotropic chemical etching through a slit-shaped opening.(a) A {100} silicon substrate. (b) Groove formation by the etching.

Anisotropic chemical etching uses that etch speed depends on the crystal orientation. Figure 6 explains how to fabricate a vertically oriented surface for the double cantilever. In single crystal silicon, the etch speed for {100} plane is higher than those of other orientation planes when aqueous solution of potassium hydrate (KOH) is used as an etchant. When a {100} silicon wafer is used as a substrate, {100} plane is parallel (e. g. (100) plane) or normal (e. g. (010) plane) to the wafer surface. By KOH etching the {100} silicon wafer, (010) and $(0\bar{1}0)$planes appear and form the normal to the wafer surface with a crystallographic accuracy. First, an etching mask film, which is typically made of silicon oxide or silicon nitride for KOH etching, is formed on the wafer. The wafer is dipped in the etchant after the slit-shaped opening is formed in the etching mask film as shown in Figure 6. The silicon substrate in this opening is exposed to the etchant and etched during the etching whereas the silicon substrate covered with the etching mask film is protected from the etchant. If we choose a silicon wafer whose crystal plane is the [001] direction and orientation flat of the lower end is $[\bar{1}00]$, the direction of the opening should be set to be parallel to the crystal direction of [100] or [010]. In the case shown in Fig. 6, the opening is along the [100] direction. As the progress of the etching, the groove is formed, whose bottom face is the (100) plane and the sidewalls are (010) and $(0\bar{1}0)$ planes. When the two parallel slit-shaped openings are formed in the etching mask, a vertically oriented thin plate is formed between the openings. By arranging the three openings so that two thin films between the three grooves remain when the substrate is fully etched, a double cantilever with a micrometer-order thickness can be formed. Thus, the thickness of the double cantilever is determined by the slit pitch and width by the thickness of the silicon wafer. Since typical wafer thickness is around 100 to 500 μm, the double cantilever with a high aspect ratio of several tens to one is possible.

It should be noted that the dimensions of target devices are partially constrained in this fabrication method. The etch rates of the crystal plane in the bottom surface and sidewalls are equal because all the planes belong to the {100} planes. In addition, the (010) and $(0\bar{1}0)$ planes are etched in the lateral direction whereas only the (100) plane is etched in the vertical direction (Figure 6). Therefore, the growth speed of the groove in the lateral direction is twice as large as that in the vertical direction. That is to say, the width is always twice as large as depth in the etched groove. In order to make the device freestanding, all the silicon substrate has to be etched. In this case, the etch depth is equal to the wafer thickness. Therefore, the width of the etched groove is larger than twice the wafer thickness. The thickness of the double cantilever is determined by the pitch of the slit-shaped openings and the width of the etched groove, that is, the wafer thickness. When the slits are arranged in the photomask, these relations should be considered carefully. In addition, the pitch of the cantilevers of the double cantilever is larger than twice the wafer thickness due to the above reason.

An example of the fabrication method of the MMP is schematically in shown in Figure 7 [11]. The directions of the wafer plane and orientation flat are set to be (001) and $(\bar{1}00)$, respectively. Considering the relationship between the width and depth of the groove, the wafer thickness should be selected, for example, around 200 μm. The larger wafer thickness causes the larger pitch of the cantilevers, which leads to the larger probe head that connects the cantilever. This decreases the natural frequency of the probe and reduces the robustness against environmental

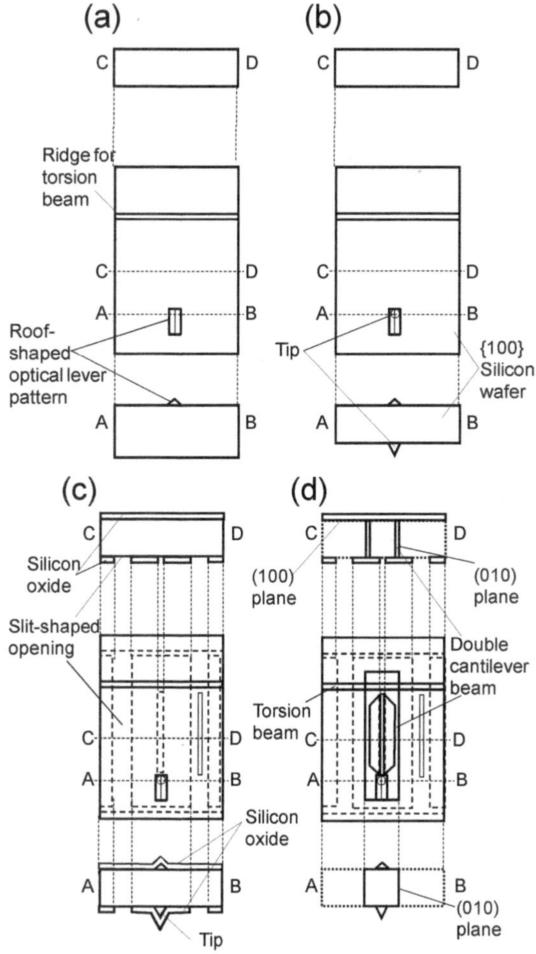

Figure 7. Fabrication process of the micro mechanical probe (MMP). (a) Formation of the optical lever pattern and ridge for the torsion beam. (b) Formation of the tip on the backside of the silicon substrate. (c) Formation of etching mask for the substrate etching. (d) Free standing probe after the substrate etching.

mechanical noises. Usually, the thickness of commercial silicon wafers has a variation of around 10 μm. Prior to the fabrication, the substrates should be uniformly etched so as to reduce the thickness variation. The fabrication processes mainly consists of three anisotropic chemical etching. The etchant is an aqueous solution of KOH for all the etchings. Silicon oxide films for etching mask can be formed by thermal oxidation method. In the first etching, the torsion beam and optical lever pattern are formed (Figure 7(a)). The optical lever pattern is formed by using the undercut etching under a rectangle-etching mask. In the second etching, the tip at the probe end is formed (Figure 7(b)). The tip is also fabricated by undercut etching. In the third etching, the double cantilever beam is formed (Figure 7(c)). Through a slit-shaped opening in the etching mask on the rear side of the wafer, the substrate silicon is etched. Finally, the freestanding double cantilever beam is formed (Figure 7(d)). The electron microscope image of a fabricated probe is shown in Figure 8. A magnified view of tip and a roof-shaped optical lever pattern are also shown. In this example, the thickness, width and length of the double cantilever were about 8 μm, 167 μm, and 1410 μm, respectively. Using the anisotropic chemical etching, the MMP with a high aspect ratio double cantilever can be formed.

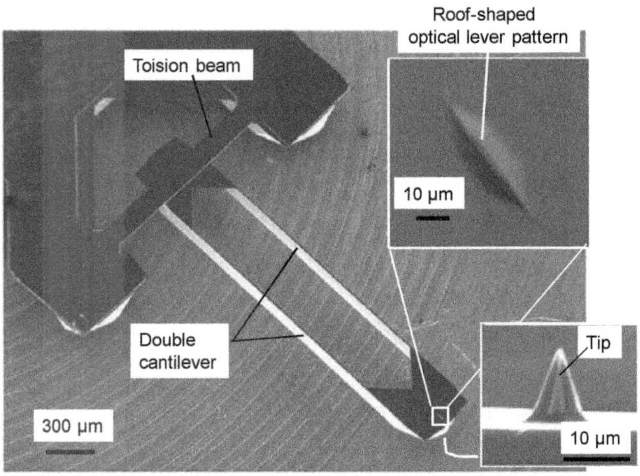

Figure 8. Fabricated micro mechanical probe (MMP).

3.4. Force calibration

The force calibration is indispensable for quantitative FFM measurements. In the MMP, the differential signals from the four-segmented PD, (I_A+I_B)-(I_C+I_D) and (I_A+I_C)-(I_B+I_D) have to be converted to the lateral deflection of the double cantilever and torsion angle of the torsion beam or vertical displacement of the probe head, respectively. After these sensitivity calibration, the deflection and torsion angle can be converted to the friction force and load, respectively.

The latter can be implemented as is done for the calibration of the vertical deflection in usual AFMs or FFMs. Using force curve measurement, the conversion coefficient can be obtained from the relationship between the piezo displacement and differential signal (I_A+I_B)-(I_C+I_D). In conventional FFMs, as described in Section 2.2, the calibration of the friction force is not easy because the torsion angle of the micro cantilever has to be measured. In contrast, the calibration of the friction force in the MMP is similar to usual calibration method for the vertical deflection in AFMs. That is is because the calibration of the friction force in the MMP can be implemented by obtaining the relationship between the differential signal, (I_A+I_C)-(I_B+I_D) and the lateral deflection of the double cantilever.

One method of the friction force calibration is shown in Figure 9. In this method, a step structure is used as a standard sample. For example, the fraction of the silicon wafer, which is glued onto the flat substrate, can be used as the step structure. The step height is set to be higher than the double cantilever width. During the probe scanning on the substrate, the probe stops at the step structure as shown in Figure 9(a). In contrast, the probe base, which is attached to the piezo scanner, continues to move. When the optical lever system of an FFM apparatus is implemented so that it moves together with the piezo scanner or probe base, the optical lever pattern appears to move in the inverse direction although the pattern stops at the step structure. This cause the lateral movement of the dark optical lever pattern on the PD of the optical lever system as showin in Figure 9(b). Since the lateral movement of the pattern is equal to the displacement of the probe base i. e. piezo scanner, the relationship between the signal (I_A+I_C)-(I_B+I_D) and the lateral deflection can be obtained.

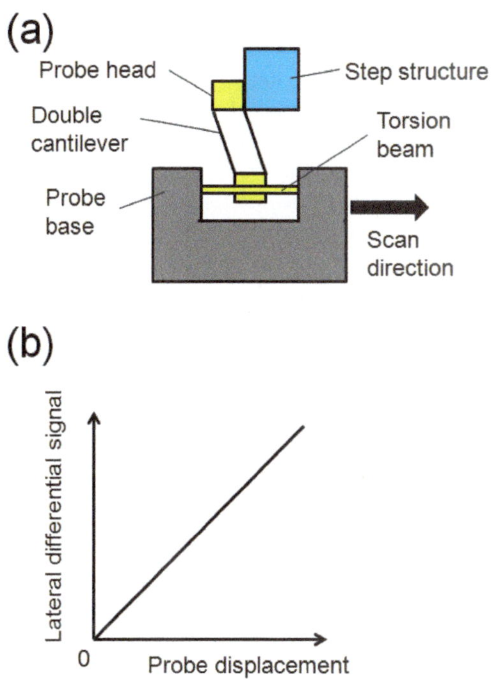

Figure 9. Calibration of friction force with a step-structure. (a) Arrangement of the micro mechanical probe and step structure. (b) Typical relationship between the differential signal for probe lateral deflection and probe displacement.

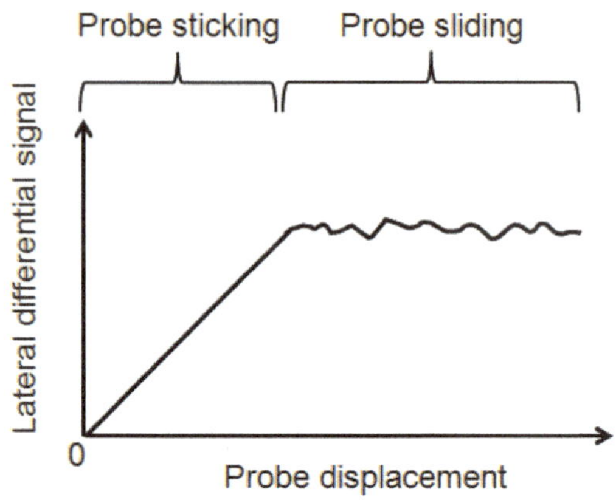

Figure 10. Calibration of the friction force by using the probe sticking due to the static friction at the beginning of the probe scan.

In the above method, the step structure that specially prepared is used. If the probe sticks to the substrate at the beginning of the probe scan by the static frictional force, the step structure is not necessary. In many cases, at the beginning of the scan, the probe remains still on the sample substrate due to static friction although the probe base or piezo scanner moves. After the elastic force overcomes the static friction, the probe start to slide on the substrate and the friction mode changes from the static to the kinetic friction. The conversion coefficient can be obtained from the relationship between the differential signal and piezo displacement before the probe sliding as shown in Fig. 10. Although this method is easily implemented, the error is possible to occur when the sample does not stick the probe enough. Therefore, the sample for the calibration has to be carefully selected.

In the previous report [13], the conversion coefficient of signal to the deflection, where the MMP was installed to the commercial AFM setup (NanoScope, Bruker), was of the order of several mV/nm. Here, this was determined by the step structure method. From this result, the minimum detection limit of the lateral deflection was obtained as the order of 0.1 nm because the typical noise level was around 1 mV. These results indicate that sub nN force can be detected when the lateral spring constant of the double cantilever is of the order of 1 N/m,

3.5. Example of MMP-based FFM measurement

Figure 11 shows an example of the FFM measurement. The sample was a nm-thick-lubricant film on a substrate. In this example, the lubricant was pepfluoropolyehter (PFPE) and the substrate was a magnetic disk. Both of the lubricant and disk are used in the lubrication of the HDDs. In addition, the lubricant film was patterned by the ultra-violet (UV) light treatment [14]. The UV light illumination bonds the lubricant molecule onto the disk surface by photo-chemical reaction. If some parts of the lubricated disk is irradiated and whole disk is rinsed by the solvent, the UV-irradiated lubricant film remains and the unirradiated lubricants not. By controlling the irradiation area with a photomask, the patterned nm-thick lubricant film can be obtained. Here, a line and space pattern whose line width and pitch were about 5 μm and 15 μm, respectively. The lubricated region showed the lower friction whereas the disk exposed region showed higher friction. In this measurement, the contact mode was used, which means that the load was constant. Therefore, the friction force distribution shown in Figure 11 shows the distribution of the friction coefficient because the coefficient is obtained by dividing the friction force by the load.

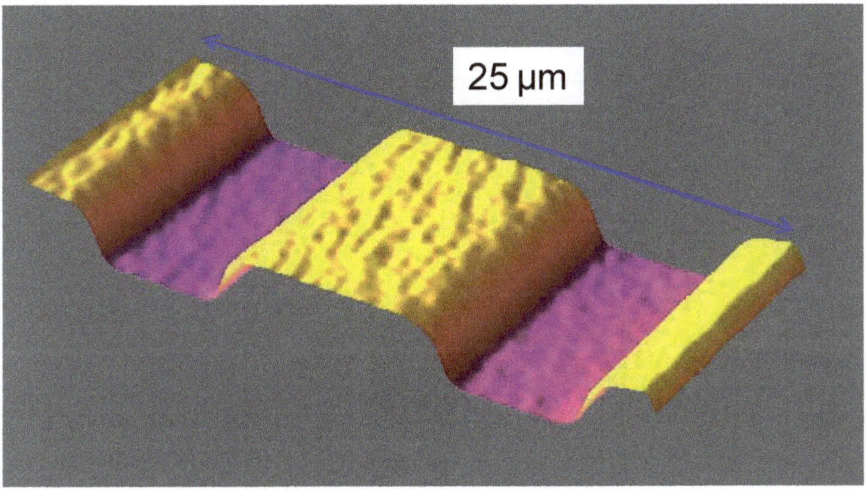

Figure 11. Example measured by a micro mechanical probe. The sample was a nanometer thick lubricant film with line-and-space-patterned by the ultra violet light radiation. The lower region is lubricant remained region and the higher is substrate magnetic disk exposed region.

Thus, the MMP can elevate FFM from the qualitative visualization method to the quantitative measurement method. In addition, the optimum conditions as a FFM probe can be selected because the MMP-based dual-axis FFM become free from the limitation due to the reduction of the mechanical interference. This can provide high sensitive measurement of both friction force and load, which is difficult in the conventional probe. The MMP is expected to be a standard probe for FFM by overcoming the problem of the conventional FFM probes.

3.6. Future directions

As described above, the significant problems of the dual-axis probe can be solved by the MMP-based FFM and the MMP can provide the friction properties on the nanometer scale. This is expected to open up the new field of FFM as a quantitative evaluation method. For further advancements, the various sliding speeds are expected. In the current dual-axis probes, the sliding speed is limited to the scan speed, typically around 1 μm/s, which is determined by the scanner piezo. In many applications, especially mechanical applications such as MEMS/ NEMS or HDDs, the friction properties at a higher sliding speed is required.

4. Summary

In this chapter, the basics and latest research results of FFM, which is one of the key measurement methods for micro/nano mechatronics, is described. FFM can provide the local friction properties on the micro/nano scale. FFM is widely used in tribology and material characterization as chemical force microscopes. Dual-axis probes are new types of FFM probes, which can overcome the significant drawbacks of the conventional cantilever probes, especially, the reduction of the mechanical interference between the lateral friction force and vertical load. However, the dual-axis probes have problems to be solved such as the lateral displacement detection method. The micro mechanical probe can overcome the problems and is possible to elevate the FFM from a visualization tool to the quantitative evaluation method for local friction properties on the micro/nano scale. This is expected to be a useful method for designing micro/nano mechatronic devices such as MEMS/NEMS or the head-disk lubrication of HDDs and improve their performance drastically.

Acknowledgements

This work was supported in part by the Japanese Ministry of Education, Culture, Sports, Science and Technology through Global Center of Excellence (GCOE) project, Grant No. 20360077, the NEDO Green-IT Project, and the Storage Research Consortium.

Author details

Kenji Fukuzawa

Department of Micro-Nano Systems Engineering, Nagoya University, Nagoya, Japan

References

[1] Binnig G, Rohrer H. Scanning tunneling microscope combined with a scanning electron-microscope. Review of Scientific Instruments 1986;57(2) 221-224.

[2] Binnig G, Rohrer H. Scanning tunneling microscopy, Scanning tunneling microscopy - from birth to adolescence. Reviews of Modern Physics 1987;59(3) 615-625.

[3] Binnig G, Quate C. F, Gerber C. Atomic Force Microscope, Physical Review Letters1986;56(9) 930-933.

[4] Mate C. M, McCelland G. M, Erlandsson R, Chiang S. Atomic-scale friction of a tungsten tip on a graphite surface. Physical Review Letters 1987;59 1942-1945.

[5] Mitsuya Y, Ohshima Y, Nonogaki T. Coupling and nonlinear effects of cantilever deflection and torsion encountered when simultaneously measuring vertical and lateral forces using the scanning probe method. Wear 1997;211 198-202.

[6] Amakawa H, Fukzuawa K, Shikida M, Zhang H, Itoh S. Dual-Axis Micro-Mechanical Probe for Highly Sensitive Friction Force Microscope. In: Proceedings of World Tribology Congress 2009 , WTC2009; 2009.

[7] Munz M. Force calibration in lateral force microscopy: a review of the experimental methods. Journal of Physics D 2010;43(6) 063001.

[8] Chui B. W, Kenny T. W, Mamin H. J, Terris B. D, Rugar D. Independent detection of vertical andlateral forces with a sidewall-implanted dual-axis peizoresistive cantilever. Applied Physics Letters 1998;72(11) 1388-1390.

[9] Ando Y, Nagashima T, Kakuta K. Using FIB-processed AFM cantilevers to detemine microtribology characteristics. Tribology Letters 2004;9 15-23;

[10] Fukuzawa K, Terada S, Shikida M, Amakawa H, Zhang H, Mitsuya Y. Dual-Axis Micro-Mechanical Probe for Independent Detection of Lateral and Vertical Forces. Applied Physics Letters 2006;89 173120.

[11] Amakawa H, Fukzuawa K, Shikida M, Zhang H, Itoh S. Improved Sensitivity of Dual-Axis Micro-Mechanical Probe for Friction Force Microscope. Tribology Online 2008;3(7) 356-360.

[12] Fukuzawa K, Terada S, Shikida M, Amakawa H, Zhang H, Mitsuya Y. Mechanical Design and Force Calibration of Dual-Axis Micro-Mechanical Probe for Friction Force Microscopy. Journal of Applied Physics 2007;101(3) 034308.

[13] Amakawa H, Fukzuawa K, Shikida M, Zhang H, Itoh S. Quantification of friction force on dual-axis micro-mechanical probe for friction force microscopy. Tribology Online 2010;5(3) (2010), 144-149.

[14] Zhang H, Mitsuya Y, Fukuoka N, Imamura M, Fukuzawa K. Self-organized patterning of molecularly thin liquid polymer films utilizing molecular flow induced by ultraviolet irradiation. Applied Physics Letters2007;90 123119.

Chapter 10

Tribology for Biological and Medical Applications

Noritsugu Umehara, Takayuki Tokoroyama and
Hiroyuki Kousaka

1. Introduction

Tribology is the science and technology of two interacting surfaces in relative motion and of related subjects and practices. The field of tribology is interdisciplinary. Until recently, it has been dominated by mechanical and chemical engineers who conducted macro-tests to predict friction and wear in machine components and developed new lubricants. After the development of biological engineering and medical engineering, tribology for various interacting surfaces of human, creatures, cells and so on has keen attention in bio-technology. Also interaction of surfaces between medical devices and bio-cells as a blood is becoming to be interested. On the basis on the old tribology, we can not solve issues for bio-medical applications, because bio and medical surfaces are basically soft and changed their mechanical and chemical properties with temperature drastically. So we should propose and develop novel technology for bio-medical applications.

In this chapter, we introduce several new technologies to reduce adhesion of thermally coagulated blood [1, 2] and fish proteins [3] to metal surface for a radio knife and grill rods respectively. Also ion beam irradiation increased the bonding strength of a probe DNA to polymer substrate for a DNP sequence tip in low price [4]. And in order to reduce the adhesion and friction of rubber, plasma and UV irradiation methods were proposed for the rubber plunger of disposal plastic syringe [5] and rubber plug of medicine bottle [6] respectively. Those proposed surface technologies for biological and medical applications will be explained in the following sections.

2. Reduction of adhesion of coagulated blood on a tip of radio knife

Radio knife is a common operation tool for surgery to avoid intense bleeding. However, there is severe trouble of severe adhesion of coagulated blood on the tip of radio knife during operation. The non-conductive coagulated blood stops electric current and the radio knife can not work. In order to avoid this issue, nurses have to remove the coagulated blood from the tip of the radio knife frequently with abrasive clothes.

In order to overcome this issue, we confirmed the adhesion mechanism of blood during the usage of a radio knife and proposed two types of new tips for radio knife as vibration-cooled type and water-cooled type. Water-cooled type showed good results [1, 2]. Figure 1 shows the experimental setup to obtain surface temperature of radio knife by thermo camera. High frequency electric voltage was applied from electric power source as shown in Figure 1 after the radio knife tip was dipped into bovine blood. During operation with electric power, temperature distribution around the tip was measured with infrared thermo camera. Also after the operation, the adhesion force of coagulated blood on the tip was measured with s scratch test. At first we observed the coagulated blood on the tip with vibration by a bimorph-typed piezoelectric actuator. Figure 2 shows the observation results of coagulated blood on the tip without and with vibration after test. It can be seen that amount of coagulation of blood decreased with increasing of amplitude. After the observation, adhesion force was measured for each coagulated blood on the tip by a scratch test of a pin stylus. The maximum scratching force was defined as an adhesion force. Figure 3 shows the effect of amplitude of vibration on the adhesion force. It can be seen from Figure 3 that the adhesion force decreased with the increasing of vibration amplitude gradually. When the amplitude was reached to 19 μm, the adhesion force with vibration was 1/3 of that without vibration. On the view point of temperature surrounding a radio tip with electric current, the vibration decreased temperature of surface of blood. We supposed that the temperature reduction is the essence to reduce the coagulation of blood. Therefore we made a proto-type of a tip with water cooling. Figure 4 shows the observation results of coagulated blood on the tip without and with water-cooling after test. Water was circulated inside a radio knife tip under various flow rate of cooling water. It can be seen that amount of coagulation of blood decreased with temperature. Figure 5 shows the relationship between adhesion force of coagulated blood and flow rate of water. It can be seen that water-cooling with 120 ml/min of flow rate provided 1/4 of adhesion force comparing to the conventional tip. On the basis of the well achievement of proposed tip, surface temperature should be essential to control the adhesion of coagulated blood.

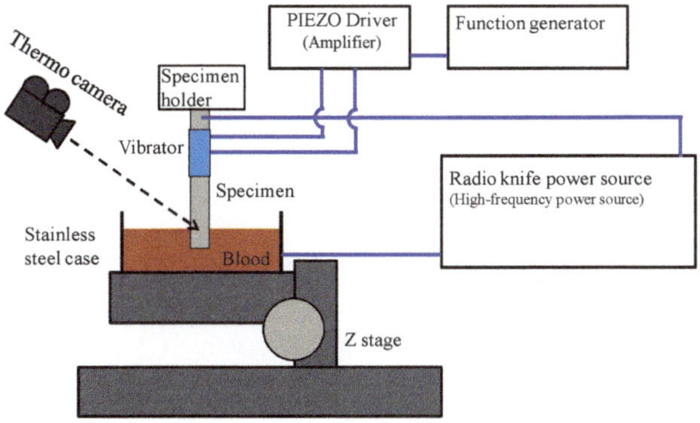

Figure 1. Schematic diagram of experimental setup for blood coagulation to a radio knife tip

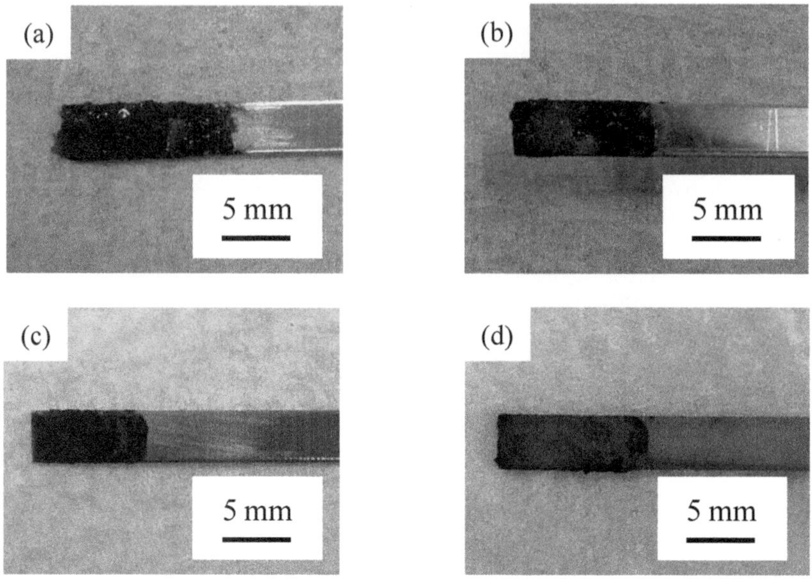

Figure 2. Observation results of coagulated blood on the tip without and with vibration (a) No vibration as conventional, (b) vibration amplitude:7 μm, (c) vibration amplitude:14 μm and (d) vibration amplitude:19 μm

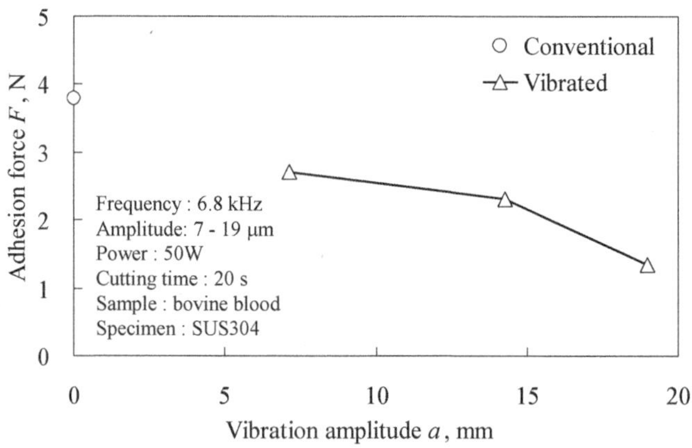

Figure 3. The relationship between vibration amplitude and adhesion force of conventional and vibrated tips

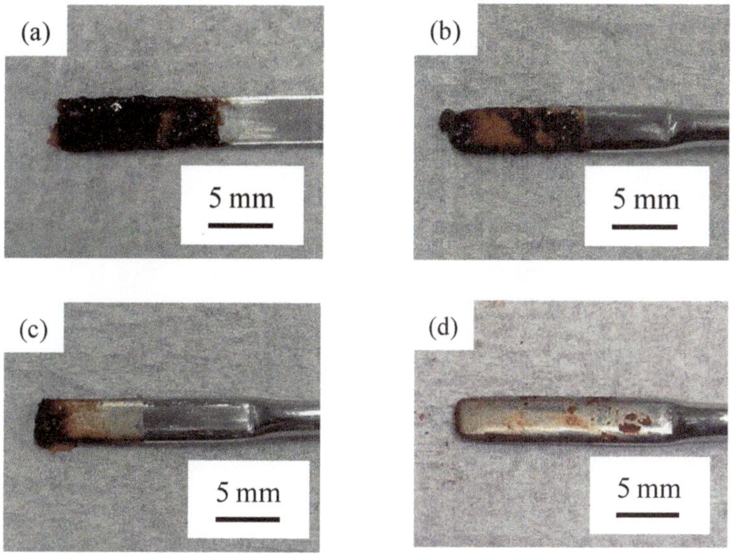

Figure 4. Observation results of coagulated blood on the tip without and with water-cooling (a) No water-cooling as conventional, (b) flow rate: 0 ml/min, (c) flow rate: 60 ml/min and (d) flow rate: 120 ml/min

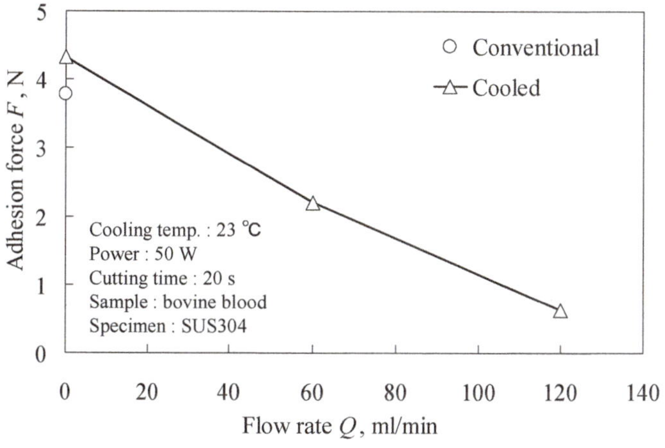

Figure 5. The relationship between flow rate and adhesion force of conventional and water cooled tips

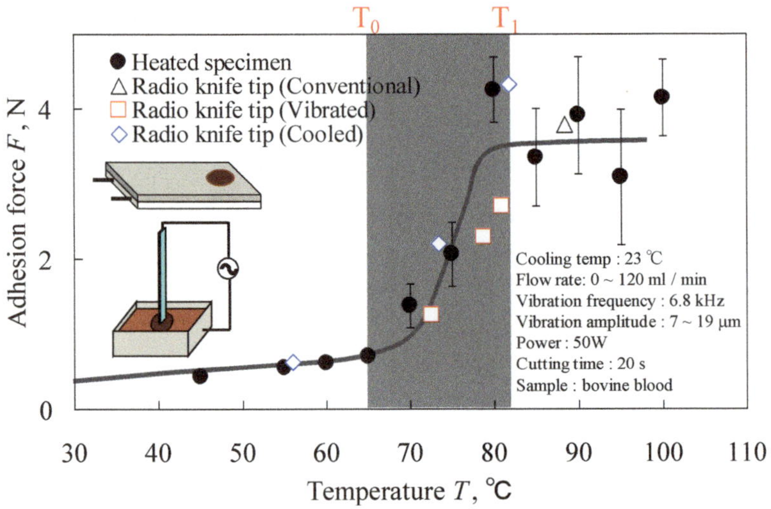

Figure 6. Relationship between adhesion force of coagulate blood and surface temparature on the tip of conventional, vibrated and water-cooled tips.

Therefore we obtained the relationship between surface temperature of tip on adhesion force as shown in Figure 6. It can be seen from this figure that adhesion force of blood was constant under 65°C, rose up and reached to the large value if the temperature of blood was more than 80 °C during temperature rising. It was clarified that this temperature dependence of adhesion force of blood was similar to the temperature dependence of blood deterioration at temperature rising. Also it was found that we can realize high temperature at the tissue surface even if the temperature of a tip is lower that sever adhesion temperature with both proposed tips.

3. Reduction of adhesion between steel and grilled fish protein with Ultra-Hydrophobic DLC

When you grill fish, you probably have experiences that skin and meat of the fish has adhered to some metal grids. This issue is still unsolved on the adhesion of the denatured fish body by heating. Conventional grids of grill are coated with Fluorine resin to avoid the severe adhesion of grilled fish to the grids. If the maximum temperature of grilling room is less that 300 °C, the coated grid with Fluorine resin can work well. However when the temperature in the grilling room rises up to 500 °C, conventional grill rods with Fluorine resin coating can not work well, because Fluorine resin should disappear by heat decomposition. The recent grilling equipment heats both from top side and bottom side, and extra heating is applied to the grilling rods that rise it up to 500 °C even if the temperature on contact part between grid and fish is at about 150 °C.

Therefore we have tried to develop new grills that have weak adhesion to grilled fish and temperature endurance at high temperature. We consider that the reason why Fluorine resin coating can reduce the adhesion is the Fluorine resin has low surface energy and hydophobicity. If we can introduce the new coating that has ultra-hydrophobicity and temperature endurance up to 500 °C, it can work well. So we focused on Ultra-Hydrophobic Diamond-like Carbon coating named as UH-DLC coating [7]. If it can keep the ultra-hydrophobic properties up to the 500 °C, it has the possibility to use for the new grill rods.

In this study, we introduced UH-DLC coating to grill rods and evaluate the possibility of its usage. In order to know the temperature endurance of UH-DLC coatings, we evaluated the effect of annealing temperature on the morphology and contact angle of water. Also we evaluated the effect of UH-DLC coating on the adhesion of real fish after grilling [3].

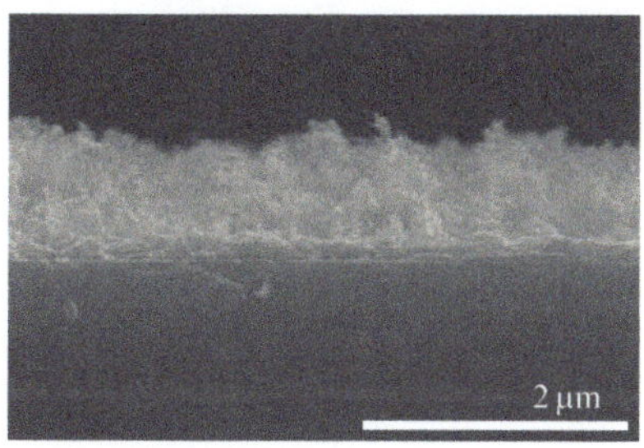

Figure 7. FE-SEM image of the needle-like structure of UH-DLC surface

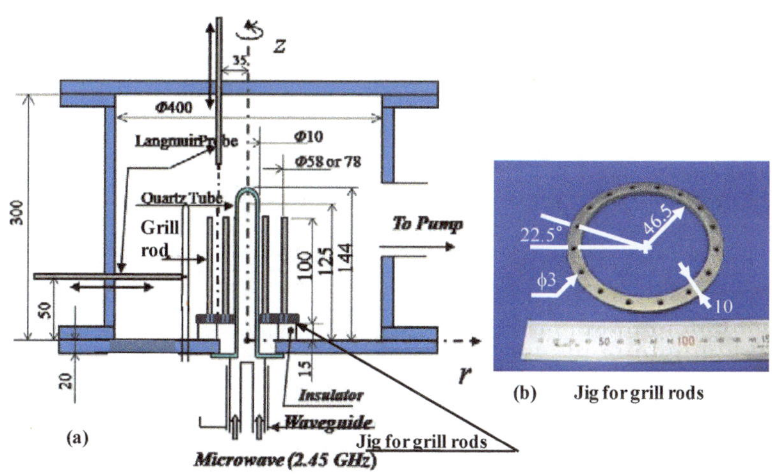

Figure 8. (a); The schematic of 2.45 GHz surface wave-excited plasma (SWP) apparatus, (b); Jig for grill rods

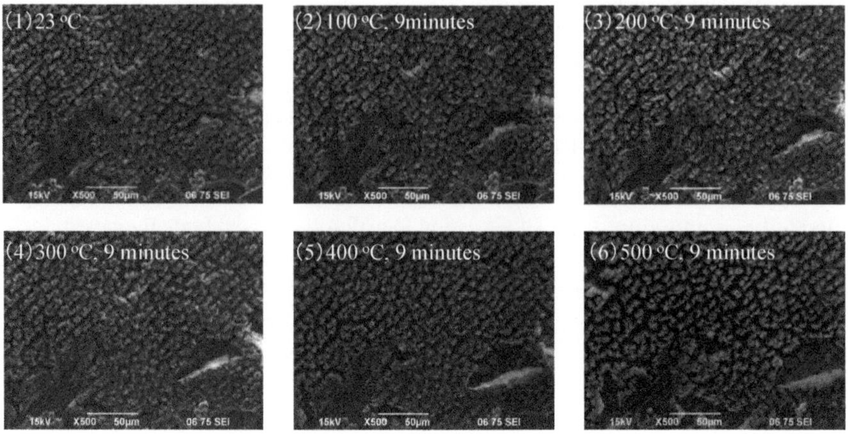

Figure 9. The status of UH-DLC film after heating at each temperature for 9 minutes

Ultra-Hydrophobic Diamond-like Carbon were coated by surface wave-excited plasma CVD that used Ar gas and TMS (Tetra Methyl silane, [$Si(CH_3)_4$]) gas as precursor gases on stainless steel block and grilling rods. Figure 7 shows the cross sectional SEM image of the UH-DLC coatings of on specimen.

Here we explain how to coat the UH-DLC film. Figure 8 shows the schematic of the PECVD equipment with a stainless-steel chamber that is 400 mm in inner diameter and 300 mm in height, presently employed for some grill rods. Figure 8 (b) shows a jig which keep grill rods parallel to the quartz tube. We set it on the insulator and attached some grill rods on its jig's 3.0 mm hole. And in case of block specimens, we only place them on the jig. Prior to deposition, gas pressure in the chamber was decreased to degrees of 10^{-4} Pa with turbo-molecular pump. The bottom end of the chamber was connected to a coaxial waveguide through which 2.45-GHz microwaves were introduced via a quartz plate coupled with a quartz tube that is 20 mm in diameter and 2 mm in thickness.

2.45-GHz microwaves introduced into the chamber propagate as surface wave along the interface between the surface of quartz tube and plasmas, generating high-density plasma column along the tube as shown in Figure 8. As precursor gases, CH_4 and Tetra Methyl silane (TMS, Si $(CH_3)_4$) are introduced into the chamber together with Ar gas.

For ultra-hydrophobic structure of DLC coating on the sample surface, an azimuthally symmetric 2.45 GHz surface wave-excited plasma (SWP) was proposed. When high working pressure at 220 Pa and TMS gas mixture, it has a profound effect on nanoscale post generation because of large TMS particle is flow inner chamber. And also etching with glow discharge of Ar gas was another effect of rough surface.

At first, we observed the block specimen coated Ultra-Hydrophobic Diamond-like Carbon by SEM. And then we heated the flat specimen coated UH-DLC in an electric furnace for 9 minutes at 100 °C. After that, we observed UH-DLC coating by SEM. Next, we heated this specimen which was heated at 100 °C in the electric furnace for 1 hour at 200 °C. The same as described above, we also heated at 300 °C, 400 °C, and 500 °C. And then observed its specimen by SEM after heating at each temperature. Figure 9 shows six SEM pictures of the UH-DLC coating after heating in the electric furnace every for 9 minutes at each temperature. We could confirm that surface of the UH-DLC coating looked a little rough as the heating with the rise of temperature, but this coating did not disappear and show big change.

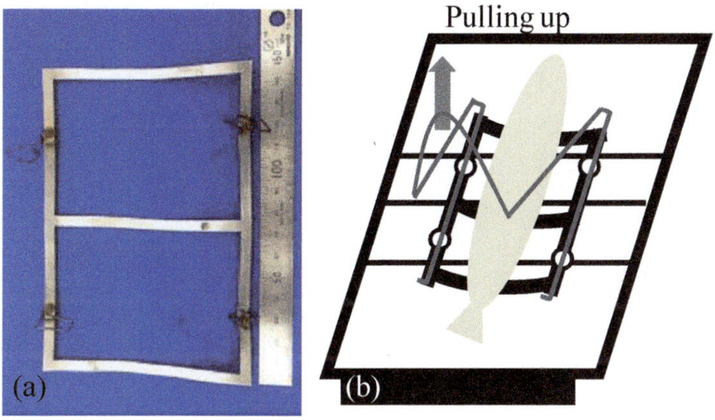

Figure 10. (a) The exclusive jig for pulling fish, (b) The schematic of pulling a fish

In order to know the reduction effect of UH-DLC and Fluorine coatings on adhesion of grilled fish against grilling rods, the adhesion was evaluated from the maximum pilling force of the fish after grilling with gas grill equipment.

We used an exclusive jig for pulling a fish as shown in Figure 10(a). We used a horse mackerel as standard fish, because a horse mackerel is one of the standard fish for the test. We put an exclusive jig and then a horse mackerel on the center of grids of grill as shown by Figure 10(b), and put them in the grilling equipment. We used only three grill rods for this experiment as shown in Figure 10(b). We heat them for 9 minutes. While we heat them, we checked the temperatures with two thermocouples in the grill equipment. Certainly, we confirmed that the temperature in the grill equipment was about 500 °C and the contact temperature between fish and the grill rod was 150 °C. After that, we take them out from the grilling room, and leave them for 1 minute in air. And then we pull the fish up with a pulling gauge and measure the maximum adhesion force.

We did the adhesion test of a real fish in the grilling equipment for different rods as no coating, UH-DLC coating and Fluorine resin coating. Figure 11 shows the adhesion result of this experiment. These results were the average adhesion force of four time experiments. The adhesion forces of UH-DLC coating and Fluorine resin coating grill rods were as about half as that of no coating grills. Also it was confirmed that the adhesion force of the UH-DLC coating grill rods was as much as those of the Fluorine resin coating grill rods.

In this section, we proposed new grill rods that were coated with Ultra-Hydrophobic Diamond-like Carbon. The adhesion of grilled fish to Ultra Hydrophobic DLC coating grill rods were as half as that of no coating grill rods and as much as that of Fluorine resin coating grill rods. Moreover, Ultra Hydrophobic Diamond-like Carbon has heat endurance at 500 °C of grilling room. On the other hand, the hydrophobic property of Fluorine resin coating disappeared at 500 °C of the grilling room. So, we believe that Ultra Hydrophobic Diamond-like Carbon coating has a possibility for the usage as new grilling grids at high temperature.

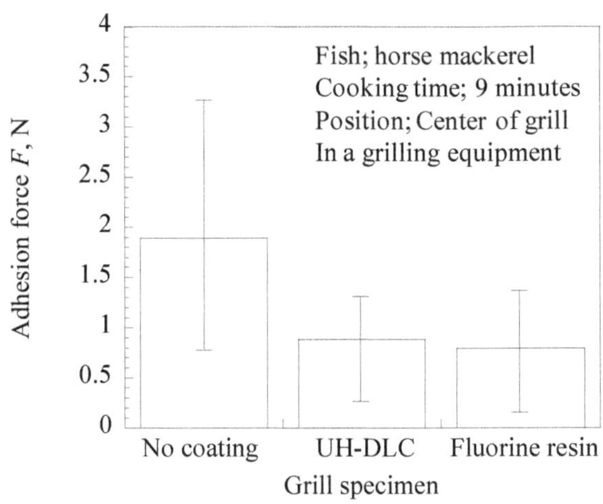

Figure 11. Experimental result of grilling fish in the gas grill equipment

4. Increasing of bonding strength between probe DNA and polymer substrate for DNA chip with ion beam irradiation

Mapping of all DNA sequence of human had finished by The International Human Genome Project in 2000. Thus difference of gene of each human can be measured. If distinctive sequence in DNA sequence of human having disease caused by gene can be found, observation of mechanism of such disease and development of new medicine moved ahead well. DNA chip is an apparatus to examine DNA, on which DNA is fixed. Existing DNA chip is "for research and development", and expensive. So, "for clinic", inexpensive, more reliable and sensitive DNA chip is needed as shown in Figure 12.

Therefore DNA chip made from a plastic was suggested [8]. For high reliability and sensitivity, affinity between DNA and plastic must be higher. But plastic is hydrophobic, affinity to DNA is not well. Therefore, it is necessary to improve affinity between DNA and plastic by modifying surface of plastic hydrophilicity using surface modification. In this study, in order to achieve DNA chip made from a plastic, it is purpose to develop technology to improve adhesion force of DNA [4]. In particular, influence of ion species and ion dose on wettability and surface roughness of PC (polycarbonate) included TiO_2 at ion irradiation is revealed. And in order to keep hydrophilicity after irradiation, influence of atmosphere in storage and irradiation times on keeping time of hydrophilicity is revealed. In addition, influence of ion irradiation on adhesion of DNA is revealed.

At first, PC included TiO_2 was set in ECR ion irradiation apparatus, and it was exhausted to $5.0\times10^{-3} \sim 3.0\times10^{-4}$ Pa in vacuum chamber. Then ionized gas in plasma source was accelerated, irradiated on the surface of specimen. In this study, Ar, N_2 and O_2 ions were irradiated on the surface of specimen controlling ion dose (cm^{-2}).

Before ion irradiation, 0.25 l distillated water was dropped on the surface of specimen, contact angle α deg was measured by optical microscope. Contact angle was average value of 7 times measurement per a specimen. And the specimen after irradiation was stored in air or water, contact angle was measured every several hours.

Then, in order to know influence of ion irradiation on wettability of the surface of plastic substrate, specimens were analyzed by SEM, AFM, and XPS.

In order to know influence of ion irradiation on adhesion force of DNA, it was measured by force curve measurement using AFM in water [9]. 0.05 M DNA solution (SSC buffered) was dropped on the substrate, and DNA was fixed by UV crosslinker after drying. DNA was 400 base length, of which molecular length was 290 nm. After fixing, the substrate held in water, AFM force curve measurement was conducted. AFM cantilever was made from silicon, of which constant of spring was 0.03 nN/m, and peak radius was about 10 nm. And contact force between cantilever and substrate was about 5 nN.

Figure 13 shows relationship between ion dose D and contact angle α of water on the PC included TiO_2. the value of α was about 75 deg at untreated, and then rapidly decreased, finally became less than 8 deg at more than 4×10^{19} cm^{-2}.From analysis by AFM and XPS, it is seemed that increase of TiO_2, which is hydrophilicity, on surface of specimen and increase of surface roughness lead to decrease of contact angle.

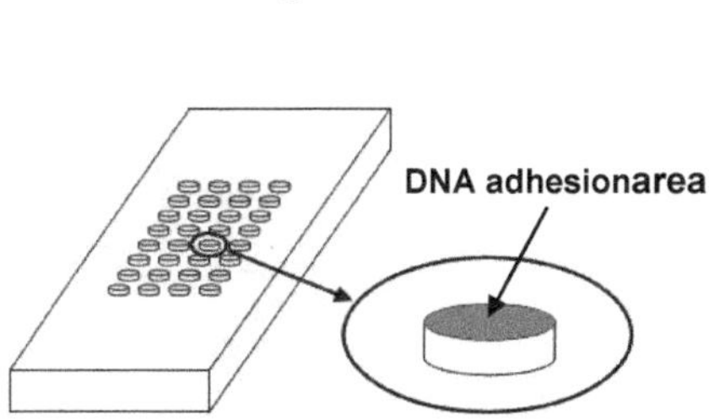

Figure 12. The proposed DNA chip made from plastic substrate

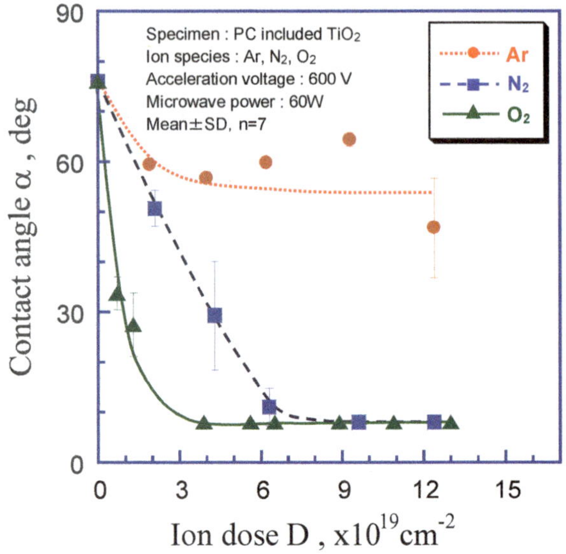

Figure 13. Relationship between ion dose and contact angle of water on the PC included TiO_2.

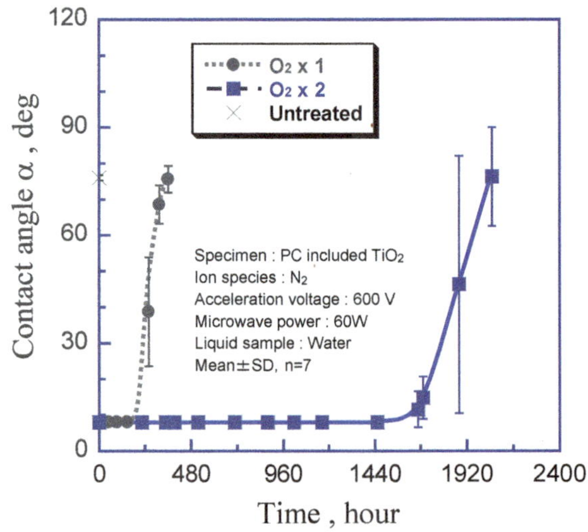

Figure 14. Relationship between contact angle and time under air after several times irradiation of O_2 ion, '× Number' is ion irradiation times.

In order to treat DNA chip substrate easily, it is necessary that hydrophilicity of substrate is keep for a long time in air. Figure 14 shows relationship between contact angle and time under air after several times irradiation of O_2 ion. Hydrophilicity of substrate was kept for about 200 hours at 1 time irradiation of O_2 ion, but for 1440 hours at 2 times irradiation. Ikada et al. reported that hydrophilic groups of surface of polymer modified by plasma turn around (hide under) as time passes, therefore hydrophilicity descend [10, 11]. In this study, similar phenomenon occurred, and it is seemed that hydrophilic groups became difficult to turn around by repeated irradiation.

Figure 15 shows relationship between adhesion force of DNA probe to the surface of specimen and several times irradiation of ions. Adhesion force of DNA was about 0.09 nN at untreated, and then increased by repeated irradiation of N_2 ion from about 0.13 nN at 1 time irradiation to about 0.62 nN, about 7 times as large as at untreated, at 4 times irradiation. In contrast, adhesion force did not increase at irradiation of O_2 ion. From analysis by XPS, it is seemed that electrostatic bond between amino group on surface of specimen formed by ion irradiation and phosphoric acid in DNA lead to increase of adhesion force of DNA.

In conclusion, in order to development of DNA chip made from plastic, ion beam was irradiated to PC include TiO_2 and contact angle of specimen was measured. The results clearly showed that contact angle decreased with increasing ion dose.

Hydrophilicity of surface of specimen formed by ion irradiation could be kept for a long time by repeated irradiation.

In order to know influence of ion irradiation on adhesion force of DNA, it was measured by force curve measurement using AFM in water. Adhesion force of DNA increased by repeated ion irradiation of N_2, but did not by ion irradiation of O_2 ion.

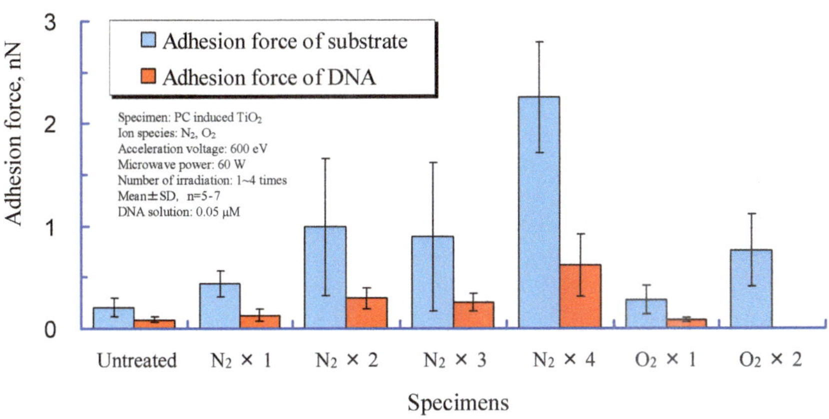

Figure 15. Relationship between adhesion force of DNA probe to the surface of specimen and several times irradiation of ions.

5. Reduction of friction between thermoplastic elastomers and plastics with photochemical fluorination

In medical field, plastic and glass syringes are widely used to insert medicines into human bodies directly. From a hygiene standpoint, they are disposed after single use. Generally, glass syringes are inferior in the accuracy of dimension and the produce cost and they also require great care at the time of disposal. Thus, replacing glass syringes with plastic ones is desired. Plastic syringes are generally used with silicone oil lubricating the sliding area between barrels and gaskets, where the barrels are typically made of plastics such as PP (Polypropylene) and the gaskets are usually made of either vulcanized rubber or TPE (Thermoplastic elastomer) as shown in Figure 16. Silicone oil is biologically and chemically inert but considered to have some demerits: possibility of accumulation in human bodies and decrease of efficacy because of adsorption of medicine's constituent [12]. These demerits are especially pronounced in prefilled-type syringes. Therefore, the development of unlubricated plastic syringes is desired for medical use. In this study, in order to decrease the friction force between barrels and gaskets under unlubricated condition, we tried to fluorinate the surfaces of PP and TPE specimens by using PFPE (Perfluoropolyether) and VUV (vacuum ultraviolet) irradiation with excimer lamp. This method has been already tried to PP [13].

First, we dropped PFPE on a specimen and put a fused silica glass on it to make thin and flat PFPE layer. Then we irradiated them with VUV to make specimen's surfaces react photochemically with PFPE as shown in Figure 17. After the irradiation, we cleaned the specimen with HFE (Hydrofluoroether) by using ultrasonic cleaner to remove residual PFPE. The effect of the photochemical treatment was evaluated by friction test, measurement of surface free energies, and FTIR (Fourier transform infrared) analysis where ATR (Attenuated total reflection) method was adopted. In the friction measurements, treated PP was slid against non-treated TPE. And treated TPE was slid against non-treated PP. Surface free energies of a specimen were calculated from the measured contact angles of water and CH_2I_2 droplets on the specimen.

In the experimental results, it was confirmed that the friction coefficient between treated TPE and non-treated PP was decreased by up to 77% as shown in Figure 18. Moreover, C-F peak, which indicates fluorination of surfaces of specimens, was detected in FTIR spectra. And surface free energies of them decreased. It is suggested that the surfaces of both PP and TPE were fluorinated and fluorination of TPE was predominantly effective for decreasing the friction coefficient between PP and TPE. We expected that the photochemical fluorination changed only chemical property. Thus, to validate possibility of other changes, we investigated the change in real contact areas between fluorinated TPE specimens and a glass plate (BK7) with contact microscope. Contrary to our expectation, decreases in real contact area of fluorinated TPE were observed with decreasing friction coefficient, indicating the change in surface mechanical properties of the fluorinated TPE specimens. Accordingly, it is indicated that the decreasing of friction coefficient between fluorinated TPE and non-treated PP is attributed to decreasing of real contact area and adhesion arisen from the reduction of surface free energy.

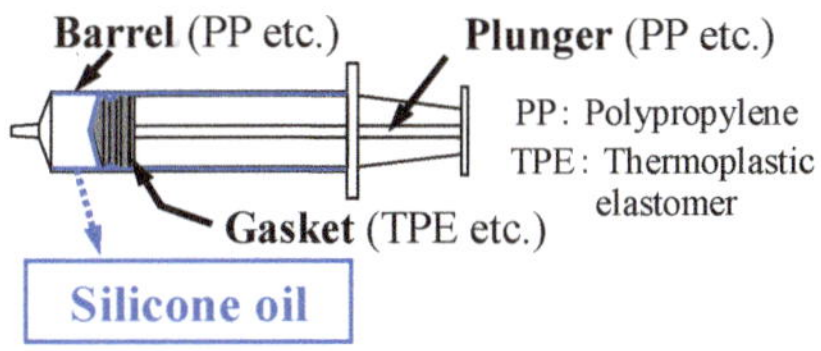

Figure 16. Schematic view of plastic syringe

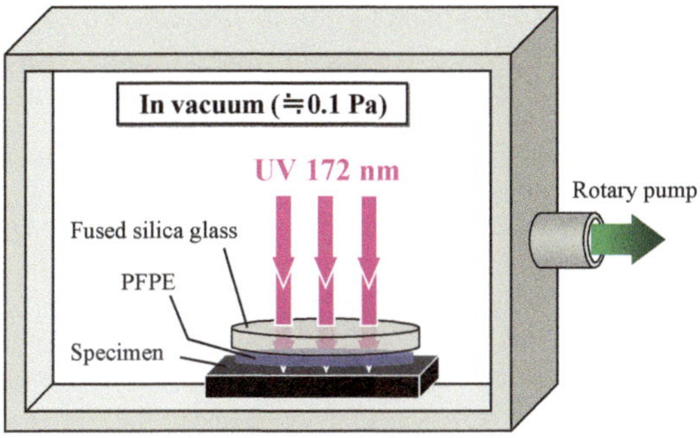

Figure 17. Schematic of photochemical treating process with UV and PFPE.in vacuum

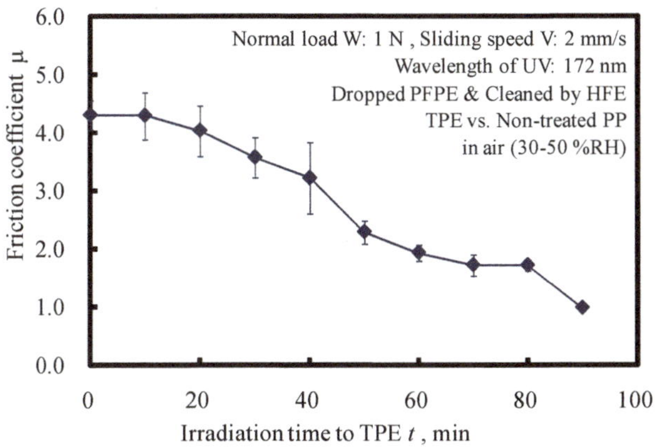

Figure 18. Relation between irradiation time to TPE and friction coefficient.

6. Reduction of adhesion of CIIR rubber to steel plate with plasma irradiation

Adhesion, or the sticking of different materials at their interface, is of general interest in many branches of technology, including micro-electronic devices, medical products and manufacturing. Adhesion between rubbers and metals is often the main source of trouble in a machine. Thus, if molded rubber products easily stick to molds, rollers, and pick-up hands made of metals, the productivity of their manufacturing line becomes low. Therefore, efficient utilization of rubber sheets demands modifications on some desirable properties of the rubber surface without affecting the bulk characteristics. Plasma treatment is one of the most employed methods to attain this goal. One of the most significant benefits of the plasma process is it offers and additional advantage that the surface modification does not affect the desirable bulk properties of the rubber.

In a previous study, we have demonstrated that the surface wave-excited plasma treatment reduced the adhesion force between a medical rubber, chloride-isobutene-isoprene rubber (CIIR) and stainless-steel ball (SUS440C) by using oxygen and argon gases.

We have also shown a decrease in the real contact area with increasing time and micro wave power, and a similar trend in the residual rates of the adhesion force and the real contact area of CIIR rubber. Therefore, it is assumed that the adhesion force is strongly subjected to the real contact area [14]. However, the main reason for the reductions in the real contact area remains unknown. Recent works have shown that plasma treatment increases the roughening of rubber surfaces [15]. The surface roughness may affect the adhesion force, which is largely dependent on the contact geometry and surface topography [16].

The objective of this research, we report on our attempts to clarify the factor to reduce the adhesion force during the surface wave excited plasma treatment process. It is also attempted to figure out the Young's modulus behavior in micro scale to measure without bulk property by using micro slicer and surface roughness changes are measured by 3D laser scanning microscope.

Results of adhesion forces between CIIR rubber and stainless-steel ball as a function of plasma treatment time at 200 W are shown in Figure 19. It is apparent from the figures that the adhesion force dramatically decreased with oxygen plasma treatment according to treatment time. Similar decreasing trend was also observed with argon plasma treatment. However, at 1 min treatment time, the adhesion force was higher with argon treatment than oxygen plasma treatment of CIIR rubber. After 1 min, argon plasma treatment was more effective than oxygen plasma treatment in decreasing the adhesion force. The adhesion force couldn't be measured after 10 min because it was lower than the measurable value of 0.001 N. In summary, the figures showed that plasma treatment time is a very important factor that decreases the adhesion force.

Load-penetration depth curves by using the nano-indenter with 50 μN maximum loads were obtained. The thickness of the prepared CIIR rubber was about 3 mm for the general thickness and about 50 μm for the cutting thickness by using the microwave slicer. Despite the same conditions, the penetration depth obtained from the 3mm thickness CIIR rubber was clearly different from that obtained from the 50 μm thickness. In particular, the unloading curve coincided with the loading curve at a penetration depth of more than 290 nm due to the thickness difference. This indicates that not only the bulk property has greatly affected the Young's modulus, but also it is affected its accurate measurement. In other words, a surface wave-excited plasma treatment increased the Young's modulus in μm thickness scale. As a result, we determined that it is possible to develop a clear difference between 3 mm and 50 μm Young's modulus of the CIIR rubber without any influence from the bulk property by using the microwave slicer.

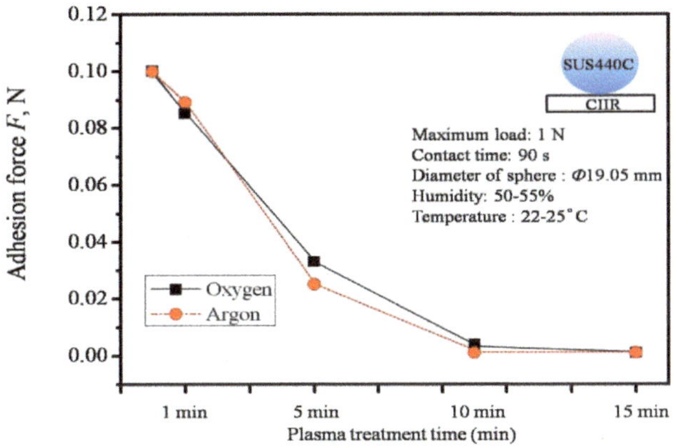

Figure 19. Adhesion force between CIIR sheet and stainless-steel ball after oxygen and argon plasma treatments at a microwave power of 200 W and a gas pressure of 30 Pa, for 0, 1, 5, 10, and 15 minutes.

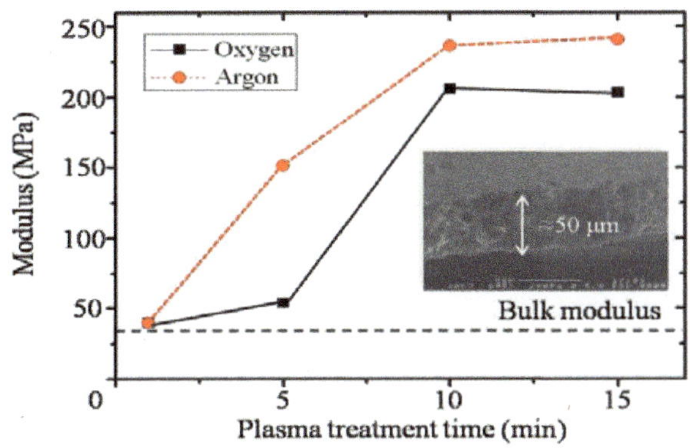

Figure 20. Effect of plasma treatment time on the elastic modulus of CIIR rubber 50 μm in thickness which was cut from the top surface of treated CIIR sheet. The elastic modulus was measured by nano-indenter for oxygen (black) and argon (red) plasma treated CIIR sheets.

Figure 20 shows the Young's modulus profile of 50 μm thickness CIIR rubber measured by nano-indenter after oxygen and argon plasma treatment with increasing time. The results treatment time (206.1 MPa), and followed by a steady state. A higher Young's modulus (236.4 MPa) was obtained with argon plasma treatment, but only after 15 min treatment with argon gas. As a result, this improved Young's modulus by the oxygen and argon are depicted with squares (oxygen plasma treatment) and circles (argon). For the oxygen plasma treatment, Young's modulus was slightly higher (39.8 MPa) than the untreated CIIR rubber (38.0 MPa). However, it increased significantly between 5 min (53.8 MPa) and 10 min surface wave-excited plasma treatment is an important factor that reduces the adhesion force.

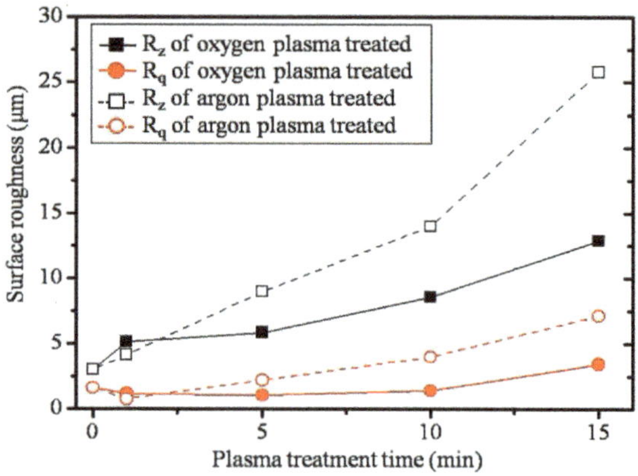

Figure 21. Surface roughness changes of CIIR sheet as a function of plasma treatment time. The measurements were done for oxygen (dotted line) and argon (solid line) plasma-treated CIIR sheets. The surface roughness parameters, R_z (maximum peak height roughness), R_q (Root mean square roughness) are defined by JIS B 0601 2001.

Figure 21 shows surface roughness as a function of plasma treatment time. The surface roughness deviations of CIIR rubber by oxygen and argon plasma treatment has changed. The entire surface roughness factors increased with increasing treatment time.

Especially, the R_z of argon plasma treated CIIR rubber (25.78 μm) is rougher than that of oxygen plasma treated rubber (12.88 μm).And these results imply that change in morphology due to surface roughness reduced the real contact area against the SUS440C ball.

Figure 22 shows 3D laser scanning microscope photographs of argon plasma treated to CIIR rubber. In the absence of argon plasma treatment at 1 min, the subsurface of the CIIR rubber looks less granular and generally has a smoother shape. However, with 5 min argon plasma treatment at 200 W, changes in the surface were visible. Nevertheless, the surface of CIIR rubber pattern was changed compared to the untreated CIIR rubber. The subsurface of the CIIR rubber was growing rougher with increasing treatment time (Figure 4c). In this study, the 200 W, 15 treatment conditions resulted in the roughest surface. As a result, this surface roughness change by etching effect might have affected the adhesion force, which is largely dependent on the contact geometry and surface topography [16].

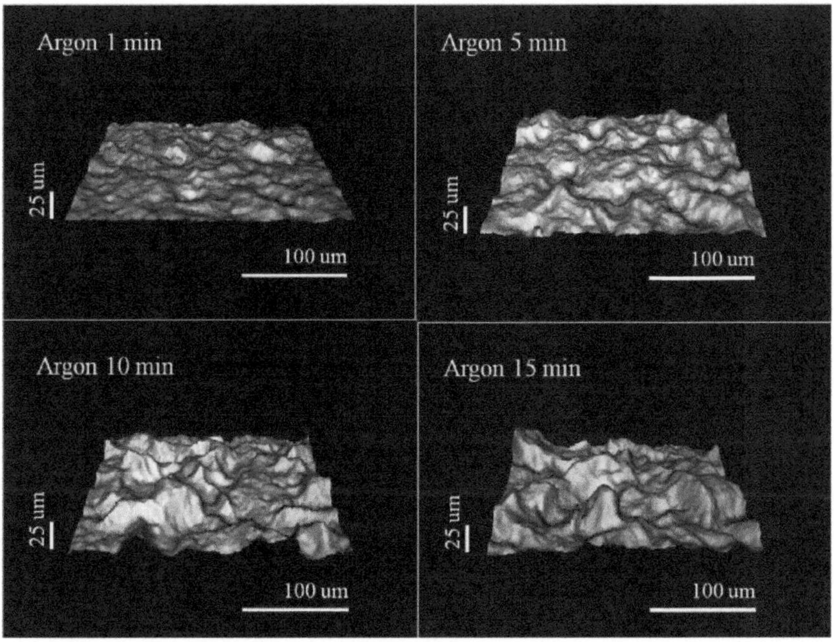

Figure 22. laser scanning microscope images of CIIR sheets after argon plasma treatment with 200 W at a gas pressure of 30 Pa for treatment times of (a) 1, (b) 5, (c) 10, and (d)15 min.

In this work, we tried to clarify the change in surface mechanical properties of CIIR sheet after the plasma treatments. In order to evaluate the change in Young's modulus of CIIR sheet surface, the top 50 μm thickness of a plasma-treated CIIR sheet was cut away to avoid the bulk property. The Young's modulus measurements with nano-indenter showed the clear difference between the surface and bulk elastic modulus of CIIR rubber after plasma treatment, indicating the success of surface modification without changing bulk property. In addition, it was shown that the plasma treatment with Ar gas increased the Young's modulus of CIIR sheet surface from 38 MPa to 236.4 MPa. And also, surface roughness of CIIR rubber has changed to rougher with both oxygen and argon gas plasma treatments. These changes in Young's modulus and roughness at the surface of CIIR sheet are considered to be the main reasons for the plasma-assisted reduction of adhesion force between stainless-steel ball (SUS 440C, JIS) and CIIR sheet.

Author details

Noritsugu Umehara, Takayuki Tokoroyama and Hiroyuki Kousaka

Department of Mechanical Science and Engineering, Nagoya University, Japan

References

[1] Nakashima Y, Tokoroyama T, Umehara N, Akagami Y. Reduction of Adhesion of Coagulation to a Radio Knife Tip with Filled Micro-Grooves. Transaction of the JSME C 2011;77(779) 2916-2921.

[2] Nakashima Y, Tokoroyama T, Umehara N, Akagami Y. Development of a Radio Knife Suppression the Adhesion of Coagulated Blood. Transaction of the JSME C 2012;78(786) 605-614.

[3] Honda N, Kajiya M, Jang Y-J, Kousaka H, Umehara H. Reduction of adhesion between steel and grilled fish protein with Ultra-Hydrophobic DLC. Journal of Advanced Mechanical Design, Systems, and Manufacturing 2010;4(1) 356-364.

[4] Demizu Y, Umehara N, Muraoka M, Imai J, Akagami Y, Kanda S. Improvement of adhering strength of DNA for polycarbonate substrate including Titaniumu Oxide with ion beam irradiation. Transaction of the JSME C 2010;76(768) 2142-2147.

[5] Usami K, Kousaka H, Umehara N, Tokoroyama T. Friction reduction between thermoplastic elastomers and plastics under unlubricated condition. Transsaction of the JSME C 2010;76(767) 1833-1837.

[6] Kim J. H, Umehara N, Kousaka H, Shimada M, Hasegawa M. Effect of micro-scale Young's modulus and surface roughness on adhesion property to plasma-treated rubber surface. Journal of Mechanical Science and Technology 2010;24 119-122.

[7] Jang Y-J, Umehara N. Wetting and Adhesion Behaviors of a-C:H Film Deposited on Nano-Scale Copper Doted Surfaces. Tribology Online 2008;3(5) 294-297.

[8] Yoshino M, Matsumura T, Umehara N, Akagami Y, Aravindan S, Ohno T. Wear 2006;274 260.

[9] Kato M, Muraoka M, Akagami Y. In: Prceeding of JSME M&M2005 Mechanics of Materials Conference; 2005.

[10] Ikada Y, Matsunaga T, Suzuki M. Journal of the Chemical Society of Japan 1985;1079 6.

[11] McCarthy T. J. In: Proceedings ACS Organic coatings and applied polymer science proceedings 1983;48 520.

[12] Yago K, Kuroyama M, Otori K, Hirayama T, Ogawa Y, Aoto H. A Study on Adsorption of Calcitonin Preparation on Syringes. Journal of the Nippon Hospital Pharmacists Association 2000; 273-279.

[13] Sato Y, Anai H, Murahara M. Functional Substitution on Polypropylene Film for Glaucoma Implant by Using Excimer Lamp. In: Proceedings of the School of Information Technology and Electronics 2003, Tokai University; 2003.

[14] Kim J. H, Nitta I, Umehara N, Kousaka H, Shimada M, Hasegawa H. Relationship between Real Contact Area and Adhesion Force of Plasma-Treated Rubber Sheets Against Stainless-Steel Ball. Tribology Online 2008;4(1) 361-365.

[15] Grythe K. F, Hansen F. K. Surface Modification of EPDM Rubber by Plasma Treatment. Langmuir 2006;22(14) 6109-6124.

[16] Burnham N. A, Kulik A. J. Handbook of Micro/Nanotribology, 2nd edition. Boca Raton, FL, USA: CRC Press; 1998.

Chapter 11

Tissue Damage and Repair Caused by Immune System and Personalized Therapy of Failed Organs by Stem Cells

Ken-ichi Isobe, Naomi Nishio,
Thanasegaran Suganya, Zhao Cheng and
Sachiko Ito

1. Introduction

Our immune system has developed to fight against invaders. Pathogenic invaders are microorganisms including virus, bacteria, fungus and amoeba.

In species development, innate immune system appeared first, then after jowed fish acquired immune system developed. Once microorganisms infects in organs, cells belonging to innate immune system such as neutrophils and macrophages move to the site of infection. Microorganisms are sensed by pattern-recognition receptors (PRRs) of the innate immune system including dendritic cells (DC). These cells phagocytize infectious agents and produce cytokines and chemokines, which induce tissue inflammation. In phagolysozome of DC, the proteins, which constitute the microorganisms, are degraded. These DC migrates to lymph node and present the microorganisms antigens to T cells, which are central of adaptive immune responses.

Similar to the eradication of pathogens, the inflammatory response is also crucial for tissue and wound repair and called sterile inflammation. Host-derived non-microbial stimuli are released following tissue injury or cell death. These endogenous molecules have been termed damage-associated molecular patterns (DAMPs). DAMPs are normally present intracellular and are therefore hidden from recognition by the immune system. Once tissues are injured these molecules are released into the extracellular environment by dying cells and trigger inflammation under sterile conditions [1] (Figure 1).

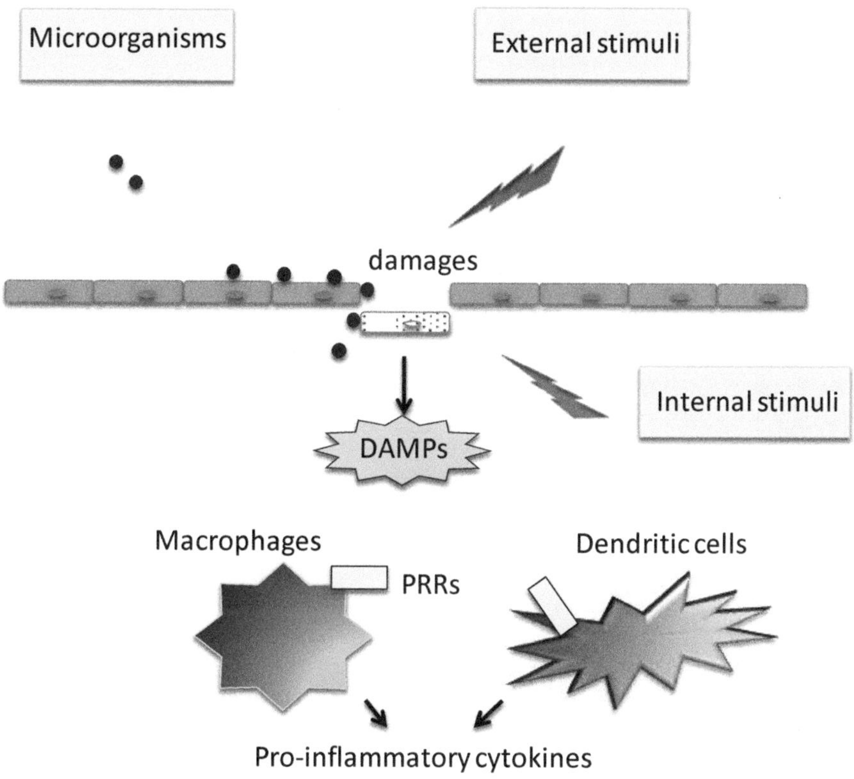

Figure 1. Damaged cells by various stimuli produce DAMPs, which activate tissue resident macrophages and dendritic cells. Activated innate cells produce pro-inflammatory cytokines.

2. Wound repair and innate and acquired immunity

2.1. Skin wound healing

Wound healing is one of the most important subjects in medicine and biology. For example accidental injuries or medical surgery initiate wound repair processes. From the practical and fundamental importance wound healing has been the subject of intense investigation for many years [2]. Recently increase of elderly person necessitates the research of wound repair further. The ability to wound repair declines with age, which induces incurable pressure ulcer in elderly patients [3]. Acute inflammatory reactions play integral roles in normal wound healing and tissue repair. Once parenchyma and blood vessel are damaged, the coagulation system is activated, which release chemical mediators that promote vascular permeability and leukocyte adhesion and recruitment. Within 30 min. neutrophils migrate to the site of wound, which phagocytize dead cells and cellular debris and clear tissue. Then neutrophils are themselves phagocytized by tissue macrophages. Although M1 macrophages damage tissue cells producing tissue-damaging proteases, which are primarily kill microorganisms, M1 macrophages also engulf damaged tissue cells. Then M2 macrophages migrate from blood vessel to repair tissue. Neutrophils have specific cell surface antigens called, Gr-1. In murine model of wound repair (Figure 2), we depleted neutrophils by anti-GR-1 antibody. Although in young mice the kinetics of wound healing was not different by the depletion of neutrophils, the depletion of neutrophils by anti-Gr-1 antibody dramatically delayed wound healing in aged mice [4]. Bone marrow has both matured neutrophils and macrophages. First, we found that in splenectomized mice, bone marrow, spleen and thymus injection accerated wound healing (Figure 3). We transplanted bone marrow cells from GFP mice (C57BL/6 background) to C57BL/6 mice. We found that bone marrow cells injection accelerated wound healing and GFP positive neutrophils and macrophages migrated to the wound tissue. Because wound healing of aged mice is relatively inefficient, we transplanted young bone marrow cells to aged mice. We found that young bone marrow cells enhanced wound healing of aged mice [5, 6]. In case of microorganisms acquired immunity help to phagocytize invader. By microorganism's infections, immunoglobulin specific for the microorganisms bind to the surface and induce phagocytosis mediated by Fc receptors. We examined whether B cells, which produce antibodies to damaged tissues, might be engaged in the process of wound healing. We found that wound healing in splenectomized nude mice was delayed and the transfer of B cells accelerated wound healing in splenectomized mice. Further we detected several autoantibodies binding to wounded tissues [7] (Fig.4). Advancing age gradually decreased the strength of IgG1 autoantibodies, which bind to wounded tissues, although the strength of IgM autoantibodies was relatively stable by advancing age [5, 6].

2.2. DSS-induced colitis

As a mucosal wound repair model we used dextran sulfate sodium (DSS). Colitis may result from DSS toxicity to colonic epithelial cells. DSS is commonly used in rodent models to chemically induce acute intestinal inflammation, and the DSS-induced colitis is characterized by weight loss, bloody diarrhea, epithelial cell damage, and immune cell infiltration, as well as an increased production of inflammatory mediators including TNF-α, IL-6, IL-12, and interferons. Colitis was induced by oral administration of DSS to two months old C57BL/6 mice at 2 % (w/v) in drinking water ad libitum for five days followed by normal drinking water. By this schedule, mice almost completely recovered. Interestingly at recovery phase of DSS-induced colitis, we observed strong up-regulation of innate immune cells having $Gr1^{+}CD11b$ $^{+}$cell surface maker in spleen and bone marrow. Transplantation of splenic DSS-derived $Gr1^{+}CD11b^{+}$ cells into DSS-treated mice improved colitis and promoted efficient colonic mucosal healing [8] (Figure 5). Anti-Gr-1 antibody treatment worsened the DSS- administered colitis. These results indicate that Gr1+CD11b+ cells induced by DSS worked to repair colon wound healing and repair colitis [9].

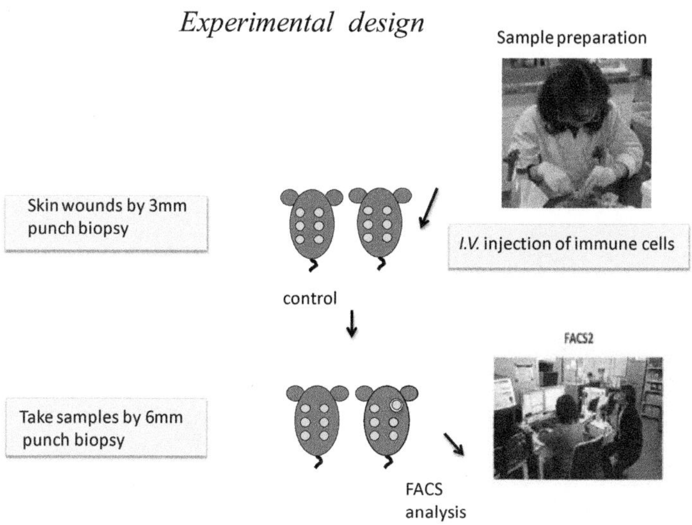

Figure 2. Murine experimental model of wound healing. The dorsal skin was punched through with a sterile disposable biopsy punch (diameter 3 mm). Separated immune cells were injected intravenously. The wounds and their surrounding areas were cut for analyses with biopsy punch with a diameter of 6 mm at each day after wounding.

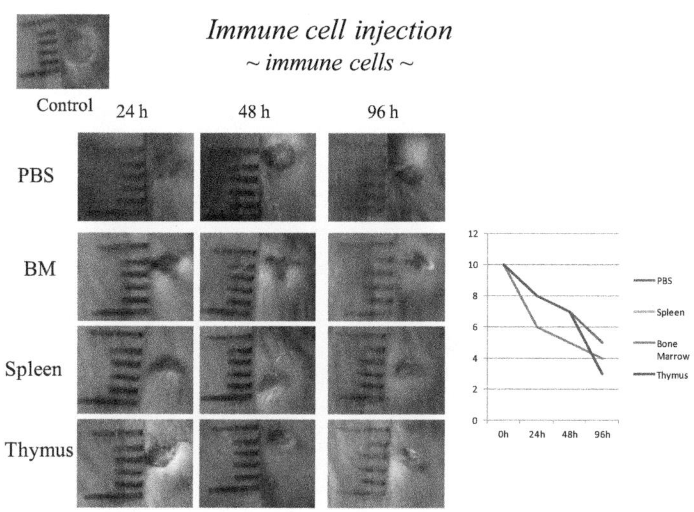

Figure 3. Examples of analysis. After 3 weeks of the removal of spleen, bone marrow, spleen or thymus cells were injected at the same time of punch biopsy.

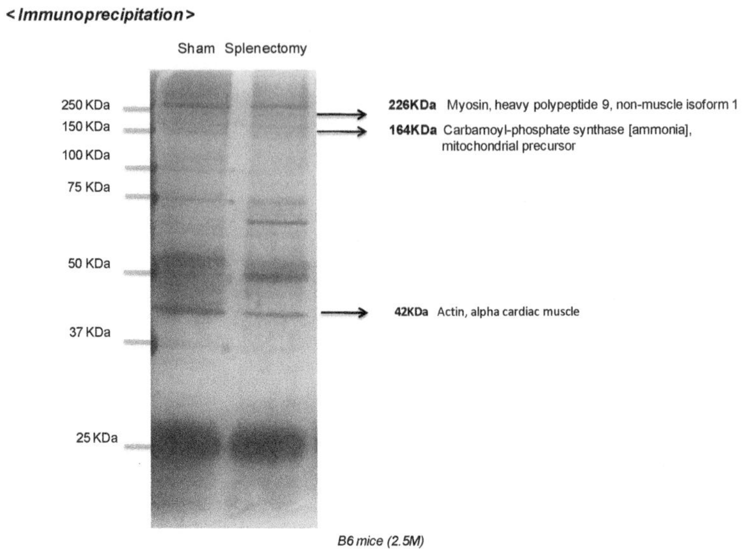

Figure 4. Immunoprecipitation of autoantibodies, which were bound to tissue cells. After separation each band was analyzed by LC/MS.

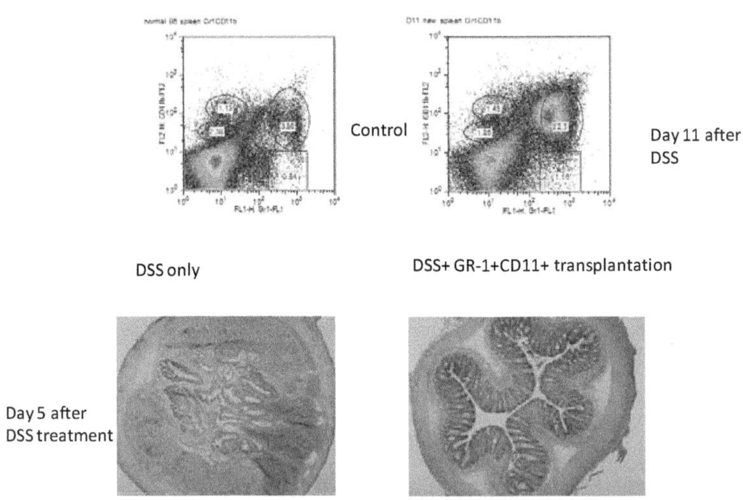

Figure 5. Colitis was induced by oral administration of DSSat 2 % (w/v) in drinking water ad libitum for five days followed by normal drinking water. Upper show the FACS analysis of spleen of DSS-treated mice. Increased GR-1+CD11b + cells were transplanted to mice of at the last day of DSS administration.

3. Sterile inflammation to tissue degeneration

3.1. Overview

Neutrophils and macrophages produce pro-inflammatory cytokines, including interleukins 1 alpha and beta (IL-1α and β) and tumor necrosis factor alpha (TNF-α), chemokines, matrix proteases. These materials help to destroy invading bacteria and virus. However, invaded tissue also be damaged by these inflammatory reactions. Once microorganisms are cleared by immune system, the invaded tissue is repaired to healthy state. However, some microorganisms such as hepatitis C virus keep staying in host tissue and repeatedly attack host tissue, which induces fatal tissue damages finally. In sterile inflammation similar phenomena occur. External stimuli, which destroy tissue cells, produce DAMPs and induce sterile inflammation. Because immune cells secrete toxic proteases to kill microorganisms, the same immune cells destroy tissue cells. If external stimuli are limited, repair system such as M2 macrophages work to repair tissue damages and returns to normal condition. However, repeated stimuli may induce next inflammation before tissue repair and will induce chronic tissue damages, which cannot be recovered. In these processes, auto reactive T cells and B cells are engaged.

3.2. Bleomycin (BLM)-induced murine model of experimental systemic sclerosis

BLM is a glycopeptide produced by the bacterium Streptomyces verticillus, which can cleave DNA, and it is widely used as an anti-tumor agent for various types of malignancies [10]. Thus BLM administration destroys tissue cells and induces sterile inflammation. We examined the effects of repeated injection of BLM to adult mice. When BLM was injected into the shaved backs of adult C3H or BALB/c mice (100 μg/mouse) 5 days per week for 3 weeks, not only skin fibrosis, but also esophageal and gastric damage related to fibrosis were observed. Injection of BLM induced innate immune inflammation, which did not resolve. Dendritic cells, which engulfed dead cells, transferred auto antigens to T cells. Transfer of CD4+ T cells from BLM-treated BALB/c mice induced the same pathological changes and antibody production in untreated-BALB/c nude mice. Th17 cells secret IL17, which stimulated innate immune cells to damage tissues [11].

3.3. Amyloid-beta peptide (Aβ) works as a DAMP

Aβ accumulation is thought to be central to the pathogenesis of Alzheimer's disease (AD) [12]. Aβ is produced by proteolysis processing of Amyloid precursor protein (APP). A balance between amyloidogenic processing of APP and the removal of soluble Aβ by clearance pathways and enzyme-mediated degradation maintains Aβ levels. Aβ is now considered to be one of DAMPs, which activate innate immune cells especially microglia. Putative sensors of Aβ are NLRP3, CD36 and RAGE [13-15]. By using microglial cells or cell line, we have shown that Aβ produce several cytokines including M-CSF by PI3K/AKT and NFκB pathway, which stimulate microglial proliferation and migration [16]. Further Aβ induces several chemokines (CCL7, CCL2, CCL3, CCL4 and CXCL2) by PI3K/AKT and Erk pathway in the microglia, which induce the migration of microglia or macrophages [17]. Aβ induced sterile inflammation also produced matrix metalloproteinase (MMP3, MMP12 and MMP13), which might degrade damaged tissue further or might work to degrade Aβ to protect against dangerous stimuli [18]. Aβ induced microglial migration may induces cytoskeletal system. We have shown that Aβ binds to IQGAP1 and actin, which are cytoskeletal components, and may stimulate Rho/Rac signaling [19].

3.4. D-galactose (D-gal) induced thymus degeneration

D-gal at high levels induces the production of reactive oxygen species (ROS) and advanced glycation end products (AGEs; [20]) and induces sterile inflammation. We administered 50 mg D-gal for 60 days to 2 months old mice. We found unorganized distributions of keratin-5 and keratin-8 proteins in the thymus of these hosts, which resembled to aged mice [21].

4. Toward personalized cell therapy

4.1. Overview

Our body has the capacity to regenerate from injury. However, damaged tissues induced by continuous microbial or sterile stimuli cannot be recovered. Continuous stimulation by TGFβ produced by macrophages induces fibroblasts to secrete matrix proteins such as fibrinogen. Damaged tissue cells are replaced by fibrous component. Many patients are suffering from these conditions. About 90% of all deaths from chronic obstructive lung diseases (COPD) are attributable to cigarette smoking. Long term Cigarette smoking induces continuous sterile inflammation. Patients who have low blood oxygen levels in their blood are given supplemental oxygen. However, oxygen supply cannot resume lung function. Long-term heavy alcoholic stimuli induce continuous sterile inflammation in liver and eventually develop cirrhosis. Adult onset diabetes mellitus is now found to be caused by repeated sterile inflammation, which is derived from obesity. Renal failure has numerous causes. The most common is diabetes mellitus. Lung, liver, and kidney failure has no treatment to cure diseases except transplantation. Transplantation to these diseases has limitation from organ shortage and immune rejection caused by HLA discrepancy. Another common degenerative diseases are Alzheimer and Parkinson disease. These diseases also are caused by continuous sterile inflammation. Because these diseases affect single organ, we may use healthy other tissue cells for regenerative medicine. Several trials are now ongoing [22].

4.2. Muscle regeneration from pluripotent Embryonic Stem (ES) cells

Aged person, who has long been in bed, has serious problem of muscle atrophy. We set up murine model of muscle regeneration. ES cells having LacZ marker were cultured on Type IV collagen dishes with 10% FCS. After 5 days of culture PDGFR-α^{+} mesodermal progenitor population was sorted and transplanted to injured muscle of KSN nude mice. We observed LacZ positive CD34+ Pax7+ cells (satellite cells) in transplanted muscle. Because of transplanted cells are satellite cells; they may autonomously proliferate *in vivo* [23]. Then we tried to differentiate skeletal muscle cells without serum. We cultured ES cells with BMP4 without serum. By this condition ES cells differentiated into osteochondrogenic cells *in vitro* and *in vivo*. Early removal of BMP4 followed by lithium chloride (LiCl) promoted the differentiation to myogenic progenitor cells and finally differentiated to skeletal muscle cells *in vitro* [24].

4.3. Thymic epithelial cells from induced pluripotent stem cells

One of the most serious problems of aged person is infection caused by age-associated immune functional decline. The age-associated atrophy of the thymus and the decline of naïve T cells output are mainly caused by degeneration of thymic epithelial cells (TEC). We tried to differentiate iPSCs to TEC. The thymus initially develops as an endodermal epithelial cell at embryonic day 10 encapsulated by mesenchyme. We follow this embryonic development to induce TEC from iPSCs. First, we try to differentiate iPSCs to mesendoderm to definitive endoderm. We added activin A and Lithium chloride (LiCl) to iPSCs in collagen IV –coated dishes. We found that endodermal epithelial cell cluster were surrounded by mesoderm cells same as embryonic development. By adding FgF8 then FgF7 and FgF10, the population of endodermal epithelial cells enlarged and matured to TEC progenitor. By the addition of RANK ligand TEC progenitor differentiated to medullary TEC like cells.

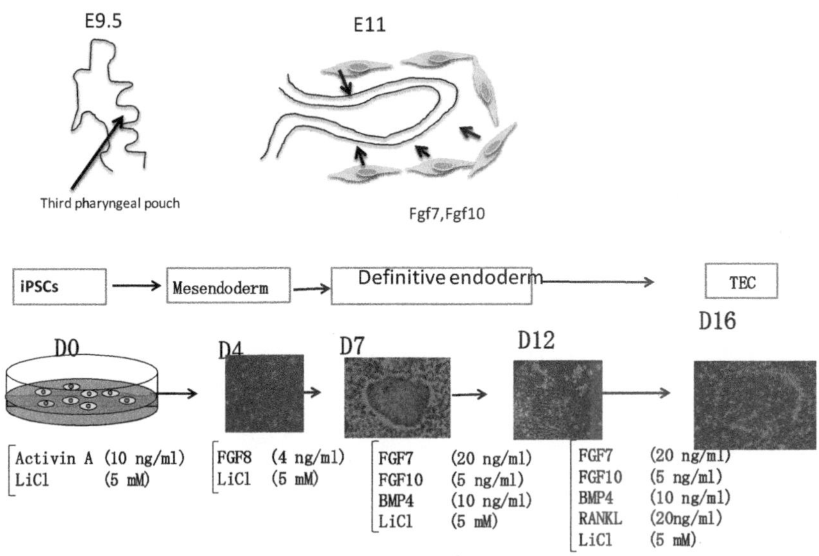

Figure 6. Upper figure shows the fetal development of TEC. Lower figure shows the protocol of TEC differentiation from iPSCs.

4.4. Establishment of induced Pluripotent Stem Cells (iPSCs) from aged mice

The generation of induced pluripotent stem cells (iPSCs) from murine and human somatic cells opens the possibility of personalized cell therapies for treating human disease and/or repairing the damaged tissues of elderly patients [25]. By the lifespan extension many developed countries has large number of elderly patients, who suffer from incurable diseases. For personalized cell therapy of elderly person, technology to establish iPSCs from aged person is necessary. Further iPSCs from aged person need to differentiate to target tissue cells. In order to establish model system to treat elderly patient by iPSCs, we designed murine model. We found that efficiency to establish iPSCs from aged mice was low. However, we could succeed to establish 2 clones of iPSCs from bone marrow derived dendritic cells of 21 month old C57BL/6 mice carrying GFP marker by retrovirus encoding four factors (Oct3/4, Sox2, Klf4 and c-Myc). Established iPSCs have pluripotent makers such as SSEA-1, Oct-4 and Pou5f1 and alkaline phosphatase activity was positive. Our Aged-iPSCs made teratoma-contained tissues of three germ layers, when they were transplanted to the dorsal flank of C57BL/6 (syngeneic) mice. They also differentiate to various tissue cells including muscle, liver and neuron by hanging drop followed by two-dimensional culture [26].

By using iPSCs we tried to differentiate into cardiovascular cells for the treatment of ischemic tissue, which might be caused by atherosclerosis. We simply cultured iPSCs in Type IV collagen dish and separated Flk1+ cells by magnetic-activated cell sorting (MACS) or FACS sorting. Flk1+ cells from iPSCs co-cultured with HUVECs on Matrigel made the network structures. Flk1+ cells were transplanted adductor muscles in the ischemic limb of nude mice. Revascularization of the ischemic hind limb was accelerated in mice that were transplanted with Flk-1^{+} cells [27].

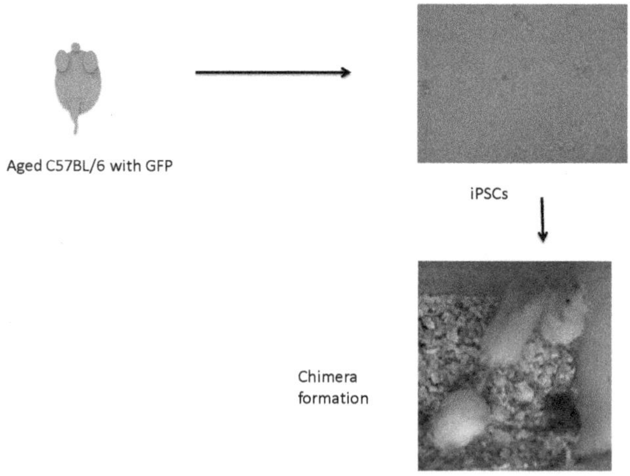

Figure 7. Establishment of aged-iPSC from GFP positive C57BL/6 mice. Chimera mice production from these iPSCs.

5. Conclusions and perspective

Human health depends on immune system. Here we present the data by using murine models, which indicate that immune cells react to self-antigens released from damaged cells by external stresses. Once microbes are cleared by immune system, invaded tissue is regenerated by tissue stem cells. Tissue recovery also occurs in single or short-term stimuli. However, continued infection or stimuli induces permanent tissue damage. Aging is one of continued stimuli caused by ROS produced from mitochondria. Recent progress of regenerative medicine especially iPS technology will enable to regenerate failed organs by using patient own tissue cells. To construct patient specific 3 D-structured organs by differentiated iPSCs with the help of computer program will open new personalized therapy.

Acknowledgements

We thank to the following researchers working in our department; Mohammad Nizam Uddin, Rong Zhang, Hirohiko Suzuki, Yuta Inami, Tohru Yoshikai, Hidetoshi Sakurai and Hideaki Ishikawa. We thank Minoru Tanaka, Ikuyo Mizuguchi and Yushikazu Fujita in the Division for Medical Research Engineering, Nagoya University Graduate School of Medicine for assistance with flow cytometry, confocal microscopy and electron microscopy respectively. We also thenk to the secretary works of Nanae Oiwa. This work was supported by the GCOE program and research grant provided by the Ministry of Education, Science, Technology, Sports and Culture, Japan.

Author details

Ken-ichi Isobe, Naomi Nishio, Thanasegaran Suganya, Zhao Cheng and Sachiko Ito

Department of Immunology, Graduate School of Medicine, Nagoya University, Nagoya, Japan

References

[1] Chen G. Y, Nuñez G. Sterile inflammation: sensing and reacting to damage. Nature Reviews Immunology 2010; 10(12) 826-37.

[2] Sonnemann K. J, Bement W. M. Wound repair: toward understanding and integration of single-cell and multicellular wound responses. Annual Review of Cell and Developmental Biology 2011;27 237-263

[3] Swift M. E, Burns A. L, Gray K. L, DiPietro L. A. Age-related alterations in the inflammatory response to dermal injury. Journal of Investigative Dermatology 2001;117(5) 1027-1035.

[4] Nishio N, Okawa Y, Sakurai H, Isobe K. Neutrophil depletion delays wound repair in aged mice. Age (Dordr). 2008;30(1) 11-19.

[5] Nishio N, Ito S, Okawa Y, Isobe K. The wound repair is control by monocyte linage cells. Proceedings of International Symposium on Micro-NanoMechatronics and Human Science, MHS2010, 7-10 November 2010; Nagoya, Japan;2010.

[6] Nishio N, Ito S, Suzuki H, Okawa Y, Isobe K. Effect of Aging on Auto-Antibodies to Wounded Tissues. The Open Geriatric Medicine Journal 2010; 3 17-24.

[7] Nishio N, Ito S, Suzuki H, Isobe K. Antibodies to wounded tissue enhance cutaneous wound healing. Immunology 2009;128(3) 369-380.

[8] Zhang R, Ito S, Nishio N, Cheng Z, Suzuki H, Isobe K. Up-regulation of Gr1+CD11b+ population in spleen of dextran sulfate sodium administered mice works to repair colitis. Inflamm Allergy Drug Targets. 2011;10(1) 39-46.

[9] Zhang R, Ito S, Nishio N, Cheng Z, Suzuki H, Isobe K. I. Dextran sulphate sodium increases splenic Gr1(+)CD11b(+) cells which accelerate recovery from colitis following intravenous transplantation. Clinical & Experimental Immunology 2011;164(3) 417-427.

[10] Povirk L. F, Austin M. J. Genotoxicity of bleomycin. Mutation Research 1991;257(2) 127-143.

[11] Ishikawa H, Takeda K, Okamoto A, Matsuo S, Isobe K. Induction of autoimmunity in a bleomycin-induced murine model of experimental systemic sclerosis: an important role for CD4+ T cells. Journal of Investigative Dermatology 2009;129(7) 1688-1695.

[12] Hardy J. Amyloid, the presenilins and Alzheimer's disease. Trends in Neurosciences 1997 ;20(4) 154-159.

[13] Yan S. D, Chen X, Fu J, Chen M, Zhu H, Roher A, Slattery T, Zhao L, Nagashima M, Morser J, Migheli A, Nawroth P, Stern D, Schmidt A. M. RAGE and amyloid-beta peptide neurotoxicity in Alzheimer's disease. Nature 1996; 382(6593) 685-691.

[14] Halle A, Hornung V, Petzold G. C, Stewart C. R, Monks B. G, Reinheckel T, Fitzgerald K. A, Latz E, Moore K. J, Golenbock D. T. The NALP3 inflammasome is involved in the innate immune response to amyloid-beta. Nature Immunology 2008;9(8) 857-865.

[15] Stewart C. R, Stuart L. M, Wilkinson K, van Gils J. M, Deng J, Halle A, Rayner K. J, Boyer L, Zhong R, Frazier W. A, Lacy-Hulbert A, El Khoury J, Golenbock D. T, Moore K. J. CD36 ligands promote sterile inflammation through assembly of a Toll-like receptor 4 and 6 heterodimer, Nature Immunology 2010;11(2) 155-161.

[16] Ito S, Sawada M, Haneda M, Fujii S, Oh-Hashi K, Kiuchi K, Takahashi M, Isobe K. Amyloid-beta peptides induce cell proliferation and macrophage colony-stimulating factor expression via the PI3-kinase/Akt pathway in cultured Ra2 microglial cells. FEBS Letters 2005;579(9) 1995-2000.

[17] Ito S, Sawada M, Haneda M, Ishida Y, Isobe K. Amyloid-beta peptides induce several chemokine mRNA expressions in the primary microglia and Ra2 cell line via the PI3K/Akt and/or ERK pathway. Neuroscience Research 2006;56(3) 294-299.

[18] Ito S, Kimura K, Haneda M, Ishida Y, Sawada M, Isobe K. Induction of matrix metalloproteinases (MMP3, MMP12 and MMP13) expression in the microglia by amyloid-beta stimulation via the PI3K/Akt pathway. Experimental Gerontology 2007;42(6): 532-537.

[19] Ito S, Nishio N, Isobe K. Analysis of Amyloid Peptide -Binding Proteins in Microglial Cells. The Open Geriatric Medicine Journal 2012;5 1-6.

[20] Lu J, Zheng Y. L, Wu D. M, Luo L, Sun D. X, Shan Q. Ursolic acid ameliorates cognition deficits and attenuates oxidative damage in the brain of senescent mice induced by D-galactose. Biochemical Pharmacology 2007;74(7):1078-1090.

[21] Uddin M. N, Nishio N, Ito S, Suzuki H, Isobe K. Toxic effects of D-galactose on thymus and spleen that resemble aging. Journal of Immunotoxicology 2010;7(3)165-173.

[22] Isobe K, Cheung H. S, Wu J. Toward Personalized Cell Therapies by Using Stem Cells. Journal of Biomedicine and Biotechnology 2012; 2012 172564.

[23] Sakurai H, Okawa Y, Inami Y, Nishio N, Isobe K. Paraxial mesodermal progenitors derived from mouse embryonic stem cells contribute to muscle regeneration via differentiation into muscle satellite cells. Stem Cells. 2008;26(7):1865-1873.

[24] Sakurai H, Inami Y, Tamamura Y, Yoshikai T, Sehara-Fujisawa A, Isobe K. Bidirectional induction toward paraxial mesodermal derivatives from mouse ES cells in chemically defined medium. Stem Cell Research 2009;3(2-3) 157-169.

[25] Takahashi K, Yamanaka S. Induction of pluripotent stem cells from mouse embryonic and adult fibroblast cultures by defined factors. Cell 2006;126(4):663-676.

[26] Cheng Z, Ito S, Nishio N, Xiao H, Zhang R, Suzuki H, Okawa Y, Murohara T, Isobe K. Establishment of induced pluripotent stem cells from aged mice using bone marrow-derived myeloid cells. Journal of Molecular Cell Biology 2011;3(2):91-98.

[27] Suzuki H, Shibata R, Kito T, Ishii M, Li P, Yoshikai T, Nishio N, Ito S, Numaguchi Y, Yamashita J. K, Murohara T, Isobe K. Therapeutic angiogenesis by transplantation of induced pluripotent stem cell-derived Flk-1 positive cells. BMC Cell Biology 2010;11(72).

[28] Tokuhashi Y, Kikkawa F, Tamakoshi K, Suganuma N, Kuzuya K, Arii Y, Kawai M, Hattori S, Kobayashi I, Furuhashi Y, Nakashima N, Tomoda Y. A randomized trial of

cisplatin, vinblastine, and bleomycin versus cyclophosphamide, aclacinomycin, and cisplatin in epithelial ovarian cancer. Oncology 1997;54(4) 281-286.

[29] Inami Y, Yoshikai T, Ito S, Nishio N, Suzuki H, Sakurai H, Isobe K. Differentiation of induced pluripotent stem cells to thymic epithelial cells by phenotype. Immunology & Cell Biology 2011;89(2):314-321.

Chapter 12

Micro-Nano Materials Characterization and Inspection

Yang Ju

1. Introduction

The individual surface atoms of flat samples could be made visible in real space until the introduction of the Scanning Tunneling Microscope (STM) in 1981 by Binnig, Rohrer, Gerber, and Weibel [1]. This powerful instrument has provided a breakthrough in our possibilities to investigate matter on the atomic scale. Within one year of its invention, the STM has helped to solve one of the most intriguing problems in surface science: the structure of the Si surface. Because of their fabulous contribution, G. Binnig and H. Rohrer were rewarded with the Nobel Prize in physics in 1986. A huge number of conductors and semiconductors have been investigated on the atomic scale and marvelous images of this world of atoms have been created within the first few years after the inception of the STM. Today, the STM is an invaluable asset in the surface scientist's toolbox.

Despite the great success of the STM, it has a serious limitation. The STM requires electrical conduction of the sample material, because the STM needs the tunneling current which is flowing between a pin contact with or very nearing the sample. Thus, the STM can only image electrically conductive samples, which limits its application to imaging metals and semiconductors. But even conductors except for a few special materials, like highly oriented pyrolytic graphite can not be studied in ambient conditions by STM but have to be investigated in an ultra-high vacuum (UHV). In ambient conditions, the surface layer of solids constantly changes by adsorption and desorption of atoms and molecules. UHV is required for clean and well defined surfaces. Taking the above condition into account, Binnig speculated the atomic force between the tip and sample, the Atomic Force Microscope (AFM) [2, 3] was invented by him in 1986. Because electrical conductivity of the sample is not required in AFM, the AFM can image virtually any solid surface without the need for surface preparation. Consequently, thousands of AFMs are in use in universities, public and industrial research laboratories all over the world.

1.1. Principle of atomic force microscope

The AFM consists of a cantilever with a sharp probe-tip at its end that is used to scan the specimen surface (see Figure 1). The cantilever is typically silicon or silicon nitride with a tip radius of curvature on the order of nanometers. When the tip is brought into proximity of a sample surface, forces between the tip and the sample lead to a deflection of the cantilever according to Hooke's law. Depending on the situation, forces that are measured in AFM include mechanical contact force, van der Waals forces, capillary forces, chemical bonding, electrostatic forces, magnetic forces, etc. Along with force, additional quantities may simultaneously be measured through the use of specialized types of probe. The deflection is measured using a laser spot reflected from the top surface of the cantilever into an array of photodiodes.

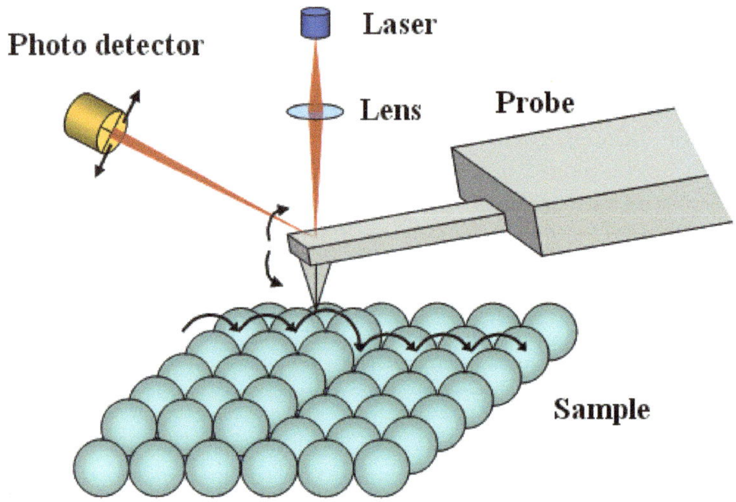

Figure 1. Atomic Force Microscope.

1.2. Microwave technique for materials characterization

The microwave methods for materials characterization generally fall into resonant methods and non-resonant methods. Resonant methods are used to get knowledge of dielectric properties at single frequency or several discrete frequencies, while non-resonant methods are often used to get a general knowledge of electromagnetic properties over a frequency range. By modifying the general knowledge of electrical properties over a certain frequency range obtained from non-resonant methods with the accurate knowledge of electrical properties at several discrete frequencies obtained from resonant methods, accurate knowledge of materials properties over a frequency range can be obtained.

1.2.1. Reflection method

In a reflection method, the properties of a sample are obtained from the reflection due to the impedance discontinuity caused by the presence of the sample in a transmission structure. The reflection method is a type of non-resonant method. From the view of transmission line, in a reflection method, the sample under test is introduced into a certain position of a transmission line, and so the impedance loading to the transmission line is changed. The properties of the sample are derived from the reflection due to the impedance discontinuity caused by the sample loading.

In a reflection method (see Figure 2), the measurement fixture made from a transmission line is usually called measurement probe or sensor. In order to increase the measurement accuracy and sensitivity, or to satisfy special measurement requirements, the measurement probes are often specially designed.

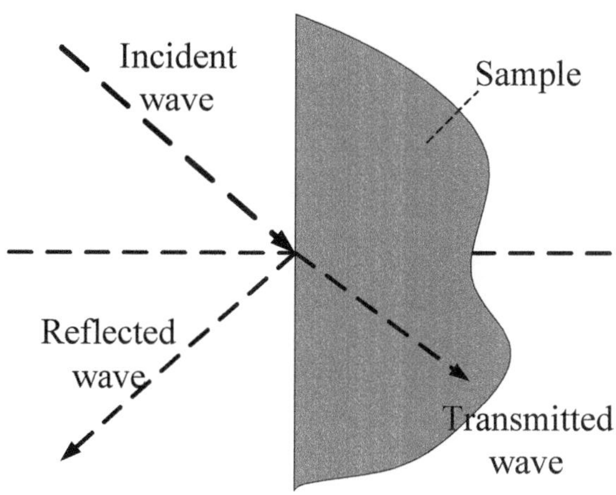

Figure 2. Boundary condition for material characterization using a non-resonant method.

1.2.2. Near-field scanning probe

In the reflection methods, there is a special method named near-field scanning probe should be introduced. Scanning techniques for local characterization of conducting and insulating films are attracting much interest. Many efforts have been made on developing microwave near-field scanning techniques, and various types of near-filed microwave microscopes have been developed for different purposed. Under the reflection method, the properties of a sample are obtained from the reflectivity due to the presence of the sample.

In principle, any type of transmission lines can be used to develop near-field microwave microscopes. In a near-field microwave microscope developed form parallel-board wave-guide, the most important part is an aperture in the form of a narrow slit (the following mentioned nano-slit plays this role in M-AFM probe). When a sample surface is in the near-filed zone of the slit, the microwave is reflected mostly from the region under the slit. Since reflection from a sample surface is determined by the resistivity, by measuring the amplitude and phase of the reflected wave while raster scanning the surface, it is possible to map the microwave resistivity of the surface. For conductive layer with thicknesses much larger that the skin depth, we can get surface impedance, while for thin layer, we can get sheet resistance. In the determination of microwave resistivity, it is necessary to measure layer thickness independently.

1.3. Developed AFM-based and microwave technique for measuring the electrical properties

Electrical properties are the most significant intrinsic characteristics of substances; they strongly affect the work functions of different materials, especially in nanometer-scale materials and devices. Thus, measuring electrical properties has become an urgent need in many areas of modern technology. For instance, in the electronics industry, critical feature sizes are becoming smaller, and it is necessary to evaluate the electrical properties of the materials with the spatial resolution on a nanometer scale to establish the knowledge to predict the behavior of materials in real devices. In addition, newly developed materials, such as conducting plastic thin films and biomaterials, which may possibly have some uncertain physical properties, will be important in the field of surface science and biological applications. Despite being intensely studied for years, their electrical properties, especially their conductivity and permittivity, are still poorly understood.

As the first section of this chapter saying, atomic force microscopy (AFM) has played an important role in nano-scale science and technology because it is one of the most versatile instruments available for imaging and manipulating structures on the nanometer scale [4-7]. Several attempts based on atomic force microscopy have been made to characterize the electrical information of materials on the nanometer scale, such as conducting atomic force microscopy (C-AFM) [8, 9], scanning capacitance microscopy (SCM) [10, 11] and electrostatic force microscopy (EFM) [12, 13]. Although C-AFM can produce a nano-scale electrical characterization of thin-films, the AFM tip must contact the conducting substrate to apply a current, so during the probing process, the AFM tip will scratch the surface of the sample. SCM can characterize electrical information by measuring the capacitance between the tip of the probe and the sample. However, it suffers from a limited spatial resolution and is sensitive to the thickness of the specimen. EFM, including Kelvin probe force microscopy (KFM) [14], scanning surface potential microscopy (SSPM) [15] and scanning Maxwell-stress microscope (SMM) [16], can measure the surface electrical potential of materials by detecting the electrostatic force between the probe tip and the sample. However, the van der Waals forces and chemical bonding forces, as well as the electrostatic forces are included in the measured data. Hence, the sample surface chemistry and atmospheric conditions greatly impact the measured electrical potential.

On the other hand, microwave measurements have been of great interest to many researchers because microwaves can propagate easily in air, and the sample response is directly related to the electrical properties of the material [17]. Thus, to obtain the microscopic electrical information, a variety of microwave microscopes have been developed [18, 19]. Steinhauer et al. developed a non-destructive and non-invasive near-field scanning microwave microscope (NSMM), which can image the local permittivity and tenability of a dielectric thin film with a spatial resolution of 1 μm [20]. Zhang and co-authors improved the NSMM to investigate the local perpendicular dielectric information of single-phase multi-ferroic thin films and single crystal materials [21]. Ferd Duewer et al. introduced scanning evanescent microwave microscopy (SEMM) [22, 23], which measures the changes of the tip-sample capacitance at the resonant frequency and the quality factor of microwave absorption. They succeeded in imaging the topography and surface resistance of metallic samples. However, to evaluate the electrical properties of materials using microwaves, it is necessary to keep the stand-off distance between the microwave probe and the sample constant because microwave signals in the near-field are extremely sensitive to this distance. Otherwise, it would be difficult to distinguish whether the changes in the signal are due to the difference of the material properties or the variation of the stand-off distance. In particular, to evaluate the electrical properties of materials with high resolution on the nanometer scale, it is indispensable to control the stand-off distance precisely on the order of nanometers.

Recently, to solve the problem of how microwave signals are affected by the stand-off distance, a technique of combining AFM with microwave microscopy has been studied [24-27]. K. Lai et al. invented a microwave impedance microscope (MIM) [24, 25], which fed a microwave signal to a silicon nitride cantilever with a Pt tip that was used to investigate the nano-scale dielectric inhomogeneity in a non-invasive manner. The Weide group combined an NSMM with an AFM (NSMM-AFM) [26, 27] by adding a microwave signal to a commercial probe. The NSMM-AFM can measure the topography and dielectric constant of thin film simultaneously. However, it is noted that MIM and NSMM-AFM do not use matched probes or cantilevers as the microwave-guide connected with the source of microwave signals. Thus, the microwave signals may not propagate along the probe and emit from the tip apex of the probe. Therefore, these techniques can only measure the changes of the probe-sample system impedance but not the intrinsic electrical properties of the measured materials.

To summarize, these AFM-based methodologies and microwave microscopy techniques can only image relative electrical properties, rather than the absolute values of the intrinsic electrical properties, such as the conductivity, permittivity, and permeability. Thus, the need remains for a microscopy technique that can provide a simultaneous measurement of topography and electrical properties on the nanometer scale.

2. The principle, fabrication and evaluation of microwave AFM

The microwave atomic force microscope (M-AFM) is a combination of the principles of the scanning probe microscope and the microwave-measurement technique [28-32]. M-AFM can maintain the constant stand-off distance between the M-AFM-probe tip and scanned sample surface, by detecting the deflection of the atomic force between them, and measure the electrical properties of materials with nanometer scale spatial resolution.

Figure 3 shows the schematic diagram of the M-AFM probe that was used to measure the electrical properties of materials in this study. Different with the normally commercial AFM probes, the M-AFM probe is having a special structure. That is, a pair of metal films was fabricated on surfaces of the M-AFM cantilever by EB (electron beam) vapor method. The metal films play a role of parallel-board microwave-signal waveguide, which can make the microwave signal propagate in the body of M-AFM cantilever and emit at the probe-tip apex. Then, by detecting the response of microwave signal reflected from the vicinity of the sample, the electrical properties can be evaluated on the nanometer scale.

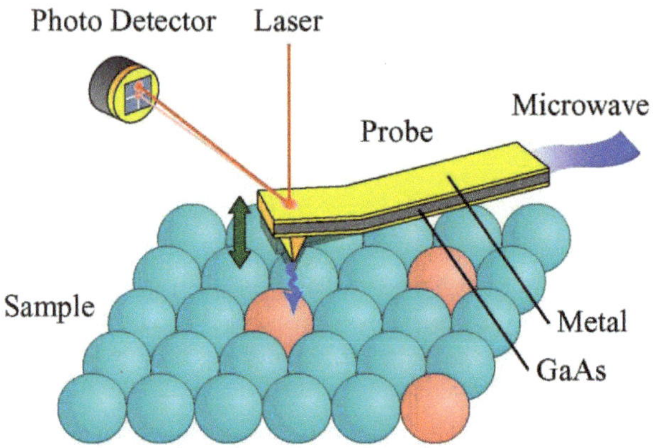

Figure 3. Schematic diagram of the M-AFM probe that was used to measure the electrical properties of materials in this study.

2.1. Fabrication of M-AFM probe

2.1.1. Fabricating the tip of M-AFM probe

To restrain the attenuation of microwave in the M-AFM probe, GaAs was used as the substrate of the probe. On the other hand, to obtain the desired structure, wet etching was used to fabricate the tip of the probe. Different with the dry etching, a side-etching will occur under the etching mask. Utilizing this property, a micro tip can be fabricated by etching a wafer, of which a small mask was introduced on the surface in advance. In the case of single crystalline wafer, such as Si and GaAs, the chemical activities are different for different crystalline planes, thereby, the etch rates are also different. Therefore, the side plane obtained at the side of the mask pattern is the most inactive plane (that is the plane having the most low etching speed) which is parallel to the side of the mask pattern. Consequently, the result of etching is strong affected by the direction of mask pattern. On the other hand, GaAs has a sphalerite structure that is more complex than that of Si, which has a similar structure as diamond. Therefore, the prediction of the etch effects is very difficult [28, 29].

Figure 4. Fabricated tip of GaAs probe.

Figure 5. GaAs probe tip with high magnification of apex part. The key component in an AFM is the tip, which should be as very sharp as possible.

In our study, it was found that only the square resist pattern can form a sharp tip (Figure 4 and Figure 5). In the case of hexagonal pattern, the reason that tip was not formed well may be due to the side of the etching mask to be too short. The reason for triangular pattern may due to that there is no crystalline plane parallel to the side of the etching mask. In addition, it was also found that one side of the square mask being 45 to the <011> direction can form a tip with a higher aspect ratio comparing with the case of one side of the resist pattern being parallel to <011> direction.

2.1.2. *Fabrication of M-AFM probe*

The process of probe fabrication is shown in Figure 6 in details: (a) Patterning the etching mask for tip generation; (b) Forming the tips by wet etching; (c) Patterning the stencil mask for the waveguide and evaporating the metal film; (d) Removing resist and film; (e) Patterning the etching mask for the beam of cantilever; (f) Forming the beam of cantilever by wet etching; (g) Patterning the etching mask on back side for the fabrication of the holder; (h) Forming the holder; (i) Evaporation of metal film on the back side; (j) Introducing slit aperture at the tip of the probe.

In the experiment, no doped semi-insulted GaAs wafer having (100) oriented surface and 350 μm thickness was used. At first, the tips were formed by etching the wafer for 100 seconds to reach the etching depth of 7.7 μm. After that, Au film used to construct the waveguide was evaporated on the substrate. The film thickness was about 50 nm (Figure 6(c)). After the deposition, the pattern of the waveguide was formed by lift-off process, where the film on the resist mask corresponding to the area without waveguide pattern was removed (Figure 6(d)). Then, in order to form the beam of the cantilever, the beam etching mask was patterned. Here, by considering the chemical activities at different crystalline planes, the length direction of the etching mask was patterned along the <011> direction. In consequence, the side-etching occurred under the resist mask, and mesa type planes appeared at the both sides of the beam (45º inclined plane). On the other hand, inverse-mesa type plane was formed at the end of the beam (60-75º inclined plane). Etching depth of the beam was about 20 μm.

In the same conditions as the beam fabrication process, holder was formed by back side etching (Figure 6(f)). Here, the etching mask was patterned on the bottom surface, and etching was carried out until the substrate was penetrated. The stirring was performed by magnetic stirrer in order to etch the sample uniformly. In the step (i) as shown in Figure 6, Au film was deposited on bottom surface of the probe to propagate a microwave signal in the probe. The thickness of the film was 50 nm, which is the same as that on the top surface of the probe. Both plane surfaces of the waveguide which were evaporated Au film are connected at the end of the beam. However, there is no Au film on the sides of the beam, since the formed inclined planes at the beam sides are not face to the direction of the evaporation. Finally, by using FIB fabrication, a slit at the tip of probe was formed to open the connection of the Au film on the two surfaces of the probe. Consequently, a homogeneous parallel plate waveguide was formed and microwaves are able to propagate along the probe and emit at the tip apex of the M-AFM probe.

It should be mentioned that dimensions of the GaAs substrate and the Au films of the M-AFM probe decide the characteristic impedance of the waveguide, in order to make certainly that microwave signals can propagate properly in the M-AFM probe for maximum sensitivity, the waveguide should have a characteristic impedance of 50 Ω (to match the characteristic impedance of a coaxial transmission line). Thus, the cantilever and the body of the M-AFM probe were designed with the dimensions of 250×30×15 μm and 2740×720×340 μm respectively.

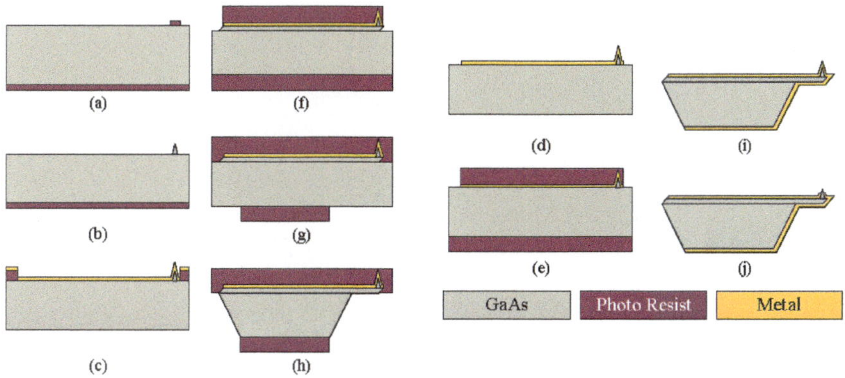

Figure 6. Fabrication processes of the M-AFM probe. (a) Patterning the etching mask for the generation of tip. (b) Forming the tip by wet etching. (c) Patterning the resist mask for the waveguide. (d) Evaporating the metal film. (e) Removing the resist and metal film. (f) Patterning the etching mask for the beam of cantilever. (g) Forming the beam of cantilever by wet etching. (h) Patterning the etching mask on back side for fabrication of the holder. (i) Forming the holder. (j) Evaporation of metal film on the back side. (k) Introducing the micro slit at the tip of probe.

2.2. SEM observation for fabricated M-AFM probes

The SEM images of the fabricated M-AFM probes are depicted in Figure 7 to Figure 10. Figure 7 shows the SEM photograph of the fabricated M-AFM probes. There 44 probes were fabricated in one process for one substrate. Figure 8 shows the as-fabricated cantilever of the M-AFM probe. The dimensions of the M-AFM probe depend on several small variations of experimental parameters, including the developing time of the resist pattern, the wet etching rate, and the EB evaporation rate. The average dimensions of the cantilever and the body of the M-AFM probes are typically 252×31×14 μm and 2742×723×339 μm, respectively. Thus, the characteristic impedance of the M-AFM probes is, on average, 49.3 Ω. Figure 9 depicts an SEM photograph of the FIB-fabricated nano-slit that has been patterned across the cantilever through the center of the probe tip. The observed tip is located near the front edge of the cantilever. As can be observed in Figure 10, the tip is approximately 7 μm high, and the nano-slit is approximately 100 nm in width.

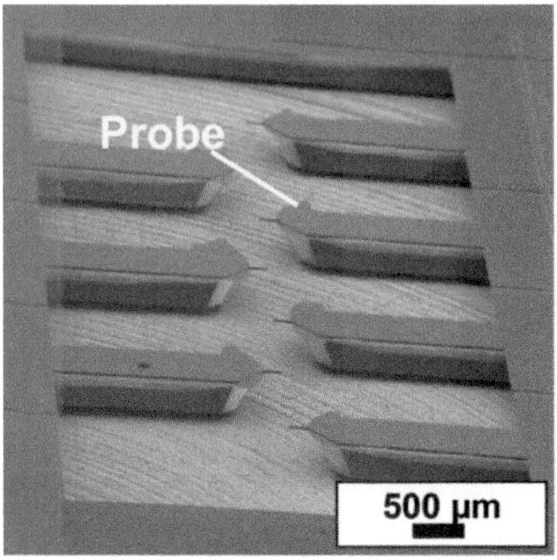

Figure 7. The fabricated M-AFM probes.

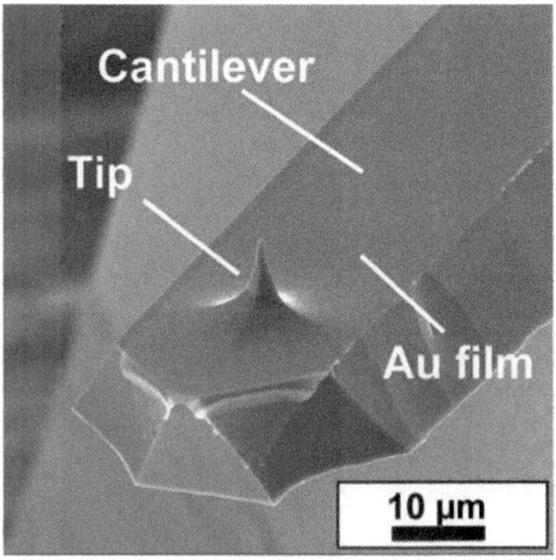

Figure 8. The cantilever of the M-AFM probe.

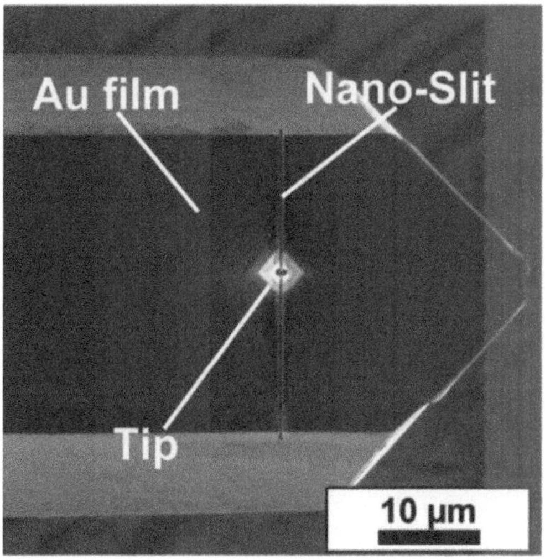

Figure 9. The 100-nm-wide-FIB-fabricated nano-slit that is across the cantilever and through the center of the tip.

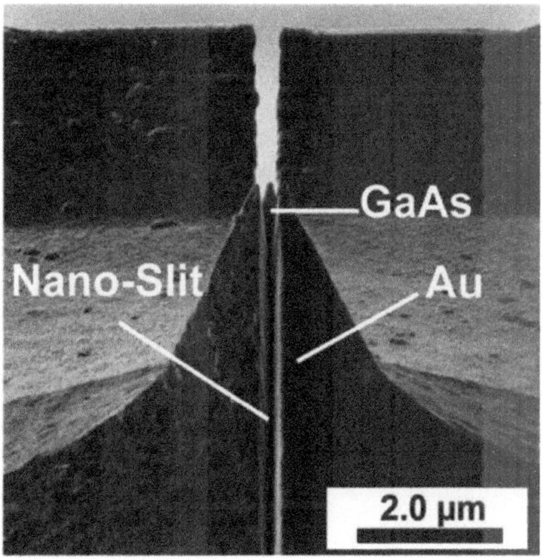

Figure 10. The high-magnification image of the M-AFM-probe tip.

2.3. Measuring topography by M-AFM probe

In order to confirm the spatial resolution of the fabricated M-AFM probes, the AFM topography of two grating samples having 2000 line/mm and 17.9 nm step height were measured by a commercial Si AFM probe, a GaAs probe without nano-slit and a M-AFM probe with nano-slit, respectively.

Probe	The resonance frequency (kHz)	Q-value	Spring constant (N/m)
A commercial Si probe	262	370	Typical value: 42
A GaAs probe without the nano-slit	185	510	Typical value: 134
A M-AFM probe with the nano-slit	201	333	Typical value: 134

Table 1. The properties of AFM probes in the atmosphere.

A JSPM-5400 was used for measurement of the sample under the noncontact mode (frequency modulation (FM) mode). The properties of three kinds of probe are given in Table 1, the resonance frequency was swept and the Q value was defined by the following relation, $Q=f_0/(f_+-f_-)$, where f_0 is the peak frequency, f_+ and f_- the shifted frequency from f_0 at 70.7% of peak intensity. The Q value indicates a resonance sharpness of the cantilever; the higher the Q value, the better stabilization of the oscillation.

Figures 11 to 13 show the topographies of the standard sample having 2000 lines/mm obtained by the commercial Si probe, GaAs probe without the nano-slit, and M-AFM probe with the nano-slit under the non-contact mode, respectively. The measurements were performed in the air, and the AFM worked in non-contact mode, with a working environment temperature of 25.0 °C and a relative humidity of 50%. The resonance frequency of Si probe and M-AFM probes (without nano-slit and with nano-slit) were 262 kHz, 185 kHz and 201 kHz, respectively, and the Q-value of them were 370, 510 and 333. The scan area was 2×2 μm^2, scanning speed was 3 μm/s, and the white spots in these figures are due to micro-dust on the sample surface. Even though the Q-value is lower than that of the GaAs probe without the nano-slit, the commercial Si probe still can obtain a little higher resolution topography due to the higher aspect ratio of the tip.

Comparing the obtained images of Figure 12, Figure 13 with the ones in Figure 11, the results illustrate that M-AFM probe has a similar capability for sensing surface topography of materials as that of commercial AFM probes.

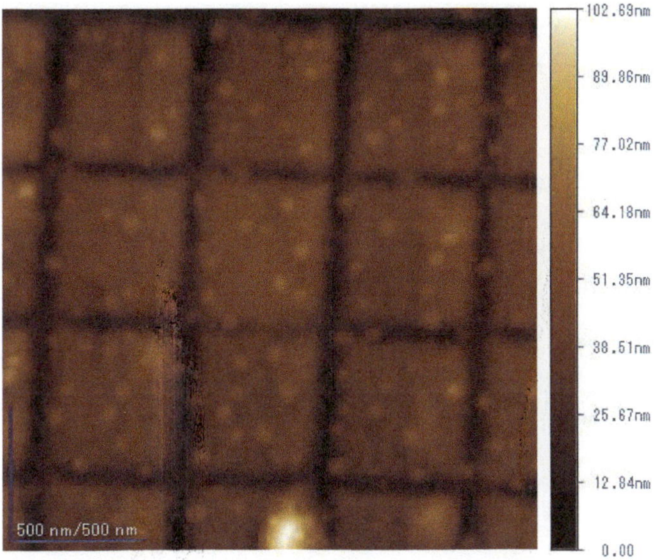

Figure 11. Surface topography of the grating sample obtained by the commercial Si probe.

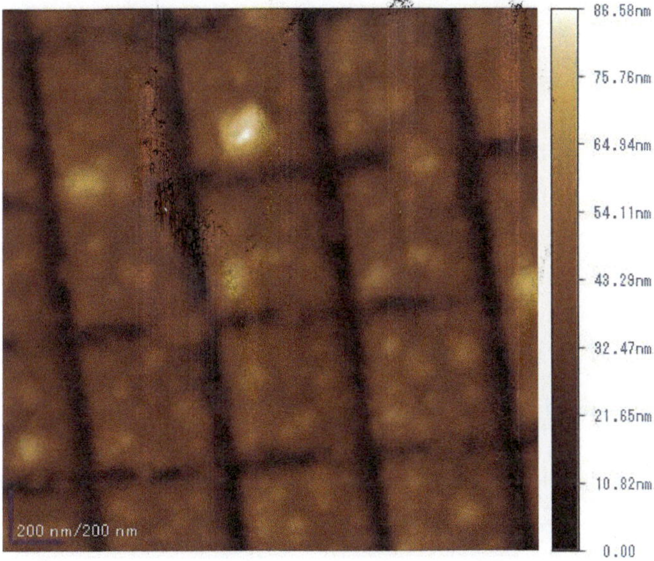

Figure 12. Surface topography of the grating sample obtained by the GaAs probe without the nano-slit.

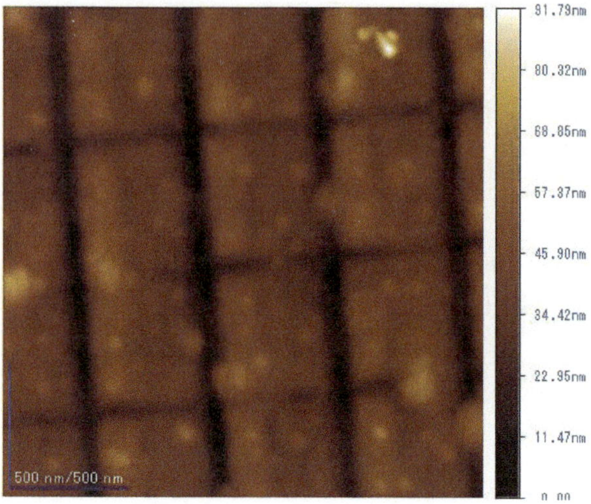

Figure 13. Surface topography of the grating sample obtained by the M-AFM probe with the nano-slit.

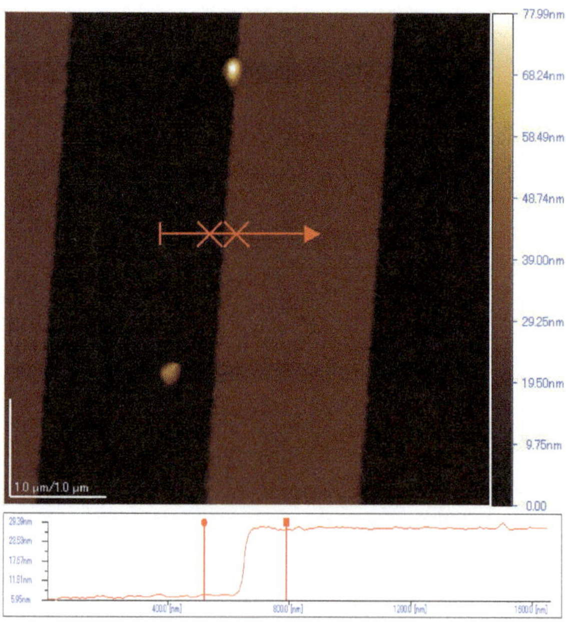

Figure 14. Topography of the grating sample obtained by the commercial Si probe.

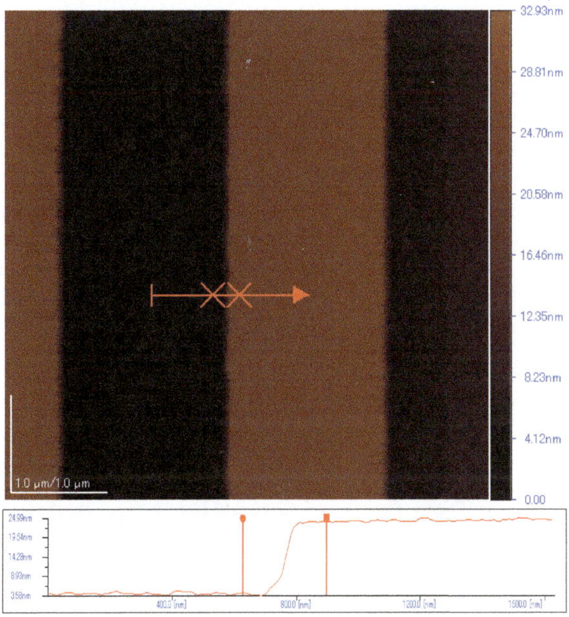

Figure 15. Topography of the grating sample obtained by the commercial Si probe.

In order to evaluate the accuracy of height measurement, a grating sample having 17.9 nm ±1nm step height was measured by using the commercial Si probe and the M-AFM probe, respectively. Figures 14 and 15 show the AFM topographies and the cross-section profiles of the grating sample obtained by the commercial Si probe and the fabricated M-AFM probe, respectively. From the slope of the step of the cross-section profile in the figures, it is confirmed that the fabricated M-AFM probes have the capability to catch the AFM topography with the resolution of nanometer order. The height of the step of the grating sample obtained by each probe was 19.17 nm and 19.67 nm, respectively. The fabricated M-AFM probe has also high resolution although the resolution was inferior as compared to the commercial Si probe. The reason why the resolution degraded with the fabricated probe is that the tip of the probes was cut by FIB fabrication. From these results, it is considered that the control of the standoff distance between the probe and the sample with high precision was achieved by the fabricated M-AFM probe.

3. Microwave imaging for materials on nanometer-scale

3.1. Experimental setup

Figure 16 schematically depicts the integrated test system of the M-AFM [31]. In our M-AFM system, the initial microwave signals, which are working at a frequency f = 16.66 GHz, are generated by a microwave generator. Next, the frequency of the microwave signals is extended by a six-frequency multiplier, which results in a stable testing frequency f = 94 GHz. The microwave signals propagate through an isolator and a circulator and then propagate into the M-AFM probe. The transmission line that connects the circulator and the probe changes from a rectangular waveguide into a coaxial line, which then changes into the parallel-plate waveguide (in the M-AFM probe). A detector is connected to the circulator, to measure the microwave signals that are received by the tip of the probe and indicate the voltage data that are converted from the reflected microwave signals. The measured signals are synchronized with positional information that is obtained from the AFM scanner, which is then used to create a microwave image. At the same time, by evaluating the output voltage data, the electrical properties of the measured materials can be determined.

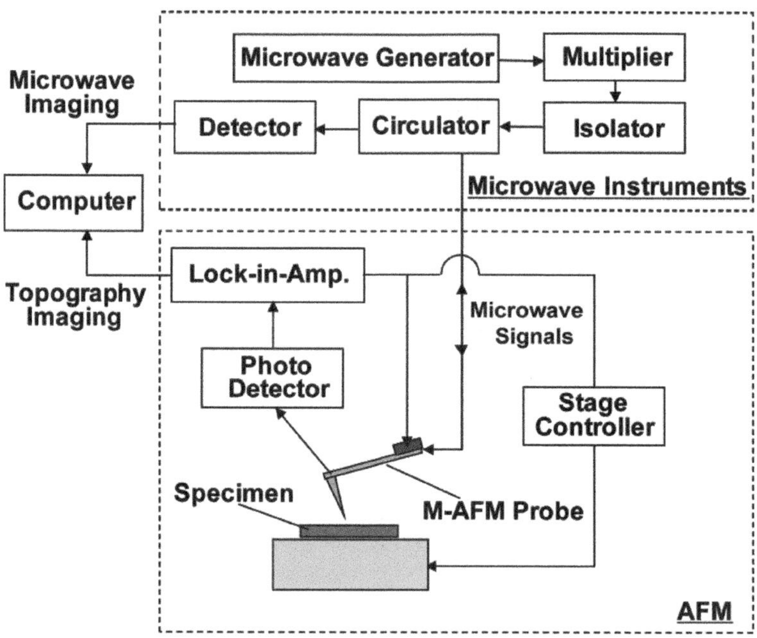

Figure 16. Diagram of the M-AFM system.

We prepared a sample for the scanning test of surface topography and microwave imaging. At first, a resist mask was patterned onto the glass substrate wafer by lithography. After developing the resist pattern, a 200-nm thick Au layer was deposited on the glass substrate by electron beam (EB) evaporation. Finally, the unexposed photo-resists were lifted off in acetone. The resulting Au and glass step structure is depicted in SEM image of Figure 17.

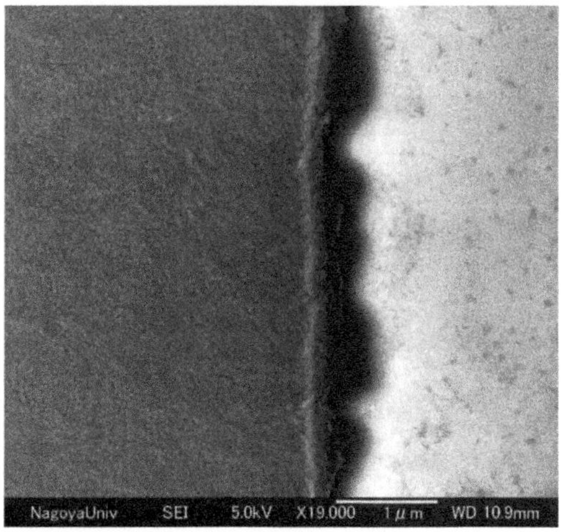

Figure 17. SEM image of measured sample.

3.2. Microwave image of Au/glass step sample

Figures 18 and 19 depict the M-AFM scanning results of the sample at the step area between the Au coating film and the glass wafer substrate. The measurements were performed in the air, and the M-AFM worked in non-contact mode, with a working environment temperature of 24.5 °C and a relative humidity of 38.4%. The resonance frequency of M-AFM probe was 133 kHz and the Q-value of it was 295. The scan area was 10×10 μm^2, scanning speed was 5 μm/s. Figure 18 depicts the surface topography of the M-AFM-measured sample. In this image, the left side represents the Au film, whereas the right side is the glass substrate. As can be seen in the scanning profile depicted in Figure 18, the thickness of the Au film was approximately 200 nm on average.

Figure 19 depicts the microwave image of the voltage that was converted from the measured microwave signals, which were simultaneously acquired by the M-AFM probe at the corresponding position depicted in Figure 18. This experimental result demonstrates that the microwave image has two spatial phases. Because the standoff distance between the tip of the M-AFM probe and the surfaces of the Au film and glass substrate is constant and controlled by the atomic force, thus, the response of the microwave signals were observed to change based on the different electrical characteristics of the measured materials. As per Figure 19, the output voltage over the glass area is larger than that over the Au area, because the scanning started from the Au area with the initial offset from the nulling operation.

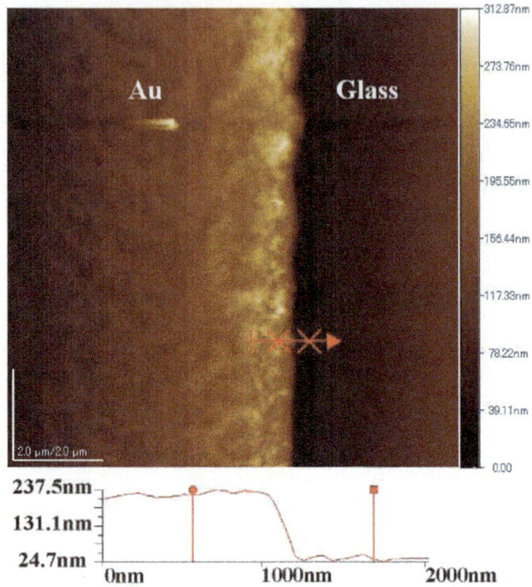

Figure 18. AFM topography image of the Au/Glass step sample.

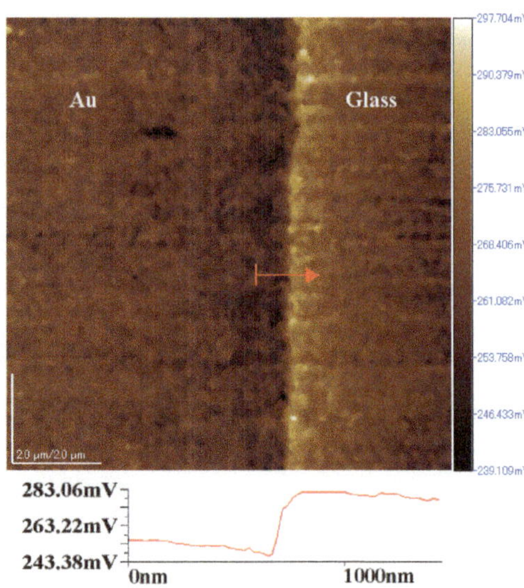

Figure 19. Microwave image of the output voltage that was converted from the measured microwave signals.

An analysis of the scanning profile depicted in Figure 19 demonstrates that the spatial resolution is higher than 120 nm, and that the output voltage measured over the Au and glass areas were 262.5 mV and 281.7 mV, respectively. The difference between the measured voltages between the Au and glass areas is 19.2 mV. Since, the stability of the measurement is high, this value is large enough for evaluating the electrical properties of other materials having the conductivity between Au and glass. As the results presented, the M-AFM should allow us to scan the electrical conductivities of other conductor materials on a nanometer scale. In addition, based on the same principle, it can also be used to measure the permittivities of dielectric materials on a nanometer scale.

4. Quantitative measurement of the electrical properties of materials on the nanometer-scale

For quantitative measurement, the operating frequency of M-AFM is set at 94 GHz. The high-frequency microwaves are easy to propagate in the waveguide and emit from the nano-slit on the probe tip. Since the width of the nano-slit is around 100 nm, the field of microwave interacting with the measured materials can be considered to be in 100 nm order. Thus, if the thickness of measured materials is larger than 100 nm, the reflection from the bottom surface of the sample can be neglected. Therefore, only the reflection from the top surface needs to be considered.

Moreover, the diode detector works in a small signal range, where it is considered to be a square-law detector. Therefore, while keeping the standoff distance between the tip of the M-AFM probe and samples constant, the output reflected voltage V, which varies only with the conductivity of the sample, has a relationship with the squared absolute value of the top surface reflection coefficient, $|\Gamma_s|^2$ as

$$V = k_0 |\Gamma_s|^2 + b_0 \tag{1}$$

The two undetermined constants k_0 and b_0 can be calibrated with two samples whose conductivities are known. For good conductors, which are used in this experiment, the surface reflection coefficient $|\Gamma_s|$ can be written as [34]

$$|\Gamma_s| = \left| \frac{1 - \sqrt{\sigma / j\omega\varepsilon_0}}{1 + \sqrt{\sigma / j\omega\varepsilon_0}} \right| \tag{2}$$

where ε_0 and σ represent permittivity of free space and the conductivity of the measured material, respectively, and ω is the angular frequency of the microwave. For semiconductor or isolating materials, similar equations can also be constructed. Then, the conductivity can be determined from Eq. (4-7) as

$$\sigma = \omega\varepsilon_0\left[4|\Gamma_s|^2 - \left(|\Gamma_s|^2+1\right)^2\right] / \left[\left(|\Gamma_s|^2+1\right)\sqrt{4|\Gamma_s|^2 - \left(|\Gamma_s|^2-1\right)^2} - 4|\Gamma_s|^2\right] \quad (3)$$

After k_0 and b_0 in Eq. (1) are calibrated using two reference samples with known conductivities, the conductivities of any samples can be calculated from the measured voltage. Therefore, the M-AFM allows us to quantitatively evaluate the electrical conductivities of materials on the nanometer scale.

It should be noted that Eq. (2) and (3) are derived under the plane wave condition, while the probe works in near-field mode. Although near-field analysis may further improve the precision of evaluation results, it requires more reference samples, which will increase the complexity of the measurement. Since the tested material was very close to the open end of the probe-tip (the standoff distance of several nanometers was extremely small as compared with the waveguide width (~100 nm) and the wavelength), this problem can be equivalent to the case that the material surface is terminated at the end of the waveguide, which can be represented by the plane wave model. Therefore, the plane wave approximation is used in this study.

There is a limitation of M-AFM technique we have to face. In the case of that the thickness of measured sample smaller than 100 nm, the reflection of microwave signal from the bottom surface of the sample and the substrate should be considered. Therefore, the M-AFM can not use the previous mentioned equations to obtain the electrical properties of measured sample quantitatively.

Five different metallic films (Cu, Pb, Al, Co and Zn) with EB fabrication were prepared for the quantitative measurement. The tested electrical conductivities by the four point probe van der Pauw method were obtained as the standard values for calibration and evaluation of M-AFM results. The tested electrical conductivities of these metal films are in the range of 4.46×10^6 S/m to 5.68×10^7 S/m. The measurements were performed in the air, and the AFM worked in non-contact mode, with a working environment temperature of 23.0 °C and a relative humidity of 50%. The resonance frequency of M-AFM probe was 107 kHz and the Q-value of it was 675. The scan area was 2×2 μm^2, scanning speed was 1 μm/s. Before scanning, we set the original voltage to be zero while maintaining a constant distance of 2.6 μm between the probe tip and the sample. During the scanning process, the standoff distance between the probe tip and samples was fixed at several nanometers by the atomic force, and the voltage corresponding to the inspected sample was measured.

Figures 20 to 24 show the topographies and microwave images of the five samples. The variations of the measured voltages for the five samples are less than ±0.46 mV, which is much smaller than the dynamic range of the M-AFM. The signal-to-noise ratio of the M-AFM measurements was evaluated to be 20.14 dB on average.

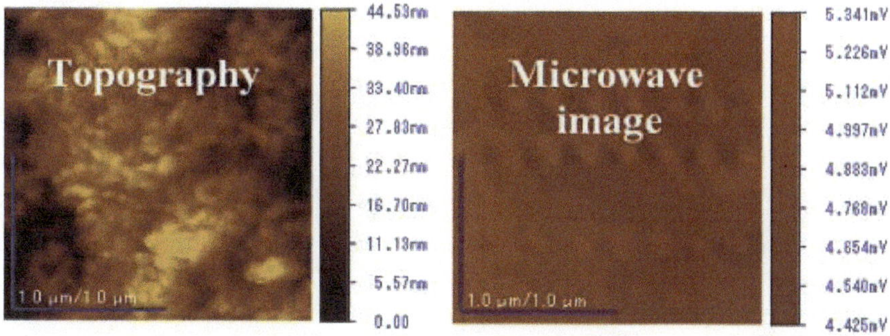

Figure 20. Topography and microwave image of measured Cu sample.

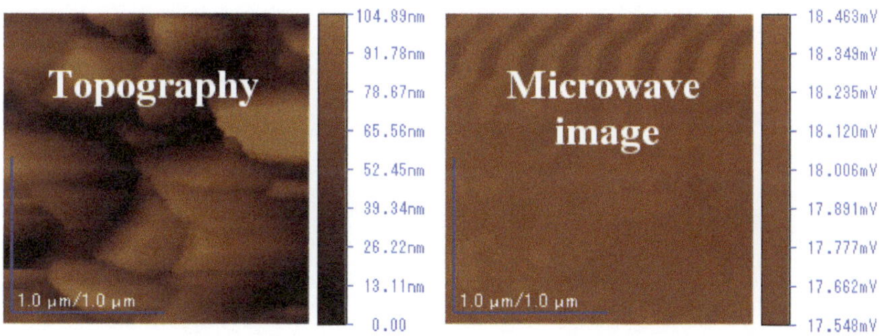

Figure 21. Topography and microwave image of measured Pb sample.

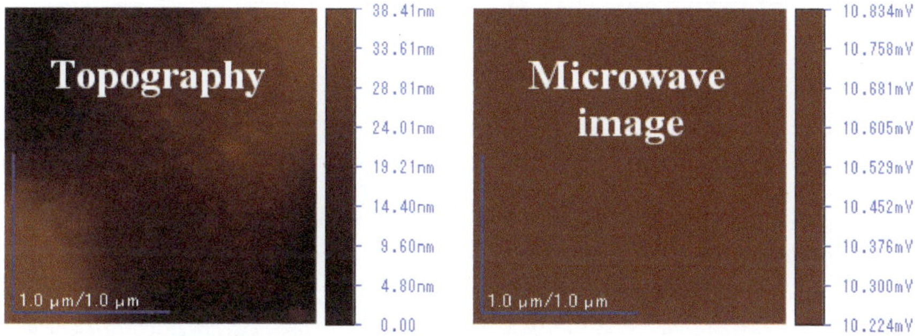

Figure 22. Topography and microwave image of measured Co sample.

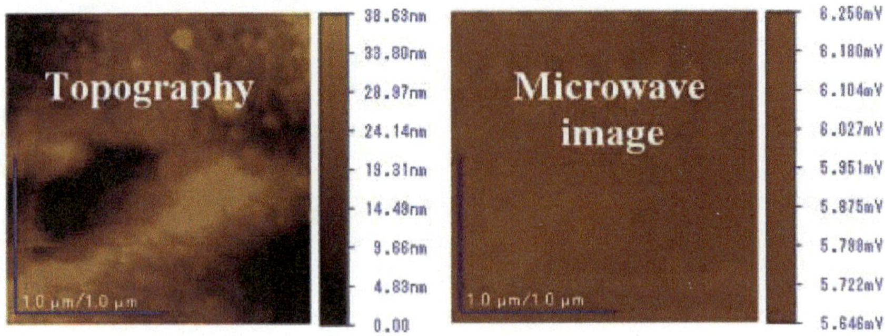

Figure 23. Topography and microwave image of measured Al sample.

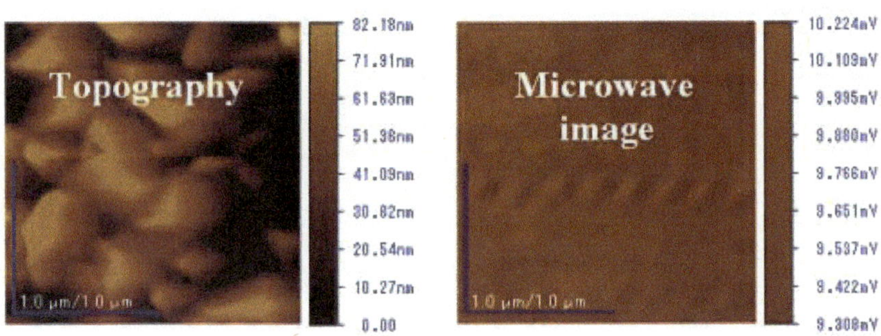

Figure 24. Topography and microwave image of measured Zn sample.

The Figure 25 shows the variation margins of measured local voltages for samples with different conductivities. Using the measured voltages of two samples obtained from Figures 20 and 21 (4.89 mV for Cu and 18.01 mV for Pb on average) and their tested conductivities (5.68×10^7 S/m for Cu and 4.46×10^6 S/m for Pb) for calibration, the two undetermined constants in Eq. (1) were calculated to be k_0 = -5.9632 and b_0= 5.9629. Then, the conductivities of Al, Co and Zn samples were evaluated with Eq. (1) and Eq. (2) by using the measured voltages obtained from Figures 20 to 24.

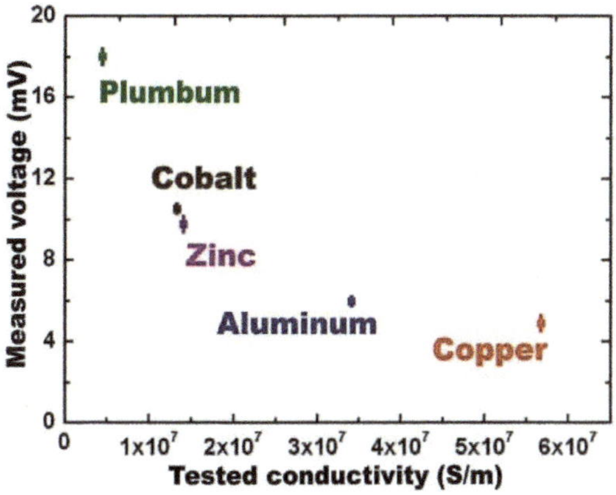

Figure 25. Variation margins of measured local voltages for samples with the different conductivities.

Figure 26 shows the variation margins of measured local voltages for samples with the square of surface reflection coefficient of them. As previous mentioned, two undetermined constants in Eq. (1) were calculated to be $k_0 = -5.9632$ and $b_0 = 5.9629$. That means the method in this work was based on a premise of that the surface reflection coefficient and measured voltage should be kept in a linear relationship. It is noted that the surface reflection coefficient and measured voltage could be provided in a linear relationship (see the fitting straight line in Figure 26). Thus, the applicability of the evaluation method in my work can be proved.

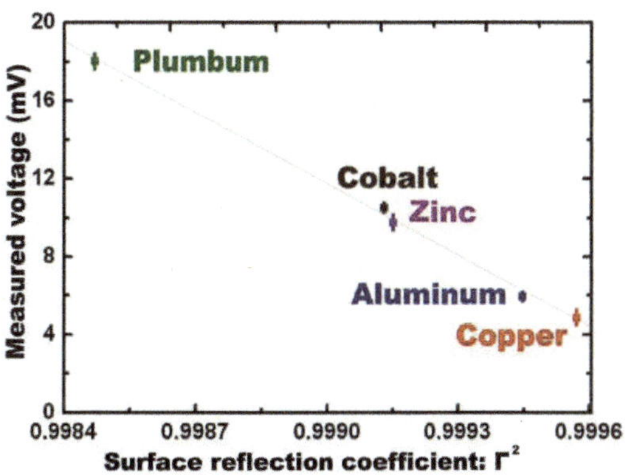

Figure 26. Variation margins of measured local voltages for samples with the square of surface reflection coefficient of them.

Figure 27 shows the evaluated results versus the tested values of Al, Co and Zn samples. It is noted from Figures 20 to 24 that no correlation can be observed between the microwave images and their corresponding geometry images. In other words, the variations of the measured local voltages are not caused by the surface morphology. The main causes of error bars of the evaluated conductivities are as follows. Firstly, the film samples prepared by EB evaporation were not homogenous in the microscopic view, and the distribution of conductivity was location-dependent (local conductivity).

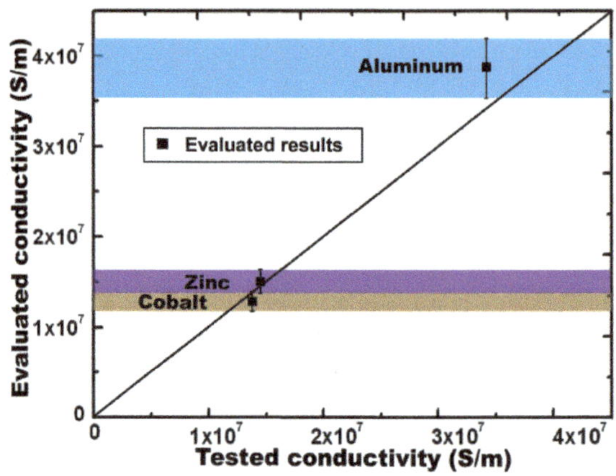

Figure 27. Evaluated conductivities of the samples in comparison with the tested conductivities of them.

It is believed that the variation margins of measured local voltages (see Figure 25) by the M-AFM caused the error bars of the evaluated conductivities. Secondly, the microwave signal for conductivity measurement was very small, which might be affected by the measurement environment. Therefore, the uncertainty of the microwave measurement may contribute to the error bars. It is also noted from Figure 27 that the deviation of evaluated conductivities from the values tested by the Van der Pauw method is 2.03%, 7.24% and 11.6% for the Zn, Co and Al, respectively. One of the causes of this deviation is that the standoff distance variation between different materials may affect the measured voltage, thereby inducing deviation of evaluated conductivity, especially for high-conductivity materials such as Al. Another cause of the deviation may be the evaluation equation which was derived under the plane wave approximation rather than the much more complicated near field analysis. The quantitative evaluation was performed three times, and the similar results as shown in Figure 25 were obtained. On the other hand, the evaluated resistivities of the five samples can be presented out as shown in Figure 28.

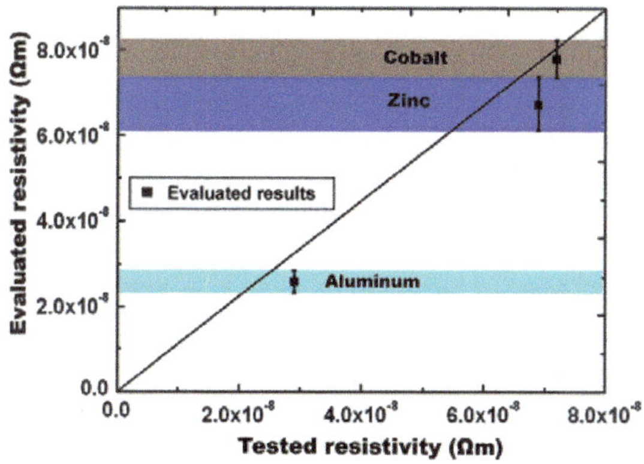

Figure 28. Evaluated resistivities of the samples in comparison with tested c resistivities of them.

5. Summary

a. We invented out a novel device named of microwave atomic force microscope (M-AFM), which is a combination of the principles of the scanning probe microscope and the microwave-measurement technique. M-AFM can maintain the constant stand-off distance between the M-AFM-probe tip and scanned sample surface, by detecting the deflection of the atomic force between them, and measure the electrical properties of materials with nanometer scale spatial resolution.

b. Microwave-AFM probes were fabricated on the GaAs wafer by using the wet etching process. A waveguide was introducing on the probe by evaporating Au film on the both surfaces of the probes. The open structure (the nano-slit) of the waveguide at the tip apex of the M-AFM probe was obtained by using FIB fabrication. SEM was used to observe the fabricated M-AFM probes. As the results, the average dimensions of the cantilever and the body of the M-AFM probes are typically 252×31×14 μm and 2742×723×339 μm, respectively. Based on these dimensions, the characteristic impedance of the M-AFM probes is, on average, 49.3 Ω. In this way, the M-AFM probe could match well with the co-axial line, which has an impedance of 50 Ω. The observed tip is located near the front edge of the cantilever and the tip is approximately 7 μm high, and the nano-slit is approximately 100 nm in width.

c. The AFM topography of the grating sample having 2000 line/mm and 18 nm step height was measured by the fabricated M-AFM probe. AFM measurements were performed by comparing with the commercial Si AFM probe. The results indicated that GaAs microwave probe has a capability to catch AFM topography of grating samples and having a high accuracy for lateral and height evaluation, similar as the commercial AFM probe.

d. We have created an M-AFM-obtained microwave image using a compact microwave instrument that was optimally synchronized with an AFM scanner. The distinguishing features of M-AFM are its ability to maintain a constant standoff distance between the probe tip and the sample surface and to measure the microwave signal interacted with the sample. Therein, both the topography and electrical-property images of the sample can be simultaneously characterized. Therefore, M-AFM is able to measure, in situ, the distribution of electrical properties on a nanometer scale. As shown in the experimental results, we successfully generated a microwave image of a 200-nm Au film coating on a glass wafer substrate with a spatial resolution of 120 nm, and, moreover, we measured the voltage difference between these two materials to be 19.2 mV. We believe that the high spatial resolution and simultaneous measurement capability of this M-AFM system will have important implications to nanotechnology characterization in the immediate future.

e. We also demonstrated a novel evaluation equation and calibration technique for the quantitative measurement of the local conductivity. Based on the analytical and explicit expressions proposed, using two reference samples with known conductivities, the conductivities of any samples can be calculated from the measured voltage. Our results demonstrate that M-AFM is able to quantitatively measure, in situ, the distribution of electrical properties on the nanometer scale.

Author details

Yang Ju

Department of Mechanical Science and Engineering, Nagoya University, Japan

References

[1] Binnig G, Rohrer H, Gerber C, Weibel E. Surface Studies by Scanning Tunneling Microscopy. Physical Review Letters 1982;49 57-61.

[2] Binnig G, Quate C. F, Gerber C. Atomic Force Microscope. Physical Review Letters 1986;56 930-933.

[3] Rugar D, Hansma P. Atomic Force Microscopy. Physics Today 1990;43 23-30.

[4] Meli M. V, Badia A, Grütter P, Lennox R. B. Self-Assembled Masks for the Transfer of Nanometer-Scale Patterns into Surfaces: Characterization by AFM and LFM. Nano letters 2002;2 131-135.

[5] Chen L. W, Cheung C. L, Ashby P. D, Lieber C. M. Single-Walled Carbon Nanotube AFM Probes: Optimal Imaging Resolution of Nanocluster and Biomolecules in Ambient and Fluid Environments. Nano letters 2004; 4 1725-1731.

[6] Wen C. K, Goh M. C. AFM Nanodissection Reveals Internal Structure Details of Single Collagen Fibrils. Nano letters 2004; 4 129-132.

[7] Wade L. A, Shapiro I. R, Ma Z. Y, Quake S. R, Collier C. P. Correlating AFM Probe Morphology to Image Resolution for Single-Wall Carbon Nanotube Tips. Nano letters 2004;4 725-731.

[8] Olbrich A, Ebersberger B, Boit C. Conducting Atomic Force Microscopy for Nanoscale Electrical Characterization of Thin SiO2. Applied Physics Letters 1998;73 3114-3116.

[9] Xu D. G, Watt G. D, Harb J. N, Davis R. C. Electrical Conductivity of Ferritin Proteins by Conductive AFM. Nano Letters 2005;5 571-577.

[10] Kopanski J. J, Marchiando J. F, Lowney J. R. Scanning Capacitance Microscopy Measurements and Modeling: Progress towards Dopant Profiling of Silicon. Journal of Vacuum Science and Technology B 1996;14 242-247.

[11] Smoliner J, Brezna W, Klang P, Andrews A. M, Strasser G. Quantitative Scanning Capacitance Microscopy on Single Subsurface InAs Quantum Dots. Applied Physics Letters 2008;92 092112.

[12] Stern J. E, Terris B. D, Mamin H. J, Rugar D. Deposition and Imaging of Localized Charge on insulator surfaces using a force microscope. Applied Physics Letters 1988; 53 2717-2719.

[13] Hu Z. H, Fischbein M. D, Drndić M. Local Charge Transport in Two-dimensional PbSe Nanocrystal Arrays Studied by Electrostatic Force Microscopy. Nano Letters 2005;5 1463-1468.

[14] Nonnenmacher M, O'Boyle M. P, Wickramasinghe H. K. Kelvin Probe Force Microscopy. Applied Physics Letters 1991;58 2921-2923.

[15] Fujihara M, Kawate H. Scanning Surface Potential Microscope for Characterization of Langmuir-Blodgett Films. Thin Solid Films 1994;242 163-169.

[16] Yokoyama H, Inoue T. Scanning Maxwell Stress Microscope for Nanometer-scale Surface Electrostatic Imaging of Thin Films. Thin Solid Films 1994;242, 33-39.

[17] Ju Y, Inoue K, Saka M, Abé H. Contactless Measurement of Electrical Conductivity of Semiconductor Wafers Using the Reflection of Millimeter Waves. Applied Physics Letters 2002;81 3585-3587.

[18] Ash E. A, Nicholls G. Super-resolution Aperture Scanning Microscope. Nature 1972;237 510-512.

[19] Rosner B. T, Van der Weide D. W. High-frequency Near-field Microscopy. Review of Scientific Instruments 2002, 73, 2505-2525.

[20] Steinhauer D. E, Vlahacos C. P, Wellstood F. C, Anlage S. M, Canedy C, Ramesh R, Stanishevsky A, Melngailis J. Imaging of Microwave Permittivity, Tunability, and

Damage Recovery in (Ba, Sr)TiO3 Thin Films. Applied Physics Letters 1999;75 3180-3182.

[21] Zhang X. Y, Wang X. C, Xu F, Ma Y. G, Ong C. K. High Frequency Dielectric Properties Distribution of BiFeO3 Thin Using Near-filed Microwave Microscopy. Review of Scientific Instruments 2009;80 114701.

[22] Duewer F, Gao C, Takeuchi I, Xiang X. D. Tip-sample Distance Feedback Control in a Scanning Evanescent Microwave Microscope. Applied Physics Letters 1999;74 2696-2698.

[23] Gao C, Hu B, Zhang P, Huang M. M, Liu W. H, Takeuchi I. Quantitative Microwave Evanescent Microscopy of Dielectric Thin Films Using a Recursive Image Charge Approach. Applied Physics Letters 2004; 84 4647-4649.

[24] Lai K, Kundhikanjana W, Peng H, Cui Y, Kelly M. A, Shen Z. X. Tapping Mode Microwave Impedance Microscopy. Review of Scientific Instruments 2009;80 043707.

[25] Kundhikanjana W, Lai K. J, Wang H. L, Dai H. J, Kelly M. A, Shen Z. X. Hierarchy of Electronic Properties of Chemically Derived and Pristine Graphene Probed by Microwave Imaging. Nano Letters 2009;9 3762-3765.

[26] Van der Weide D. W. Localized Picosecond Resolution with a Near-field Microwave/ Scanning-force Microscope. Applied Physics Letters 1996;70 677-679.

[27] Karbassi A, Ruf D, Bettermann A. D, Paulson C. A, Van der Weide D. W, Tanbakuchi H, Stancliff R. Quantitative Scanning Near-field Microwave Microscopy for Thin Film Dielectric Constant Measurement. Review of Scientific Instruments 2008;79 094706.

[28] Ju Y, Kobayashi T, Soyama H. Development of a Nanostructural Microwave Probe Based on GaAs. Microsystem Technologies 2008;14 1021-1025.

[29] Ju Y, Hamada M, Kobayashi T, Soyama H. A Microwave Probe Nanostructure for Atomic Force Microscopy. Microsystem Technologies 2009;15 1195-1199.

[30] Hosoi A, Hamada M, Fujimoto A, Ju Y. Properties of M-AFM Probe Affected by Nanostructural Metal Coatings. Microsystem Technologies 2010;16 1233-1237.

[31] Zhang L, Ju Y, Hosoi A, Fujimoto A. Microwave Atomic Force Microscopy Imaging for Nanometer-scale Electrical Property Characterization. Review of Scientific Instruments 2010;81 123708.

[32] Fujimoto A, Zhang L, Hosoi A, Ju Y. Structure Modification of M-AFM Probe for the Measurement of Local Conductivity. Microsystem Technologies 2011;17 715–720.

[33] Giessibl F. J. Advances in Atomic Force Microscopy. Reviews of Modern Physics 2003;75 949.

[34] Pozer D. M. Microwave Engineering 2nd ed. New York: Jone Wiley and Sons; 1998.

[35] Lin C. C, Juo T. J, Chen Y. J, Chiou C. H, Wang H. W, Liu Y. L. Enhanced Cyclic Voltammetry Using 1-D Gold Nanorods Synthesized via AAO Template Electrochemical Deposition. Desalination 2008;233 113-119.

[36] Yue Y. M, Chen M. J, Ju Y, Zhang L. Stress-induced Growth of Well Aligned Cu2O Nanowire Arrays and Their Photovoltaic Effect. Scripta Materialia 2012;66 81-84.

Chapter 13

Aerospace Application

Akihiro Sasoh

1. Introduction

Usually, Micro-nano technology refers to mechatronics in micro-meter to nano-meter length scale, in other words, to spatial micro-nano-meter technology. It is also possible that temporally-localized mechatronics is effective to improve macro-scale mechatronics. This concept can be referred to as 'temporal micro-nano technology,' in which mechanical/electrical inputs are done during a short time of the order of mirco- or nano-second. In this chapter, we demonstrate this technology in aerospace application.

2. Improvement of supersonic aerodynamics using repetitive nano-second laser pulses

In the aspect of the supersonic aerodynamic performance, shock wave formation leads inevitably to serious problems preventing the development of high-performance supersonic vehicle. A representative problem is sonic boom which is an impulsive noise induced by a supersonic aircraft. Wave drag force induced by shock wave is another serious problem against improvement of the aerodynamic performance. This study considers further technology to reduce a wave drag force. Wave drag reduction over a 20-mm-dia. cylinder with a truncated cone nose in a Mach 1.94 flow is done by depositing laser pulse energies at repetition frequencies up to 80 kHz and average input power of 400 W at a maximum. In actual application of energy deposition scheme [1-8] to reduce the drag, it should be taken into account both a drag coefficient and efficiency of energy deposition. The purpose of this chapter is to investigate the impacts of nose shape on supersonic drag reduction performance with repetitive energy depositions.

2.1. Trade-off in truncated cone shape

Figure 1 indicates why truncated cone shape is important to apply energy deposition scheme. In order to realize the energy depositions for improvement of drag reduction, two conditions should be satisfied. The first condition is that the magnitude of a drag force or a drag coefficient has a lower value than the one of a baseline value, for example obtained by a conical body. Another condition is to obtain a large value of an 'efficiency of energy deposition' [1] (or 'power gain'), which is defined by the ratio of a saved propulsion power to a power deposited into the flow. As shown in Figure 1, with a flat-faced cylinder body, a decrement in drag is largest although its baseline drag is largest. Without energy deposition, a conical nose leads to a smallest drag, yet the efficiency of energy deposition is worst. By truncating the cone, this tradeoff problem can be solved. Although the drag reduction performance with truncated cone model is weaker, it is easily accomplished the target drag force because its base drag force has smaller value compared to blunt body. In this case, the effective time, for which the effect of the energy deposition on a blunt body lasts, is a key parameter to know what shape of a body is suitable for obtaining a lower drag.

Sakai [8] reported that the effective time is evaluated during the interaction of the low density core created by a single laser pulse using an Nd:YAG laser with the bow shock wave over a blunt body. The evaluated effective time on a flat-faced cylinder is longer than on a hemisphere under the same energy deposition condition. The longer effective time for the flat-faced cylinder is due to the fact that the recirculation zone with vortices, which are produced due to baroclinic interaction, keeps for longer time in the forebody region. This behavior results in the reduction of the drag for a longer period. He also presented that the modulated drag for the flat-faced cylinder is nearly the same with that for the hemisphere and that the drag value is higher than that for a sharp cone with the same base diameter. It should be noted that the efficiencies for the flat-faced cylinder are typically higher than for the hemisphere under the same energy deposition condition. Thus, it is believed that a flat-faced geometry has a potential advantage to be used in the drag reduction with energy deposition.

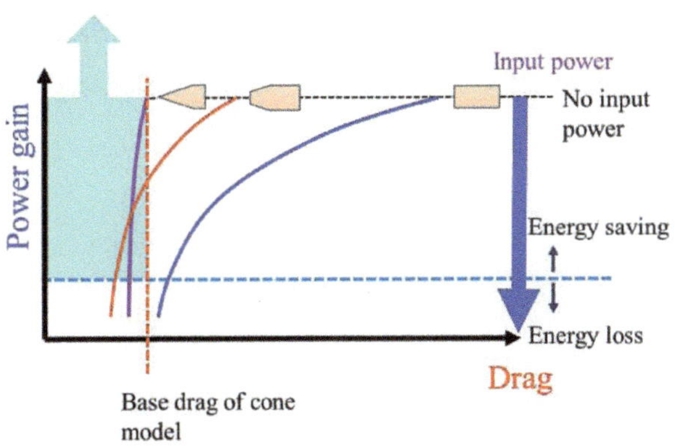

Figure 1. Trade-off in truncated cone model for improving the drag reduction performance.

Sakai [9] then proposed to employ a truncated cone and estimated its drag reduction performance using computational fluid dynamics method with Euler equations. In accordance with his results, while the magnitude of drag force is reduced with front face area of truncated cone model decreasing, the efficiency of energy deposition is higher as the front face area is increased. A quasi-steady state flowfield over truncated conical geometry is established with the higher repetitive frequency of pulse energy, typically higher than 50 kHz. In the quasi-steady state flowfield, recirculation zone composed of several vortices makes a virtual spike in front of the truncated cone body.

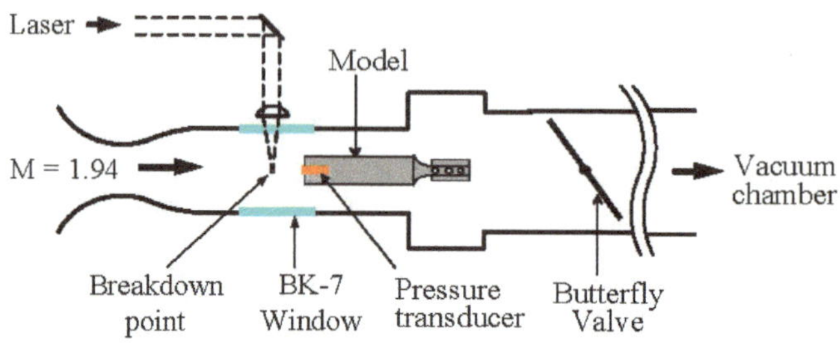

(a) Schematic diagram

(b) Photograph of in-draft wind tunnel

(c) Photograph of test section

Figure 2. Schematic and photographs of experimental apparatus.

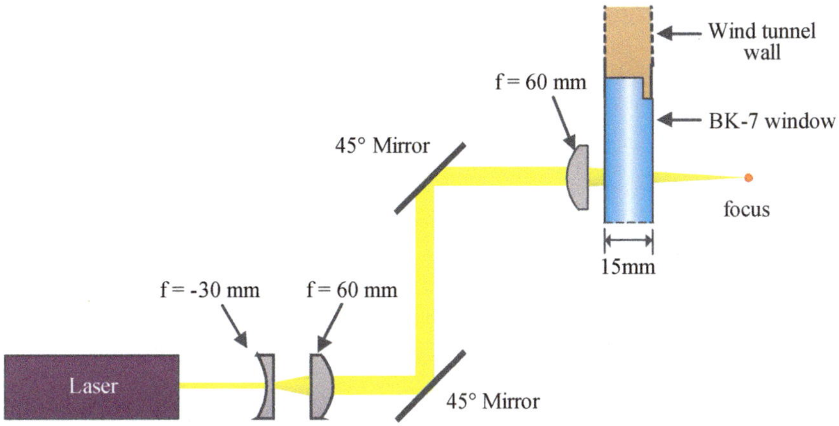

Figure 3. Schematic of laser optics system

2.2. Apparatus

Overall experimental apparatus is the same as of used in Ref. 6. The experimental facility comprises of a supersonic wind tunnel, laser optic system and diagnostic system. While the supersonic wind tunnel is operating, measurement system of stagnation pressure, drag force and visualization system is operated. Figure 2 shows the schematic and photographs of apparatus including the supersonic wind tunnel. Window diameter for visualization is 90mm. Drag reduction performance with constant pulse energy is estimated as a function of laser frequency. Nd:YVO_4 is used only to deposit the repetitive pulse energy. In our laser optic system (Figure 3), laser frequency can be applied up to 50 kHz under allowable pulse energy, E, is 7.2 mJ. Laser pulses up to 8 kHz is deposited with E=5.0 mJ.

The drag force is measured by using force balance system introduced in Ref. 5 (Figure 4). The flowfiled over the model is visualized via schlieren system including two 300-mm-dia. concave mirrors and a circular knife edge. In a single run of the wind tunnel one hundred frames of schlieren images are captured into a high-speed framing camera (Shimadzu HPV-1) with a framing interval of 4 μs, 1/4 of which is an exposure period. To measure the time-dependent stagnation pressure, a piezoelectric pressure transducer (H112A21, PCB Inc., rise time of 1 μs, sensitivity of 7.015mV/Pa) is flush-mounted at head of the model.

The data acquisition system is shown in Figure 5.

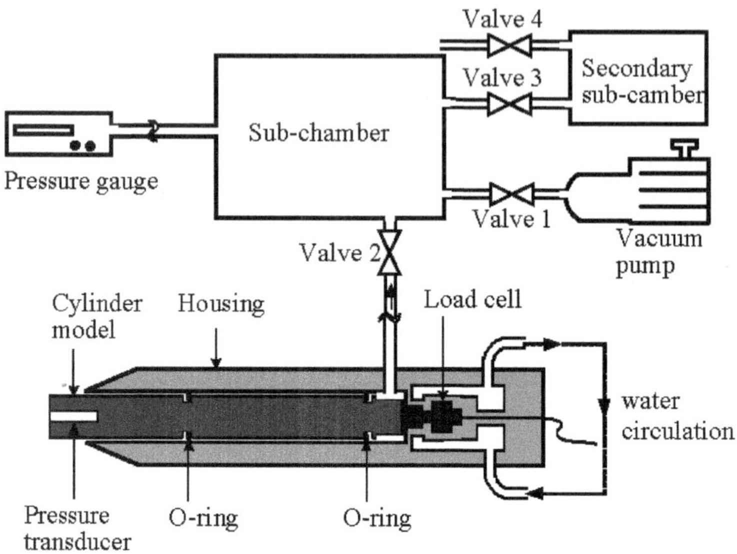

(a) Schematic of drag balance

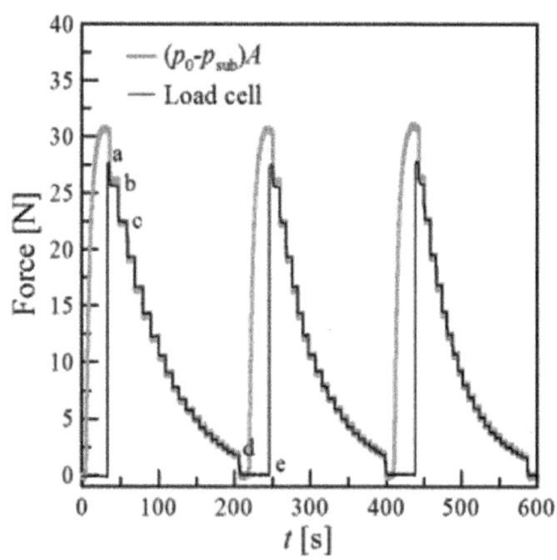

(b) Example of calibration output

Figure 4. Drag measurement system and example of calibration.

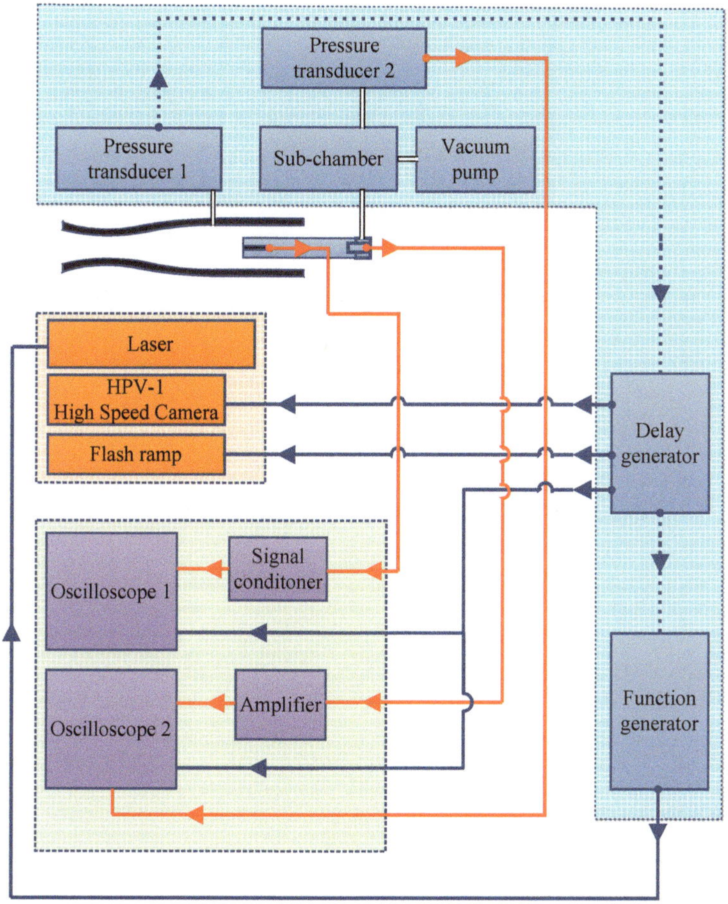

Figure 5. Data acquisition system.

2.3. Drag reduction performance of flat-faced truncated cone model

First, the drag reduction performance with flat-faced truncated conical body is investigated. Figures 6 shows the schematic illustration of truncated cone models used in the experiments. The base diameter of body, d, is 20 mm. The front face diameter is defined as d_f. The diameter ratio, d_f/d, is varied from 1.0 (cylinder model) to 0(cone model) with half apex angle of 15 degree. The stagnation pressure is measured only on 3-types (d_f / d=1.0, 0.75 and 0.5) flat-faced truncated con model, because diameter of pressure transducer, 5.56 mm, is comparable to the nose dimension.

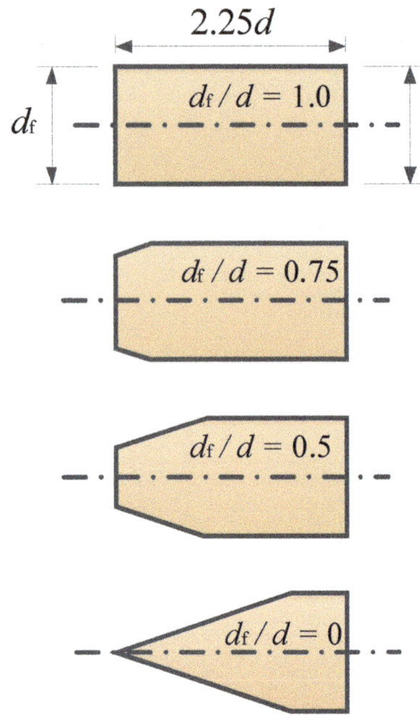

Figure 6. Truncated cone models for improving the drag reduction performance.

d_f/d	*E [mJ]*	*f [kHz]*	Shape of front face
1.0	5.0	10 ~ 80	Flat face
0.75	5.0	10 ~ 80	
0.75	7.2	10 ~ 50	
0.5	5.0	10 ~ 80	
0.5	7.2	10 ~ 50	
0	5.0	10 ~ 80	
0	7.2	10 ~ 50	
0.5	5.0	10 ~ 80	Concave face
0.5	7.2	10 ~ 50	
0.5	5.0	10 ~ 80	
0.5	7.2	10 ~ 50	

Table 1. Experimental conditions.

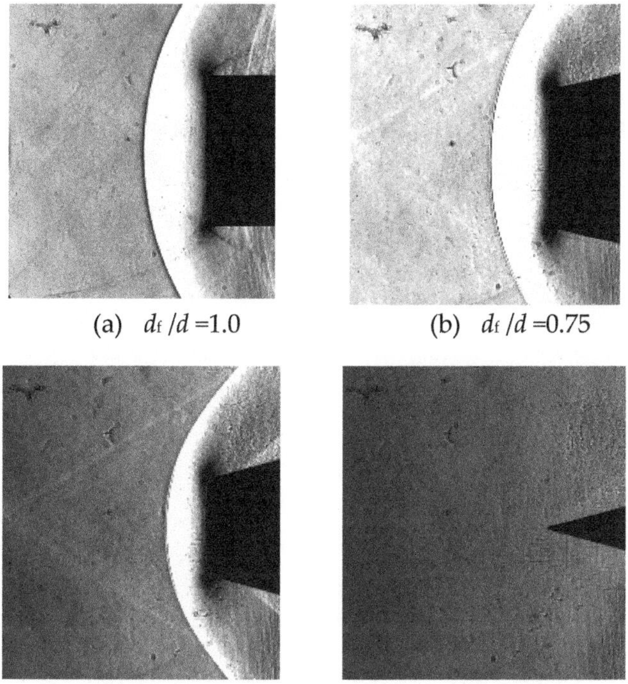

(a) d_f/d =1.0 (b) d_f/d =0.75

(c) d_f/d =0.5 (d) d_f/d =0

Figure 7. Shock layer without laser pulses.

Experimental conditions are shown in Table 1. Drag reduction performance with constant pulse energy is estimated as a function of laser frequency. In our laser optic system, laser frequency can be applied up to 50 kHz under allowable pulse energy, E, is 7.2 mJ. On the other hand, laser pulses up to 80 kHz is deposited with E=5.0 mJ.

For the absence of laser pulses, the shock layer over the truncated cone model is compared with a different d_f / d value in Figure 7. Although all of shock layer have bow shock shape except for the case of the conical model(d_f / d = 0.0), those shock stand-off distances are decreased with decreasing front face area; shock stand-off distance is 0.45 d for d_f / d = 1.0, 0.31d for d_f / d = 0.75 and 0.25d for d_f / d = 0.5.

Figures 8 presents schlieren images with laser pulse energy depositions (f=80 kHz, E=5.0 mJ). With energy depositions, the effective apex angle of distorted shock layer becomes smaller with d_f/d increasing. In particular, shock layer shape of d_f / d=1.0 is similar to oblique shock. As the residence time of vortex rings is longer, the virtual spike composed of several vortices becomes more sharply. For the d_f / d=0.0, the baroclinical vortex ring is not observed because laser-heated gas interacts with attached oblique shock wave.

The effective residence time of vortex ring [10, 11] can be seen from stagnation pressure histories in Figure 9. With $d_f / d = 1.0$, pulse-to-pulse interaction is significant at f=10 kHz. Hence, stagnation pressure history shows the almost quasi-steady state behavior. However, stagnation pressure decrement caused by vortex ring can be found for $d_f / d = 0.75$. In the case of $d_f / d = 0.5$, stagnation pressure is almost recovered into the former state, and then affected by blast wave. Even if pulse-to-pulse interaction is somewhat occurred, the effect on stagnation pressure is very weak.

Figure 10 shows examples of the time variation of the drag. Figures 11 and 12 show the drag reduction performance of flat-faced truncated cone model as a function of f. When the pulse energy of 5.0 mJ is deposited, drag force almost linearly reduces with f. For d_f / d=1.0, $\Delta D/D_0$ is obtained up to 21%, and the efficiency of energy deposition is nearly constant value of about 7. Under the same condition, amount of drag reduction and efficiency of energy deposition are decreased with decreasing of d_f / d. With E=5.0mJ, propulsion energy saving is not realized if d_f / d is smaller than 0.5.

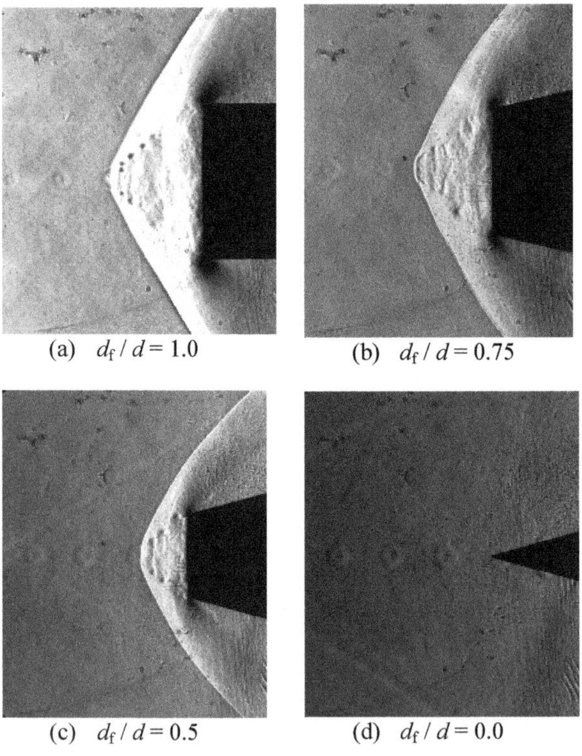

Figure 8. Instantaneous schlieren images at f= 80kHz, E=5.0mJ.

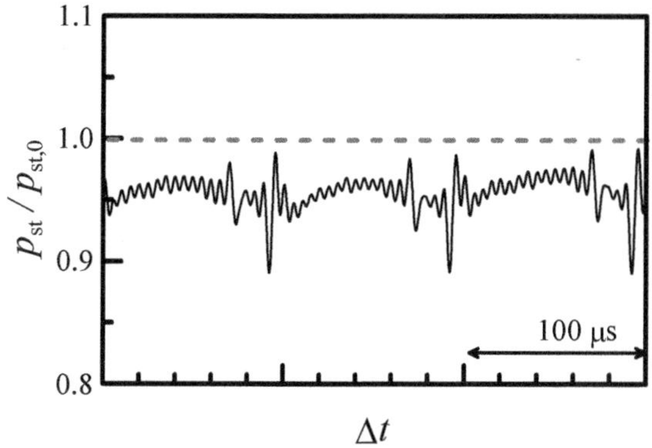

(a)d_f / d =1.0

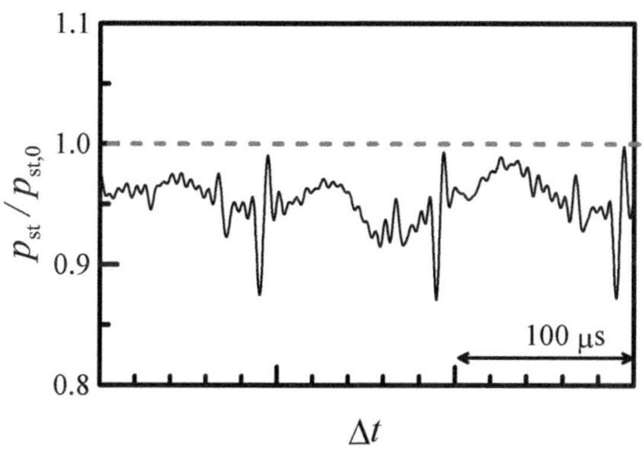

(b)d_f / d =0.75

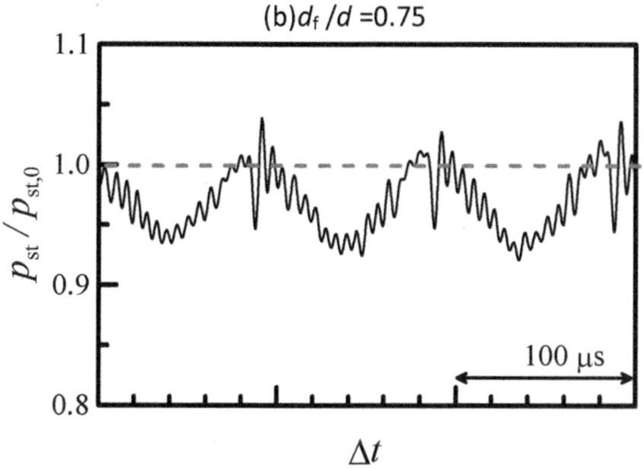

(c)d_f / d =0.5

Figure 9. Stagnation pressure histories, f=10kHz, E=5.0mJ.

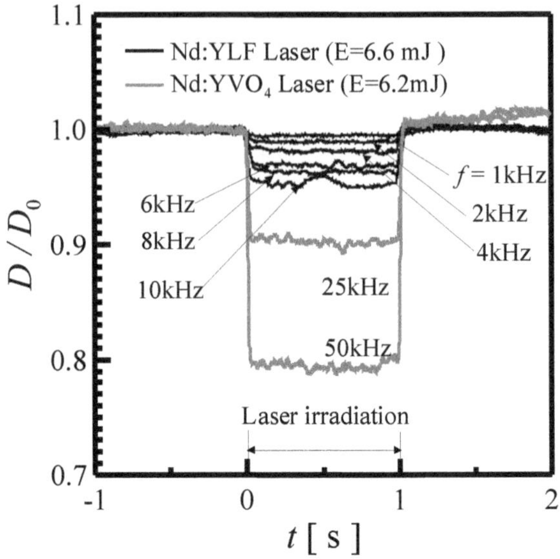

Figure 10. Example of drag variation

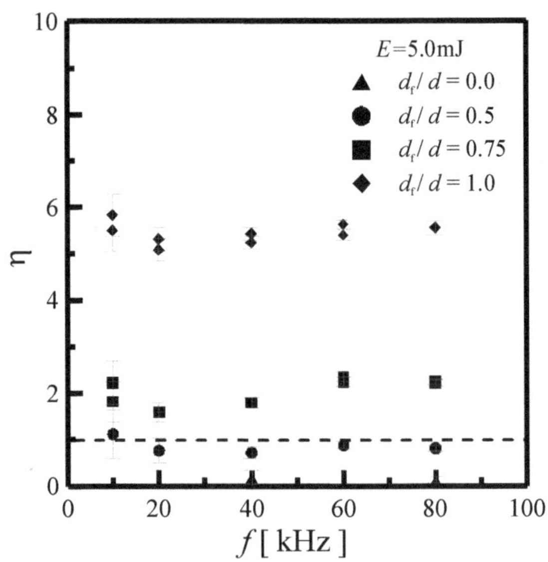

Figure 11. Efficiency of energy deposition of flat-faced truncated cone model, *E*=5.0mJ/pulse.

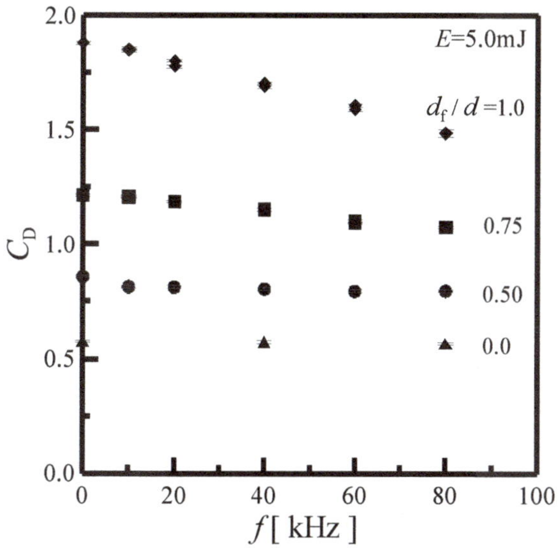

Figure 12. Variation of drag coefficient with d_f/d, E=5.0mJ/pulse.

The drag coefficient, C_D, is plotted against f in Figure 12. As repetitive laser frequency is increased, the drag coefficient is decreased. Without energy depositions, drag coefficient of d_f / d=1.0 is 1.8. Although the drag coefficient of d_f / d=1.0 is decreased down to 1.46 with f=80 kHz and E=5.0 mJ, that is still higher than the base drag(=0.55) of conical model(d_f / d=0.0).

2.4. Drag reduction performance of concave-faced truncated cone model

In order to improve the drag reduction performance, experimental studies are conducted on concave-faced truncated cone models with d_f / d = 0.5 as seen in Figure 13. The radius curvatures of front face are R/d= 0.5 and 1.0, respectively. All of truncated cone models have same length, and location of depositing pulse energy is 2d ahead of the model.

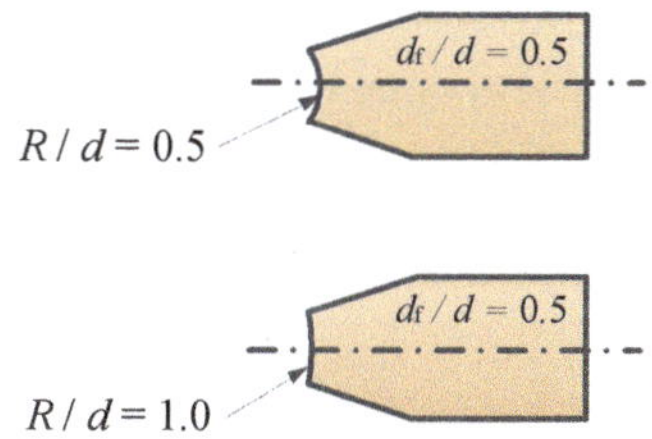

Figure 13. Schematic diagram of concave-faced truncated cone model with d_f/d = 0.5.

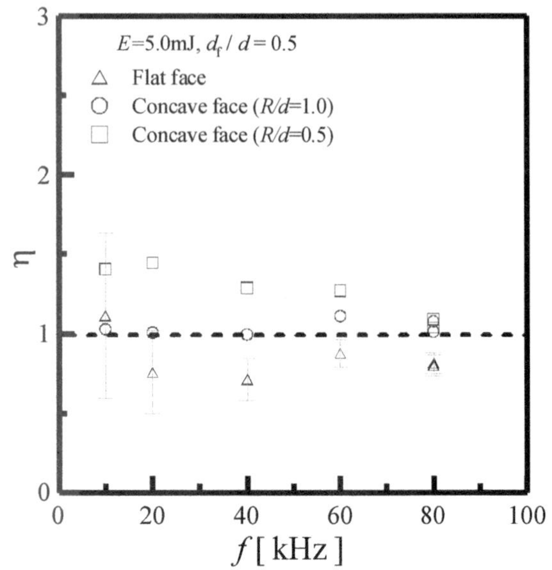

Figure 14. Power gain of concave-faced truncated cone model, d_f/d=0.50, E=5.0mJ/pulse.

From the above results, it was concluded that drag reduction performance of truncated cone become poor since the effective residence time of vortices become shorter with d_f / d decreasing. Therefore, concave-faced truncated cone shape is considered to improve the drag reduction performance of truncated cone in this section. Truncated cone model with d_f / d=0.5 is used, and radius curvature (R/d) of concave face is varied from 0.5 to 1.0.

Figure 14 presents the effect of concave radius curvature on the drag reduction performance. It is interesting that $\Delta D/D_0$ with concave-faced truncated cone becomes higher comparing with flat-faced truncated cone. $\Delta D/D_0$ is increased with R/d=1.0 for E=5.0mJ and f=80 kHz. As shown in Figure 10, the efficiency of energy deposition is slightly enhanced with concave-faced truncated cone. This implies that concave-faced truncated cone is useful to improve the drag reduction performance. From these results, it is confirmed that drag coefficient of concave-faced truncated cone is slightly decreased. In the case of R/d=0.5, drag reduction performance is almost same with one of R/d=1.0. However, drag coefficient of R/d=0.5 becomes higher than that of R/d=1.0 because small radius curvature leads to increase of base drag force.

2.5. Summary of this chapter

Although a blunt body shape brings the better drag reduction performance due to energy depositions, a truncated cone shape has considerable advantage to satisfy the necessary conditions in actual application of energy deposition scheme; the magnitude of drag force should be lower than the base drag force of a sharp conical body and the efficiency of energy depositions should be higher than unity. From these demands, drag reduction performance over truncated cone model was experimentally estimated.

In the experiments, a truncated cone model with half angle of 15 degree is used. The diameter ratio of flat-faced truncated cone is varied from 1.0 to 0.0. From the present results on flat-faced truncated cone, the effective residence time of vortices resulted by laser-heated gases interaction with shock wave plays an important role in drag reduction performance of truncated cone model. In order to improve the drag reduction performance of the truncated cones, the drag reduction performance of a concave-faced truncated cone model is evaluated. In the concave-faced truncated cone experiments, the radius curvature of concave face is varied from 0.5 to 1.0 with diameter ratio of 0.5. From the comparison with flat-faced truncated cone, the concave-faced truncated cone has more effective drag reduction performance.

3. Conclusion

As is demonstrated in this chapter, micro-nano technology is applicable also to mechanical manipulation in the temporal manner. An example of micro-nano-second mechanics is demonstrated by repetitively depositing nano-second laser pulses over a supersonic object. These short period energy deposition induce nonlinear flow mechanics, thereby improving the supersonic aerodynamics performance. It follows from these results that mico-nano technology should refer to spatiotemporal mechatronics, which will lead to further technological advancements.

Acknowledgements

The authors would like to thank Messrs. A. Saito, N. Shiraki and K. Kumazawa, Technical Division, Nagoya University for their valuable technical assistances. This research was supported by Japan Society for Promotion of Science as Grant-in-Aid for Scientific Research (S), No. 22226014.

Author details

Akihiro Sasoh

Department of Aerospace Engineering, Nagoya University, Nagoya, Japan

References

[1] Knight D. Survey of Aerodynamic Drag Reduction at High Speed by Energy Deposition. Journal of Propulsion and Power 2008;24 1153-1167.

[2] Tret'yakov P. K, Garanin A. F, Grachev G. N, Krainev V. L, Ponomarenko A. G, Tishchenko V. N, Yakovlev V. I. Control of Supersonic Flow around Bodies by Means of High-Power Recurrent Optical Breakdown. Physics-Doklady 1996; 41 566-567.

[3] Sasoh A. Is 'Fly By Light Power' possible ? In: Proceedings of 46th Autumn Joint Conference of Kansai & Chubu Branch, Japan Society for Aeronautical and Space Sciences (in Japanese); 2009.

[4] Takaki R, Liou M.-S. Parametric Study of Heat Release Preceding a Blunt Body in Hypersonic Flow. AIAA Journal 2002;40 501-509.

[5] Sasoh A, Sekiya Y, Sakai T, Kim J-H, Matsuda A. Wave Drag Reduction over a Blunt Nose with Repetitive Laser Energy Depositions. AIAA Journal 2010;48 2811-2817.

[6] Jae-Hyung Kim J. H, Matsuda A, Sakai T, Sasoh A. Wave Drag Reduction with Acting Spike Induced by Laser-Pulse Energy Depositions, AIAA Journal; in press.

[7] Adelgren R. A, Yan H, Elliott G. S, Knight D. D, Beutner T. J, Zheltovodov A. A. Control of Edney IV Interaction by Pulsed Laser Energy Deposition. AIAA Journal 2005;43 256-269.

[8] Sakai T, Sekiya Y, Mori K, Sasoh A. Interaction Between Laser-Induced Plasma and Shock Wave Over a Blunt Body in a supersonic flow. In: Proceedings of Institution of Mechanical Engineers Journal of Aerospace Engineering Part G 2008;222 605-617.

[9] Sakai T. Supersonic Drag Performance of Truncated Cones with Repetitive Energy Depositions. International Journal of Aerospace Innovation 2009;1 31-43.

[10] Ogino Y, Ohishi N, Taguchi S, Sawada K. Baroclinic Vortex Influence on Wave Drag Reduction induced by Pulse Energy Deposition. Physics of Fluids 2009;21 066102.

[11] Kim J.-H, Matsuda A, Sasoh A. Formation of a virtual spike built-up with vortex rings generated by repetitive energy depositions over a bow shock layer. Physics of Fluids 2011;23(1) 021703.

Chapter 14

Hydroxyapatite Coating on Titanium Implants Using Hydroprocessing and Evaluation of Their Osteoconductivity

Kensuke Kuroda and Masazumi Okido

1. Introduction

Titanium (Ti) and its alloys are used as artificial joints and teeth roots in orthopedic and dental settings because they have the advantage that their mechanical properties are closer to those of bone than are those of stainless steel or cobalt–chromium alloys. However, the difference in mechanical properties between Ti and natural bone leads to negative effects, such as stress shielding. To mitigate these effects, many new Ti alloys have been developed for hard tissue implants, with a focus on controlling the alloy element and its content, phase, and other characteristics.

When implants do not undergo surface modification to enhance the osteoconductivity, it takes a relatively long time to fix the metallic implant to bone such that it is stable. There are many approaches for improving the osteoconductivity of Ti and its alloys. These approaches can be classified into the following two techniques: (1) bioactive compounds that accelerate bone formation are coated on metallic implants, and (2) a rough surface at the macro-level is formed on the metallic implants, and the ingrowth of bone results in anchorage of the implants. These techniques have achieved a certain level of success, and the surface-modified implants have been used clinically. However, there are still weaknesses with the coating that need resolution, as well as unclear points regarding the effect of the surface properties on the osteoconductivity. Since hydroprocessing can be used to prepare the coating on complex-shaped substrates with complex topography, which many implants have, we focus on the use of hydroprocessing in many techniques for coating the bioactive compound, especially hydroxyapatite (HAp), and expound on the characteristics of the techniques and issues. Moreover, we describe in detail the evaluation of the osteoconductivity of implants coated with HAp, using *in vivo* testing in rat tibiae.

2. HAp coating

HAp ($Ca_{10}(PO_4)_6(OH)_2$), which is the main inorganic component in the mammal bone or tooth [1], has attracted attention as a surface-coating compound because of its high osteoconductivity. Many pyro methods of forming HAp and other calcium phosphate coatings on metallic substrates have been reported (e.g., plasma spraying [2, 3], sol-gel method [4,5], electron beam sputtering method [6], and ion beam sputtering method [7]). However, all have weak points in relation to coating with HAp on complex-shaped implants. Plasma spraying remains the most commonly used technique for HAp coating on a Ti or Ti alloy substrate in the fabrication of artificial joint replacements [1] and in endosseous dental implants [2]. On the other hand, many hydro coating techniques (e.g., cathodic electrolysis method [8-10], electrophoretic method [11, 12], and thermal substrate method [13-18]) have been proposed as approaches to forming thin film coatings on metallic substrates. The cathodic electrolysis and thermal substrate methods are single-step coating techniques in an aqueous solution, and they coat the HAp directly from the solution. The electrophoretic method is omitted from this review because it uses HAp formed by other methods in advance, despite the hydroprocessing. Therefore, in this paper, we describe the cathodic electrolysis and thermal substrate methods.

2.1. Theory of HAp coating using hydroprocessing

It is known that the solubility of HAp in an aqueous solution decreases with increasing temperature and that the relationship between the HAp solubility product, $K_{SP}/(\mathrm{mol\ dm^{-3}})^9$, and the temperature, T/K, is given by [19]:

$$\log K_{SP} = -8219.41/T - 1.6657 - 0.098215T. \tag{1}$$

Therefore, heating an aqueous solution containing Ca^{2+} and PO_4^{3-} ions results in the precipitation of calcium phosphates, such as HAp, in the solution.

The ionic product of HAp, $K_{IP}/(\mathrm{mol\ L^{-1}})^9$, is expressed as follows:

$$K_{IP} = \left[Ca^{2+}\right]^5\left[PO_4^{3-}\right]^3\left[OH^-\right], \tag{2}$$

where [X] indicates the molar concentration ($\mathrm{mol\ L^{-1}}$) of ionic species X. The increase in $[Ca^{2+}]$ or $[PO_4^{3-}]$ content or pH value in the solution initiates the precipitation of HAp because K_{IP} achieves K_{SP}. Moreover, $[PO_4^{3-}]$ increases with increasing pH value (Figure 1). Therefore, the increase in pH directly accelerates the precipitation of HAp, which indirectly increases the $[PO_4^{3-}]$ content.

Figure 2 shows the solubility curves of various compounds on a calcium orthophosphate [19]; as shown, there are many compounds other than HAp. This figure indicates that $CaHPO_4$ (DCPA) is the most stable compound at pH < 5, with HAp the most stable at pH > 5. Therefore,

HAp can be easily obtained in a solution where pH > 5 and where the ion content and temperature are controlled. However, HAp cannot precipitate in the pH < 5 solution, and hydroprocessing using the precipitation phenomenon in the aqueous solution cannot give β-$Ca_3(PO_4)_2$ (β-TCP), a bioactive compound.

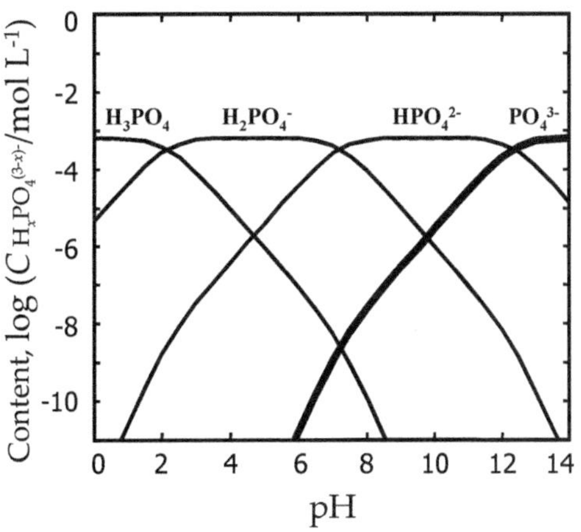

Figure 1. Logarithmic concentration diagram for orthophosphoric acid.

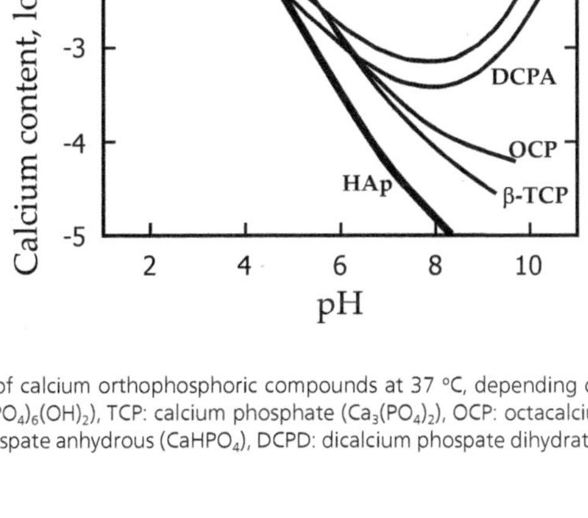

Figure 2. Solubility curves of calcium orthophosphoric compounds at 37 °C, depending on pH in aqueous solution. HAp: hydroxyapatite ($Ca_{10}(PO_4)_6(OH)_2$), TCP: calcium phosphate ($Ca_3(PO_4)_2$), OCP: octacalcium phosphate ($Ca_8H_2(PO_4)_6$ $5H_2O$), DCPA: dicalcium phospate anhydrous ($CaHPO_4$), DCPD: dicalcium phospate dihydrate ($CaHPO_4$ $2H_2O$).

2.2. Thermal substrate method in aqueous solution [13]

This process involves passing an alternating current through a metallic sample immersed in an aqueous solution. The immersed metallic sample heats up to more than 100 ºC by Joule heating, even though the hydroprocessing occurs at atmospheric pressure (Figure 3). Therefore, this method can produce the special reaction conditions (>100 ºC in an aqueous solution) by not using the pressure vessel.

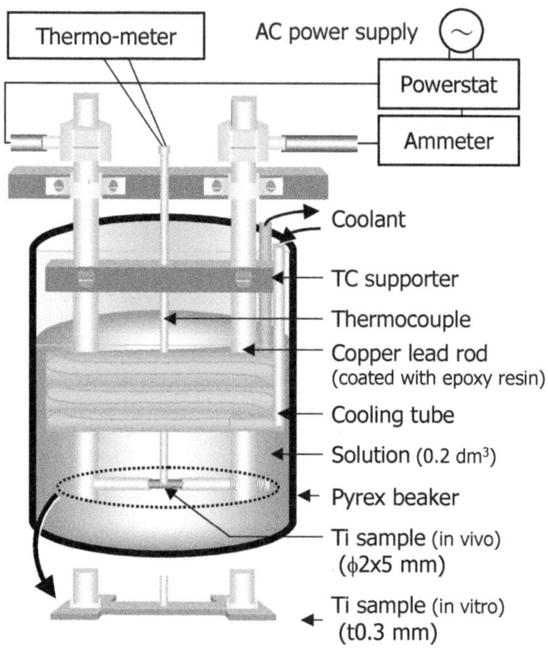

Figure 3. Experimental apparatus for HAp coating.

When the thermal substrate method in an aqueous solution is used, the fact that the solubility of HAp decreases with increasing temperature means that HAp precipitation occurs only on the substrate. It is important to coat HAp while controlling the concentration of the solute and the pH value of the solution and temperature, because they affect the degree of supersaturation of HAp in the solution (Eqs (1) and (2)). Figure 4 (a)–(d) shows the change in the surface morphology of the samples coated with HAp under controlled pH and temperature, whose foctors determined the degree of the supersaturation with respect to HAp [14, 15, 17, 18]. The precipitate at pH = 4.0 (Figure 4 (a)) appeared to pile up like bricks and was identified as DCPA, which is a stable compound. On the other hand, in the solution at pH = 8.0, the precipitate was

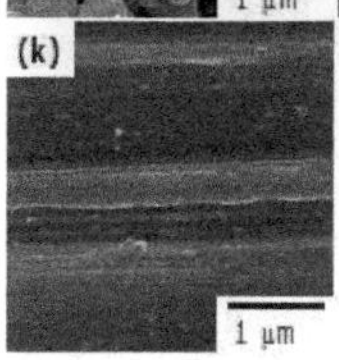

#	comp.		technique	conditions (CC: $CaCl_2$, CP: $Ca(H_2PO_4)_2$)				Fig. 6
(a)	DCPA		thermal substrate	0.7 mM CC 0.3 mM CP	pH 4	150 °C	15 min.	-
(b)	HAp	net			pH 8	40 °C	30min.	①
(c)		plate				60 °C		②
(d)		needle						③
(e)	CO_3Ap	needle		0.7 mM CC 0.3 mM CP pH 8	0.5 mM $NaHCO_3$	140 °C	15min.	④
(f)	CO_3Ap	spherical			10 mM $NaHCO_3$			⑤
(g)	HAp/gel.	needle			72 mg L^{-1} col. (Type I)			⑥
(h)	HAp/col.	plate				40 °C	30min.	⑦
(i)	HAp	net	cathodic electrolysis	7 mM CC 3 mM CP pH 5.5	6 mass% H_2O_2 -0.756 V vs. Ag/AgCl	35 °C	60 min.	-
(j)		needle			-1.1V vs. Ag/AgCl	120 °C	30 min.	-
(k)	control	as polished		#400 SiC				⑧

Figure 4. SEM photographs of the surface of the samples treated by various methods.

composed of HAp (Figure 4 (b)–(d)). From the EDX analysis, the molar ratio of calcium to phosphorous (Ca/P) of HAp was 1.41-1.43. This shows the coated HAp was calcium deficient. The pH dependence of the solubility of the calcium phosphate compounds explains why the precipitate changed with increasing pH of the solution, i.e., the solubility curves of DCPA and HAp cross at approximately pH = 5 for various compounds of calcium phosphate [19]. The surface morphology of the precipitated HAp strongly depends on coating temperature: low temperature (40 ºC) gave net-like HAp (Figure 4 (b)); high temperature (140 ºC) gave needle-like HAp (Figure 4 (d)); and mid temperature (60 ºC) gave plate-like HAp (Figure 4 (c)). That is, by using hydroprocessing, we can control the crystalline form, which could not have been achieved using traditional methods. Figure 5 shows the scanning electron microscopy (SEM) photographs of the HAp-coated samples on porous Ti alloy surfaces formed by sintering Ti6Al4V particles (ca. 100 μm in diameter) on cpTi substrates [16]. Heating at 100 ºC for 15 min. in a pH=7 solution led to HAp precipitation over the entire surface of the Ti6Al4V sintered particles (on both front and back faces) and on the base cpTi substrate of the experimental samples. In particular, it was found that HAp precipitate was also detected at the sinter neck regions of adjacent particles and on the base substrate, while the original open-pored geometry was maintained. Therefore, this method can be used to apply the HAp coating to a substrate with complex topography.

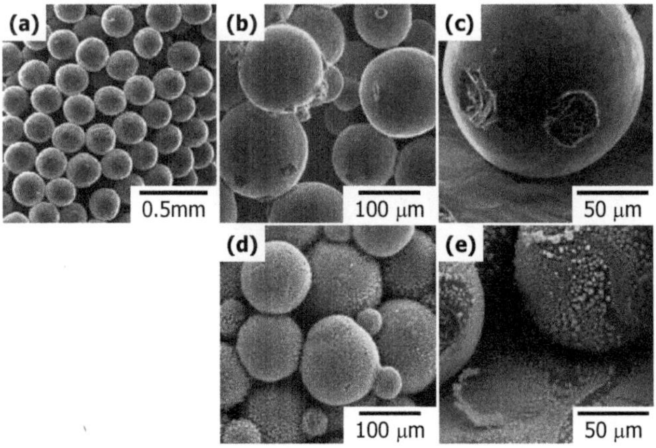

Figure 5. and/or cross-sectional views of the samples with surface roughness. (a)-(c) beads-sintered porous samples (as sintered), (d)(e) HAp coated on beads-sintered porous samples (thermal substrate method, 0.7 mM $CaCl_2$+0.3 mM $Ca(H_2PO_4)_2$, pH 7, 100 °C, 15 min.).

Biological apatite in natural bone does not appear in the form of pure HAp and it contains a considerable amount of carbonate ions [20] (about 7.4 mass% with respect to total bone and 11.4 mass% with respect to the inorganic component in natural bone [21]). Carbonate apatite (CO_3–Ap), which replaces PO_4^{3-} and/or OH^- ions with CO_3^{2-} ions, is similar to the inorganic component of bone, and it seems to be a more promising bioactive material than stoichiometric HAp, because CO_3–Ap has greater solubility than pure HAp [20]. In addition, it has been reported that $CaCO_3$ displays bioactivity, such as cell compatibility and hard tissue compatibility [22, 23]; that is, CO_3^{2-} is expected to influence biological reactivity and osteoconductive properties. It is also well known that the solubility of $CaCO_3$ in an aqueous solution decreases with increasing temperature [24]. In the solution, when CO_3^{2-} ions are added, CO_3–Ap or CO_3–Ap/$CaCO_3$ composite films are easily obtained on a substrate. Typical SEM photographs of the surface of the samples are shown in Figure 4 (e)–(f), coated in the pH = 8 solution with <0.5 mM $NaHCO_3$ added at 140 ºC for a period of 15 min., and after the steam autoclaving treatment (5 mass% CO_3 in this film) [25, 26]. The precipitates coated from the solution with >0.5 mM $NaHCO_3$ added contained CO_3–Ap and $CaCO_3$ at all temperatures, and the X-ray diffraction spectra showed a mixture of calcite, vaterite, and aragonite. The crystalline form of CO_3–Ap was changed, depending on the added $NaHCO_3$ content, as well as coating temperature. In particular, adding a significant amount of $NaHCO_3$ (>5 mM) brought about sphere-like-shaped CO_3–Ap (Figure 4 (f)) in the 140 ºC coating. In CO_3–Ap films, FT–IR analysis revealed that CO_3^{2-} was substituted for PO_4^{3-} (Type B CO_3–Ap) in advance, which was similar to biological apatite [27], and adding more CO_3^{2-} to the solution gave the substitution for OH^- (Type A). Therefore, in the samples with <0.5 mM $NaHCO_3$ added, Type B CO_3–Ap was obtained, and in the samples with >5 mM $NaHCO_3$ added (i.e., having the binary phase of CO_3–Ap/$CaCO_3$), Type AB CO_3–Ap was formed.

Natural bone contains CO_3–Ap and a considerable amount of organic components, such as collagen (about 23 mass% [21]). It is known that the hybrid organic–inorganic structure initiates pliable bone. Some researchers have reported the preparation of nanocomposites of HAp/collagen and HAp/gelatin [28–30], as natural bone is considered a nanocomposite of mineral and proteins. Moreover, immobilization of collagen on implants displays a tighter fixation with the surrounding tissue, since the collagen behaves as an adhesive protein with cells because of the amino groups in the collagen molecules [31, 32]. From the viewpoint of osteoconductivity, we expected that preparing the HAp/collagen composite coating would be a more promising approach than using an individual coating of either CO_3–Ap or HAp. In the solution to which acid-soluble collagen is added, HAp/collagen or HAp/gelatin composite films are easily obtained on a substrate, depending on coating temperature. In general, as mammalian collagen rapidly denatures to gelatin at >45 ºC, HAp/collagen composite can be obtained at <40 ºC and HAp/gelatin composite at >50 ºC. Figure 4 (g)–(h) shows the surface of the samples coated in the pH = 8 solution with 72 mg L^{-1} collagen, derived from calf, at 140 ºC and 40 ºC (10–15 mass% collagen or gelatin in the film) [33]. The surface morphologies of HAp/collagen and HAp/gelatin significantly depend on the coating temperature and are not affected by whether the composite film contains collagen or gelatin. That is, collagen and gelatin have only a small effect on the HAp crystal growth of the adsorption onto HAp. Hydroprocessing can be used to form HAp/collagen and HAp/gelatin composite films, which could not be formed using high-temperature processing, and the content of collagen and gelatin in the films can be controlled up to 60 mass%.

2.3. Cathodic electrolysis in aqueous solution [8-10]

In the electrochemical technique, a redox reaction produces supersaturation of OH^- ions near the electrode in the aqueous solution containing Ca^{2+} and PO_4^{3-} in the same manner as in the thermal substrate method. This local effect induces heterogeneous nucleation on the metal substrate serving as the electrode. The addition of hydrogen peroxide to the solution prevents H_2 gas generation at the cathodic electrode and promotes nucleation and growth of the HAp coating. Adding H_2O_2 to electrolytes enhances the formation of OH^- ions at the solution–electrode interface at a lower cathodic potential, as described in the following reaction (3) [10]:

$$H_2O_2 + 2e^- = 2OH^-. \quad (3)$$

In this method, the surface morphology of the precipitated HAp greatly depends on coating temperature [34] in the same manner as in the thermal substrate method. The effect of temperature on the surface morphology of coated samples is shown in Figure 4 (i)–(j). The HAp crystals had a similar shape to those formed using the thermal substrate method, although the size of HAp crystals differed between the cathodic electrolysis and the thermal substrate methods. The molar ratio (Ca/P) of HAp was almost same as that using the thermal substrate method. The coatings at >100 ºC were conducted in the pressure vessel. When using the electrolysis solution, to which CO_3^{2-} or collagen are added, CO_3–Ap, HAp/collagen, or HAp/gelatin composite films are formed on a substrate, depending on coating temperature.

3. Evaluation of osteoconductivity

The evaluation methods for the bioactivity of the implants are classified into *in vitro* and *in vivo* methods. In this review, the *in vivo* evaluation method is described. In *in vivo* evaluation, many types of animals at different ages were used in various studies, and different researchers used a different implanted part of the animals. Moreover, a unified quantification criterion has not yet been established, and the criteria used in various studies are not compatible with one another. Therefore, we use the bone–implant contact ratio, R_{B-I}, as an osteoconductive index based on the observation of body tissue on the implants. Bone–implant contact was determined by linear measurement of direct bone contact with the implant surface. The sum of the length of the bone formation on the implant surface was measured and expressed as a percentage of the total implant length (bone–implant contact ratio) in the cancellous bone and the cortical bone parts [17, 18, 25, 26, 33].

$$R_{B\text{-}I}(\%) = \frac{\text{sum of the length of the part of bone formation on the implant surface}}{\text{total implant length}} \times 100 \qquad (4)$$

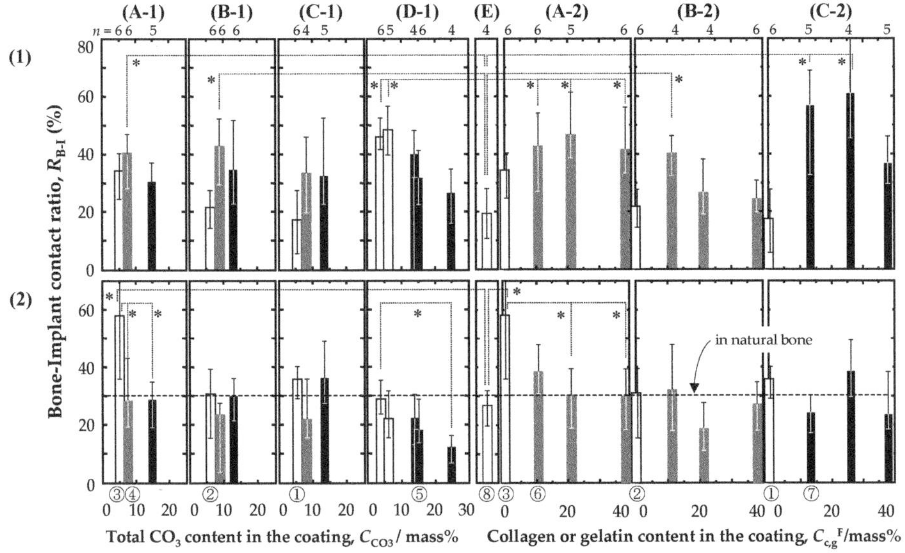

Figure 6. Bone-implant contact ratio, R_{B-I}, for the various surface coated samples (rat's tibia, 14 days). *$p < 0.05$, (1) cortical bone part, (2) cancellous bone part. (A) needle-like, (B) plate-like, (C) net-like, (D) spherical-like, (E) as-polished Ti. □ Hap ■ CO_3Ap (A-1, B-1, C-1) or HAp/gel. composite (A-2, B-2) ■ $CO_3Ap/CaCO_3$ composite (A-1, B-1, C-1, D-1) or HAp/col. composite (C-2)

Figure 6 shows the bone–implant contact ratios, R_{B-I}, of the samples coated under the various conditions mentioned above and classified based on the following four surface morphologies: (A) needle-like, (B) plate-like, (C) net-like, and (D) sphere-like. The samples are then compared with the control implant ((E) noncoated Ti). In Fig. 6, the samples are distinguished according to color based on whether or not the coating contained $CaCO_3$ and collagen or gelatin (white: HAp; gray: CO_3–Ap or HAp/gelatin; black: CO_3–Ap/$CaCO_3$ or HAp/collagen). The R_{B-I} value of HAp-coated samples (white bar) is the same as or higher than that of the as-polished one (E). In particular, R_{B-I} in the cancellous bone part is highest in the sample coated with the needle-like HAp (A-1). The influence of the different surface morphologies on R_{B-I} is apparent [17, 18]. A small amount of CO_3 included in CO_3–Ap does not influence osteoconductivity, and an increased amount of CO_3 (>15 mass%CO_3), including that in CO_3–Ap/$CaCO_3$, has a negative effect on (black bar in (A-1), (B-1), (C-1), and (D-1)) [25, 26]. The R_{B-I} value of HAp/gelatin-coated samples is the same as that of HAp (gray and white bars in (A-2) and (B-2)), and we did not find a positive effect of the addition to HAp on the osteoconductivity, or any negative effects within the limit of gelatin content used. In the HAp/collagen films (C-2), osteoconductivity was improved, and maximum R_{B-I} was obtained when the collagen content was the same as that in natural bone. The addition of too much collagen, exceeding that amount of collagen content in natural bone, inhibited the improvement of the osteoconductivity [33].

4. Conclusion

The inside of the human body is equivalent to a water environment at room temperature, since the water content in the body is about 60%. It is thought that hydroformed HAp has greater osteoconductivity than HAp synthesized using pyroprocessing, because synthesized HAp in the aqueous solution at neutral pH and room temperature is similar to that formed in the body. In addition, titanium dioxide, TiO_2, which does not exist in the human body, is a remarkable compound with respect to its osteoconductivity. It is important to research and improve the osteoconductivity of substances such as HAp, TiO_2, and $CaTiO_3$. However, we need to pay attention to the properties of their compounds, such as surface roughness [35], crystallinity, and corrosivity, all of which influence osteoconductivity. Furthermore, the evaluation criterion for osteoconductivity has not been adequately established.

The development of implants with high functionality is an important problem that urgently needs to be solved, instead of merely making progress in medical technology. It is thought that nothing can compete with such implants in the progress and development of the individual technology. We hope that these important problems can be solved using the combination of the discovery of new bioactive compounds (organic and inorganic) and their coating techniques, alloy designs for the implants, and/or the growth of related surrounding techniques for them.

Author details

Kensuke Kuroda and Masazumi Okido

Department of Materials Science and Engineering, Nagoya University, Nagoya, Japan

References

[1] Hench L. L, Wilson J. An introduction to bioceramics. Singapore, World Scientific Publishing; 1993.

[2] Pilliar R. M, Deporter D. A, Watson P. A, Pharoah M, Chipman M, Valiquette N, Carter S, DeGroot K. The Effect of Partial Coating with Hydroxyapatite on Bone Remodeling in Relation to Porous-Coated Titanium-Alloy Dental Implants in the Dog. Journal of Dental Research 1991;70(10) 1338-1345.

[3] Kweh S. W. K, Khor X. A, Cheang P. Plasma-sprayed hydroxyapatite (HA) coatings with flame-spheroidized feedstock: microstructure and mechanical properties. Biomaterials 2000;21(12) 1223-1234.

[4] Mavis B, Taş A. C. Dip coating of calcium hydroxyapatite on Ti-6Al-4V substrate. Journal of American Ceramic Society 2000;83(4) 989-991.

[5] Langstaff S, Sayer M, Smith T. J. N, Pugh S. M, Hesp S. A. M, Thompson W. T. Resorbable bioceramics based on stabilized calcium phosphates. Part I: rational design, sample preparation and material characterization. Biomaterials 1999;20(18) 1727-1741.

[6] Kim D. –H, Kong Y. –M, Lee S. –H, Lee I. –S, Kim H. –E, Heo S. –J, Koak J. -Y. Composition and crystallization of hydroxyapatite coating layer formed by electron beam deposition. Journal of American Ceramic Society 2003;86(1) 186-188.

[7] Chen T. S, Lacefield W. R. Crystallization of ion beam deposited calcium phosphate coatings. Journal of Materials Research 1994;9(5) 1284-1290.

[8] Momma H. Thin hydroxyapatite layers formed on porous titanium using electrochemical and hydrothermal reaction. Journal of Materials Science 1994;29(4) 949-953.

[9] Okido M, Kuroda K, Ishikawa M, Ichino R, Takai O. Hydroxyapatite coating on titanium by means of thermal substrate method in aqueous solutions. Solid State Ionics 2002;151(1-4) 42-57.

[10] Okido M, Nishikawa K, Kuroda K, Ichino R, Zhao Z, Takai O. Evaluation of the hydroxyapatite film coating on titanium cathode by QCM. Materials Transactions 2002;43(12) 3010-3014.

[11] Nie X, Leyland A, Matthews A. Deposition of layered bioceramic hydroxyapatite/ TiO2 coatings on titanium alloys using a hybrid technique of micro-arc oxidation and electrophoresis. Surface and Coatings Technology 2000;125(1-3) 407-414.

[12] Sena L. A, Andrade M. C, Rossi A. M, Soares G. A. Hydroxyapatite deposition by electrophoresis on titanium sheets with different surface finishing. Journal of Biomedical Materials Research 2002;60(1) 1-7.

[13] Kuroda K, Ichino R, Okido M, Takai O. Hydroxyapatite coating on titanium by thermal substrate method in aqueous solution. Journal of Biomedical Materials Research 2002;59(2) 390-397.

[14] Kuroda K, Ichino R, Okido M, Takai O. Effects of ion concentration and pH on hydroxyapatite deposition from aqueous solution onto titanium by the thermal substrate method. Journal of Biomedical Materials Research 2002;61(3) 354-359.

[15] Kuroda K, Miyashita Y, Ichino R, Okido M, Takai O. Preparation of calcium phosphate coating on titanium by the thermal substrate method and their in vitro evaluation. Materials Transactions 2002;43(2) 3015-3019.

[16] Kuroda K, Nakamoto S, Ichino R, Okido M, Pilliar R. M. Hydroxyapatite coatings on a 3D porous surface using thermal substrate method. Materials Transactions 2005;46(7) 1633-1635.

[17] Kuroda K, Nakamoto S, Miyashita Y, Ichino R, Okido M. Osteoinductivity of HAp films with different surface morphology coated by the thermal substrate method in aqueous solutions. Materials Transactions 2006;47(5) 1931-1934.

[18] Kuroda K, Nakamoto S, Miyashita Y, Ichino R, Okido M. Osteoinductivity of HAp films with different surface morphology coated by the thermal substrate method in aqueous solutions. Journal of Japan Institute of Metals 2007;71(3) 342-345.

[19] Elliot J. C. Studied in inorganic chemistry 18, Structure and chemistry of the apatites and other calcium orthophosphates. NewYork: Elsevier; 1994.

[20] Tang R, Henneman Z. J, Nancollas G. H. Constant composition kinetics study of carbonatedapatite dissolution. Journal of Crystal Growth 2003;249(3-4) 614-624.

[21] Aoki H. Marvelous Biomaterial, Apatite (in Japanese). Tokyo: ISHIYAKU PUBLISHERS, INC.; 1999.

[22] Hanein D, Sabanary H, Addadi L, Geiger B. Selective interactions of cells with crystal surfaces Implications for the mechanism of cell adhesion. Journal of Cell Science 1993;104(2) 275-288.

[23] Ohgushi H, Okumura M, Yoshikawa T, Inoue K, Senpuku N, Shors E. C. Bone formation process in porous calcium carbonate and hydroxyapatite. Journal of Biomedical Materials Research 1992;26(7) 885-895.

[24] Nihon Kagaku Kai. Kagaku Binran, Kiso-Hen (in Japanese). Tokyo: Maruzen; 2004.

[25] Kuroda K, Moriyama M, Ichino R, Okido M, Seki A. Formation and osteoconductivity of carbonate apatite and carbonate apatite/CaCO3 composite films by the thermal substrate method in aqueous solutions. Materials Transactions 2008;49(6) 1434-1440.

[26] Kuroda K, Moriyama M, Ichino R, Okido M, Seki A. Formation and osteoconductivity of carbonate apatite and carbonate apatite/CaCO3 composite films by the thermal substrate method in aqueous solutions. Journal of Japan Institute of Metals 2009;73(5) 346-353.

[27] Zhang Q, Chen J, Feng J, Cao Y, Deng C, Zhang X. Dissolution and mineralization behaviors of HA coatings. Biomaterials 2003;24(26) 4741-4748.

[28] Chang M. C, Ikoma T, Kikuchi M, Tanaka J. Preparation of a porous hydroxyapatite/collagen nanocomposite using glutaraldehyde as a crosslinkage agent. Journal of Materials Science Letters 2001;20(13) 1199-1201.

[29] Wang R. Z, Cui F. Z, Lu H. B, Wen H. B, Ma C. L, Li H. D. Synthesis of nanophase hydroxyapatite/collagen composite. Journal of Materials Science Letters 1995;14(7) 490-492.

[30] Chang M. C, Ko C, Douglas W. Preparation of hydroxyapatite-gelatin nanocomposite. Biomaterials 2003;24(17) 2853-2862.

[31] Ishii T, Koishi M, Tsunoda T. Nure-gijutsu handbook (in Japanese). Tokyo: Techno-system; 2001.

[32] Okada T, Ikada Y. Surface modification of silicone for percutaneous implantation. Journal of Biomaterials Science, Polymer Edition 1995;7(171-180).

[33] Kuroda K, Moriyama M, Ichino R, Okido M, Seki A. Formation and osteoconductivity of hydroxyapatite / collagen composite films by the thermal substrate method in aqueous solutions. Materials Transactions 2009;50(5) 1190-1995.

[34] Ban S, Maruno S. Hydrothermal-electrochemical deposition of hydroxyapatite. Journal of Biomedical Materials Research 1998;42(3) 387-395.

[35] Yamamoto D, Kawai I, Kuroda K, Ichino R, Okido M, Seki A. Osteoconductivity of Anodized Titanium with Controlled Micron-Level Surface Roughness. Materials Transactions 2011;52(8) 1650-1654

Tissue Engineering and Regenerative Medicine for Bone Regeneration

Hideharu Hibi and Minoru Ueda

1. Introduction

The terms *tissue engineering* and *regenerative medicine* originally described the concepts for producing tissue *in vitro* and *in vivo*, respectively. However, the goal of both approaches is to promote healing and tissue regeneration and function more predictably, more quickly, and less invasively than allowed by previous techniques. Regardless of these concepts, successful outcomes depend on the strategic employment of elements for tissue regeneration and may lead to the development of alternative techniques to whole organ and tissue transplantation for diseased, failed, or malfunctioning organs. Our research projects, including those introduced in chapter 8, range over autologous stem cells, intelligent scaffolds, and various cell signaling pathways with regard to tissue engineering and regenerative medicine, from basic principles to clinical applications. This chapter presents a landscape of tissue engineering and regenerative medicine for bone regeneration.

2. Principles of tissue regeneration

Tissue regeneration requires 3 key elements: 1) cells that are harvested and dissociated from the donor tissue, 2) scaffold substrates as biomaterials in which cells are attached and cultured resulting in the implantation at the desired site of the functioning tissue, and 3) cell signals that promote or prevent cell adhesion, proliferation, migration, and differentiation by upregulating or downregulating the synthesis of proteins, growth factors, and receptors (Figure 1). Combined with these 3 key elements at different rates, timings, and so on, the clinical strategy of tissue regeneration mainly comprises cell transplantation, implantation of bioartificial tissue constructs, and chemical induction of regeneration from tissue at the site of injury. According to the nature of the target tissue and the extent of the damage to be repaired, the treatment strategy needs to be modified. For bone tissue regeneration, mesenchymal cell transplantation and osteogenic chemical induction are preferred for small tissue deficiencies, while bioartificial mineral construct implantation is considered suitable for large deficiencies.

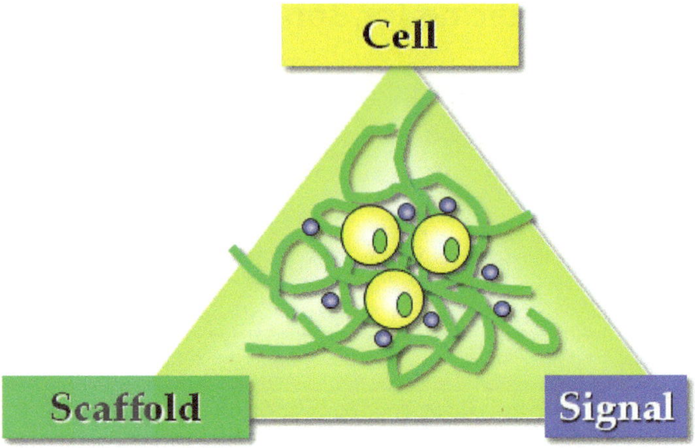

Figure 1. Elements required for tissue regeneration

3. Research projects

The following research projects are introduced in the present chapter: 1) clinical research on tissue-engineered osteogenic material (TEOM), 2) regulation of cell signaling pathways, 3) cell transplantation for improving bone quality, and 4) TEOM for promoting bone formation.

3.1. Clinical research on TEOM

TEOM, ie, injectable tissue-engineered bone, is a gel comprising *in vitro* expanded autologous bone marrow stromal cells (BMSCs) and platelet-rich plasma (PRP). Alpha granules in the platelets release various growth factors and the plasma forms a fibrin network, which serves as cell scaffold. Translational researches on TEOM have produced positive data and TEOM is considered a promising material for bone formation [1-4]. Yamada et al. reported a clinical study on injectable tissue-engineered bone for maxillary sinus floor augmentation and simultaneous placement of dental implants (Figure 2). Their study included 16 sinus augmentations in 12 patients whose alveolar crest bone height was 2 to 10 mm. All 41 dental implants installed were clinically stable at the time of second-stage surgery. Radiography at 2 years after implant installation showed increased mineralized tissue (mean, 8.8 ± 1.6 mm) compared with the preoperative bone height, and neither adverse effects nor remarkable bone absorption were seen in the 2- to 6-year follow-up periods. These results support the feasibility of using injectable tissue-engineered bone for successful bone formation and dental implants.

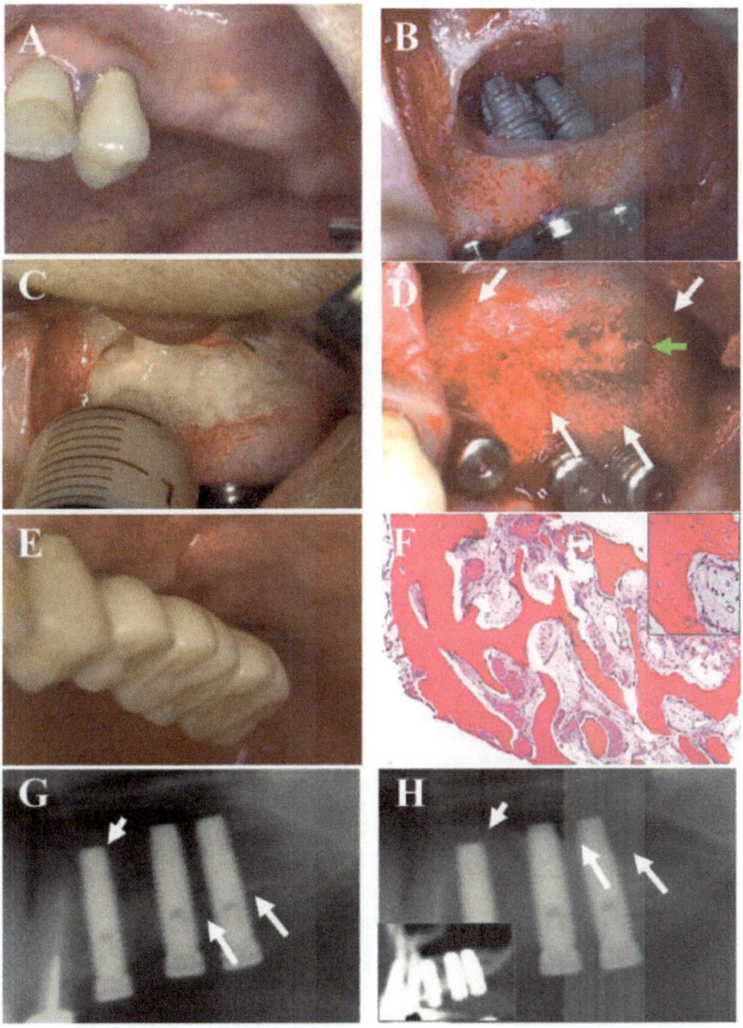

Figure 2. Treatment procedures with tissue-engineered bone A: Preoperative view. B: Magnified view of dental implants placed into the area of maxillary sinus floor augmentation and dental implant thread exposure. C: Tissue-engineered bone injection. D: Magnified view during second-stage surgery at 6 months after implant placement. The exposed thread is surrounded by newly formed bone (white arrow). The green arrow indicates the line of bone biopsy. E: Provisional prosthesis. F: Micrograph of the regenerated bone area (×4); right-upper panel (×20). G: Orthopantomogram taken immediately after first-stage surgery. Line of substratum of the maxillary sinus mucosa (arrow). H: Orthopantomogram taken at 6 months after first-stage surgery. Line of maxillary sinus mucosa after sinus augmentation (arrow). The left lower panel is a computed tomography (CT) image indicating radiodensity. (From [1]. Reprinted with permission).

3.2. Regulation of cell signaling pathways

Improvements in cell culture efficiency may contribute to a shorter treatment time and decrease its cost. Katagiri et al. examined the cultivation process for human mesenchymal stem cells (MSCs) by regulating the Wnt signaling pathway [5]. Wnt signaling was activated and inhibited with LiCl and secreted frizzled-related protein-3 (sFRP-3), respectively. Human MSCs were examined for proliferation (cell counting and BrdU assays), osteogenic differentiation (alizarin red staining), and osteogenic gene expression on days 7 and 14 after induction (reverse-transcription polymerase chain reaction [RT-PCR] and quantitative real-time RT-PCR analyses). Cell counting and BrdU assays showed higher proliferation rate of LiCl-treated MSCs than of untreated MSCs. Alizarin red staining showed that sFRP-3-treated MSCs mineralized earlier (on day 7) than untreated MSCs (Figure 3). Both RT-PCR analyses (on days 7 and 14) demonstrated that sFRP-3-treated MSCs expressed higher levels of the alkaline phosphatase gene than untreated MSCs. These results suggest that regulation of the Wnt signaling pathway contributes to cell proliferation and osteogenic differentiation of MSCs. This study implies that effective and efficient bone regeneration technologies can be developed by regulating the Wnt signaling pathway.

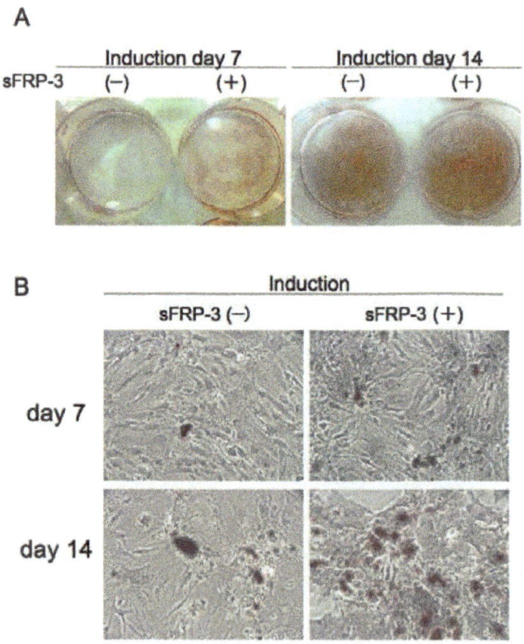

Figure 3. Alizarin red staining of sFRP-3-treated human mesenchymal stem cells (MSCs) during osteogenic differentiation A: During osteogenic induction of MSCs, more mineralization occurred in MSCs treated with secreted frizzled-related protein-3 (sFRP-3) than in untreated MSCs on days 7 and 14. B: Phase-contrast macroscopic views (×40). (From [5]. Reprinted with permission).

3.3. Cell transplantation for improving bone quality

Low bone quality is a risk factor for successful osseointegration, a prerequisite for dental implants, and typically occurs in osteoporosis. Osteoporosis is a systemic skeletal disease leading to fragile bones with decreased microstructures due to the postmenopausal decrease in estrogen secretion. Okamoto et al. investigated whether BMSCs can promote bone healing around titanium implants in rat osteoporosis models [6]. Sprague-Dawley rats were divided into 3 groups: the first group in which the ovaries were removed (OVX group), the second group in which a sham surgery was performed (SHAM group), and the third group in which BMSCs were transplanted to an OVX group (OVX-BMSCs group) (Figure 4). In the OVX-BMSCs group, 1×10^5 BMSCs were transplanted into the femur with implant. Each value of the bone-to-implant contact and the bone area of each cortical bone and cancellous bone were obtained histomorphometrically. Bone density was measured over 500 μm distally and proximally from the implants. Each ratio of bone-to-implant contact, bone area, and bone density in the OVX-BMSCs group was significantly higher than those of the OVX group as compared to the cancellous bone. BMSC transplantation therapy improved local bone healing in the cancellous bone surrounding implants and also significantly improved bone binding with implants.

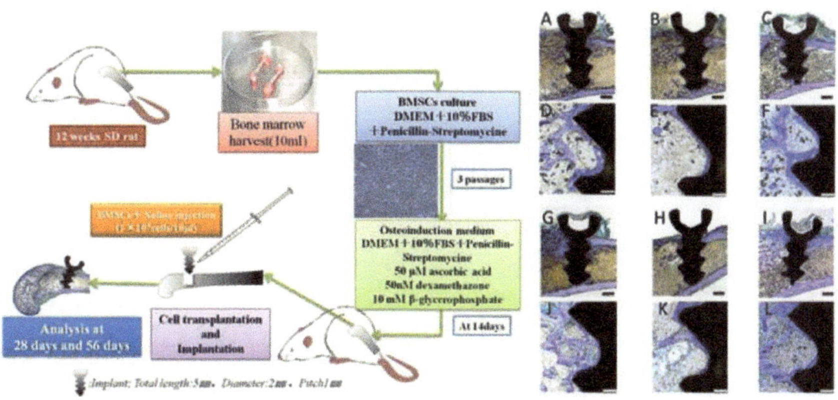

Figure 4. Experimental design (left) and histological evaluation of the distal femur at 28 and 56 days after implantation (right) A-F: Micrographs of the distal femur at 28 days after implantation. A and D: SHAM; B and E: OVX (ovaries had been removed); C and F: OVX-BMSCs (ovaries had been removed and bone marrow stromal cells had been transplanted), (toluidine blue stain, A-C ×1.25, bar = 1.1 mm; D-F ×10, bar = 200 μm). In the cortical bone area, most of the implant surface was in direct contact with newly generated bones, and the thread was widely filled with bone tissues in all groups. No remarkable difference was observed among the groups. Slightly more osteogenesis was observed in the OVX-BMSCs group than in the OVX group. The osteogenetic area was remarkably smaller in the OVX group than in the SHAM group outside the thread. In addition, osteogenesis was observed at a higher rate in the OVX-BMSCs group than in the OVX group outside the thread. G-L: Micrographs of the distal femur at 56 days after implantation. G and J: SHAM; H and K: OVX; I and L: OVX-BMSCs, (toluidine blue stain, G-I ×1.25, bar = 1.1 mm; J-L ×10, bar = 200 μm). In the cortical bone area, the area on the implant surface that was in contact with newly generated bones was greater than that observed at 28 days, and the amount of bone tissue inside the thread had increased in all groups. In the cancellous bone area, the amount of bone that was in contact with the implant surface had increased in all groups. There was no remarkable difference between the SHAM group and the OVX-BMSCs group. In the OVX-BMSCs group, trabecular structures were observed, presenting as web-like pattern. (From [6]. Reprinted with permission).

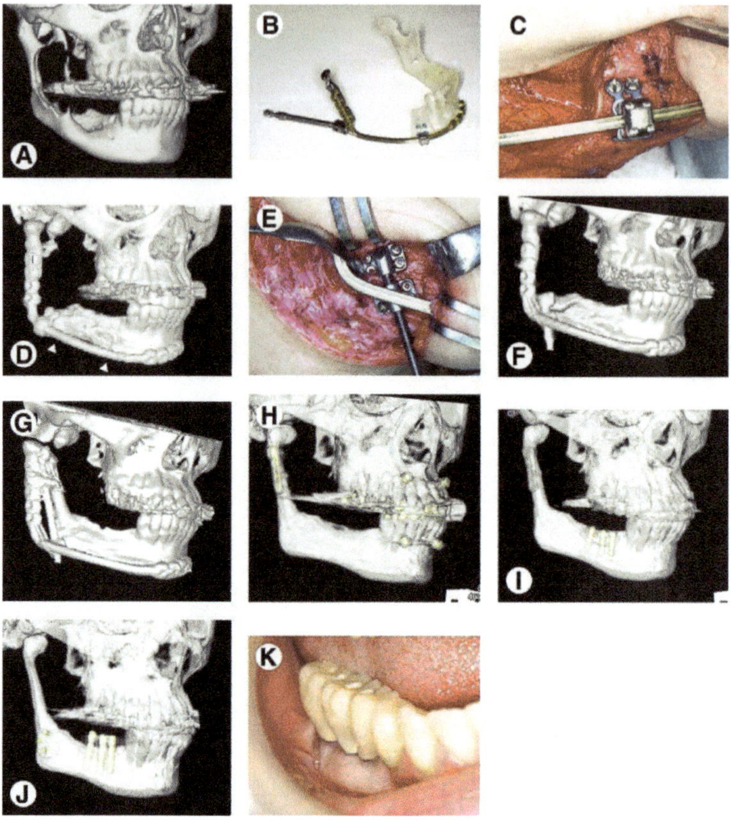

Figure 5. Two-step transport distraction osteogenesis (DO) for the posterior body and ramus of the mandible A: Mandible with recurrent ameloblastoma. Computed tomography (CT) revealed a multilocular radiolucent area in the ramus and posterior body of the mandible. B: Custom-made distraction device with an artificial condyle on the patient's stereolithographic model. Locking plates and screws enabled supraperiosteal fixation of the transport segment. C: Supraperiosteal fixation of the distraction device. A short buccal periosteal flap of the transport segment was repositioned and sutured to cover the vertical osteotomy line. The transport segment was fixed with locking plates and screws supraperiosteally. D: Regenerated posterior mandibular body. Eight months after distraction, CT showed that the newly formed 50-mm-long bone in the distraction gap (arrowheads) progressed to continuous buccal and lingual cortical surfaces. E: Secondary vertical DO for reconstructing the ascending ramus. A short buccal periosteal flap of the secondary transport segment was repositioned and sutured to cover the horizontal osteotomy line. The transport segment was fixed with another distraction device positioned supraperiosteally. F: Another transport segment. Immediately after surgery, CT showed fixation of the secondary transport segment with the distraction device. G: Regenerated ascending ramus. Nine months after the secondary distraction, CT showed that newly formed bone in the distraction gap progressed to continuous buccal and lingual cortical surfaces of the ramus. H: Device removal and mandibular reconstruction. CT showed that the regenerated ramus was connected to the preserved condylar segment with a lag screw and plate. I: Regenerated mandibular body containing implants. CT showed 3 screw-type implants installed in the right second premolar and molar regions with a buccal veneer bone graft fixed with 2 screws. J: Reconstructed mandible and osseointegrated implants for occlusal function. CT showed that the mandibular angle was adjusted with osteotomy and screw fixation in the angle region. K: Provisional implant-supported prosthesis on the reconstructed mandible. The ridge supporting the implants was surrounded by the attached mucosa, the buccal part of which originated from the palatal mucosa. (From [8]. Reprinted with permission).

3.4. TEOM for promoting bone formation

Distraction osteogenesis (DO) is considered *in vivo* tissue engineering and is a technique that provides autologous and predictable bone formation. Hibi & Ueda developed an internal device with which they were able to reconstruct a mandibular segmental bony defect of more than 10 cm [7,8]. (Figure 5).

DO is a useful technique for reconstructing bony defects without performing grafting procedures; however, it requires a long treatment time that includes latent, lengthening, and consolidation periods. To shorten these periods, attempts of applying tissue engineering technologies for DO have been made [9, 10]. Kinoshita et al. investigated whether locally injected TEOM can promote bone regeneration in a rabbit high-rate DO model (Figure 6). Bilateral osteotomies were performed in the maxilla; distraction devices were activated at a rate of 2.0 mm once daily for 4 days after a 5-day latency period. Twelve rabbits were divided into 2 groups. At the end of distraction, the experimental group of rabbits received an injection of TEOM into the distracted tissue on one side, whereas saline solution was injected into the distracted tissue on the contralateral side (as internal control). An additional control group received an injection of PRP or saline solution into the distracted tissue in the same way as the experimental group. The distraction regenerates were assessed by radiological and histomorphometric analyses. The radiodensity of the distraction gap injected with TEOM was significantly higher than that injected with PRP or saline solution at 2, 3, and 4 weeks postdistraction. The histomorphometric analysis also showed that both new bone zone and bony content in the distraction gap injected with TEOM were significantly increased when compared with PRP or saline solution. These results demonstrated that the distraction gap injected with TEOM showed significant new bone formation. Therefore, injections of TEOM may be able to compensate for insufficient distraction gaps.

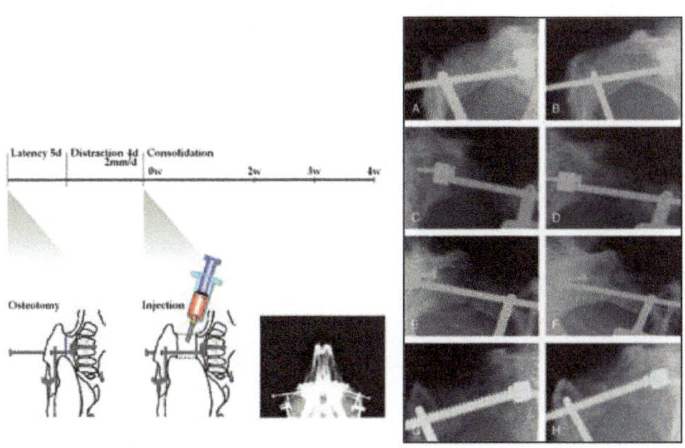

Figure 6. Experimental design (left) and radiographic view of the distracted maxilla (right) A: Two weeks after tissue-engineered osteogenic material (TEOM) injection. B: Four weeks after TEOM injection. C: Two weeks after injection of saline solution. D: Four weeks after saline solution injection. E: Two weeks after platelet-rich plasma (PRP) injection. F: Four weeks after PRP injection. G: Two weeks after injection of saline solution. H: Four weeks after injection of saline solution. (From [9]. Reprinted with permission).

4. Conclusion and future perspective

The research projects presented in this chapter and chapter 8 cover autologous stem cells, intelligent scaffolds, and various cell signaling pathways with regard to tissue engineering and regenerative medicine, from basic principles to clinical applications. Future research should further clarify the importance of functions and mutual relationships of this triad on various levels. In addition to the elements of this triad, efficient tissue regeneration and maintenance of the regenerated tissue require sufficient blood supply and mechanical stress, respectively (Figure 7). These additional elements have been also approached in our ongoing research projects; all combined will produce dramatic advances in our ability to regenerate bony tissues, multilaterally improve the quality of life for all, and provide brighter future perspectives in the field of tissue engineering and regenerative medicine.

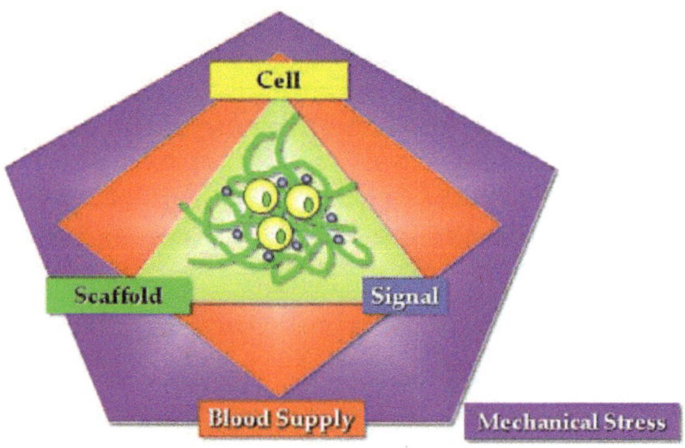

Figure 7. Elements required for efficient tissue regeneration and maintenance of the regenerated tissue

Author details

Hideharu Hibi and Minoru Ueda

Department of Oral & Maxillofacial Surgery, Graduate School of Medicine, Nagoya University, Nagoya, Japan

References

[1] Yamada Y, Nakamura S, Ito K, Kohgo T, Hibi H, Nagasaka T, Ueda M. Injectable tissue-engineered bone using autogenous bone marrow-derived stromal cells for maxil-

lary sinus augmentation: clinical application report from a 2–6-year follow-up. Tissue Engineering Part A 2008;14(10) 1699-1707.

[2] Ueda M, Yamada Y, Kagami H, Hibi H. Injectable bone applied for ridge augmentation and dental implant placement: human progress study. Implant Dentistry 2008;17(1) 82–90.

[3] Ueda M, Yamada Y, Hibi H. Osteoperiosteal tissue-engineered injectable bone. In: The Osteoperiosteal Flap: A simplified approach to alveolar bone reconstruction. Jensen O.T. (Ed.), Illinois, USA: Quintessence Publishing; 2010.

[4] Ito K, Yamada Y, Naiki T, Usami K, Mizuno H, Okada K, Narita Y, Aoki M, Kondo T, Mizuno D, Mase J, Nishiguchi H, Kagami H, Kinoshita K, Hibi H, Nagasaka T, Ueda M. Bone. In: Applied Tissue Engineering, Ueda M., (Ed.), pp.21–45, Rijeka: InTech; 2011.

[5] Katagiri W, Yamada Y, Nakamura S, Ito K, Hara K, Hibi H, Ueda M. Regulation of the Wnt signaling pathways during cell culture of human mesenchymal stem cells for efficient bone regeneration. Oral Science International 2010;7(2) 37-46.

[6] Okamoto Y, Tateishi H, Kinoshita K, Tsuchiya S, Hibi H, Ueda M. An experimental study of bone healing around the titanium screw implants in ovariectomized rats: Enhancement of bone healing by bone marrow stromal cells transplantation. Implant Dentistry 2011;20(3) 236-245.

[7] Hibi H, Ueda M. Supraperiosteal transport distraction osteogenesis. In: The Osteoperiosteal Flap: A simplified approach to alveolar bone reconstruction. Jensen O.T., (Ed.), Illinois, USA: Quintessence Publishing; 2010.

[8] Hibi H, Ueda M. Supraperiosteal transport distraction osteogenesis for reconstructing a segmental defect of the mandible. Journal of Oral and Maxillofacial Surgery 2011;69(3) 742–746.

[9] Kinoshita K, Hibi H, Yamada Y, Ueda M. Promoted new bone formation in maxillary distraction osteogenesis using a tissue-engineered osteogenic material. Journal of Craniofacial Surgery 2008;19(1) 80-87.

[10] Jensen O. T, Laster Z, Hibi H, Yamada Y, Ueda M. Alveolar distraction osteogenesis and tissue engineering. In: Tissue Engineering: Applications in oral and maxillofacial surgery and periodontics, 2nd edition, Lynch S. E, Marx R. E, Nevins M, Wisner-Lynch L. A. (Eds.), Illinois, USA: Quintessence Publishing; 2010.

Chapter 16

System Integration of a Novel Cell Interrogation Platform

Masaru Takeuchi, Gauvain Haulot and
Chih-Ming Ho

1. Introduction

Microfluidics [1, 2] becomes the backbone technology for bio-medical diagnoses and therapeutics due to the length scale matching between cells and fluidic devices. The micro-electro-mechanical-system (MEMS) based molecular sensors have reached the sensitivity of detecting a few nucleic acid molecules and several pg/ml of proteins [3, 4]. During the last decade, integration of microfluidic devices into a system for interacting with cellular system becomes possible and greatly advances the capability of controlling the cell fates.

While the number of microfluidic devices increasing with the desired functionalities of the microfluidic system, large number of interconnections are not just difficult to manufacture and can be a major road block. The Optoelectronic Reconfigurable Microchannel (OERM) [5] has basically alleviated the difficult design and fabrication problem. OERM is a new technology providing light-controlled creation, annihilation and reconfiguration of microchannels within seconds. A light pattern is projected on a reconfigurable chip resulting in the formation of a corresponding microchannel network through a low power optoelectronic effect. When the light pattern is modified, the microchannel network reconfigures accordingly. Manufacturing and reconfiguration of microchannels have been demonstrated in ice, frozen cyclohexane, frozen DMSO and frozen hexadecane. As a consequence, it is expected that the technology can be used with many more materials with reasonable freezing point and latent heat of fusion. Hexadecane is immiscible with water and is widely used in microfluidic emulsions. Its use as a reconfiguration template paves the way towards aqueous droplet manipulation on the platform. Fabrication and reconfiguration happen in seconds.

The re-configurability by defreezing and refreezing fluid certainly provide an innovative way of greatly simplifying the interconnect problem. On the other hand, most cells need to be in fluids above frozen point for proper physiological functions. For example, human cells need to be in 37°C environment. Therefore, an additional microfluidic system operated at 37°C needs to be designed and integrated seamlessly with the OERM system for accomplishing the biological studies. In this article, we will introduce the design and manufacturing methods of the two-temperature fluidic interrogative platform.

2. Opto-electro-reconfigurable-microchannel (OERM)

The first issue with designing OERM are the processes of melting and freezing channels of micron size. The problem of phase transition belongs to the Stefan problem category. It involves a moving boundary that represents the phase change location [6]. Such problems have been extensively studied [7]. Their resolution is often strenuous and requires numerical methods [8, 9]. It is beyond the scope of this chapter. Nevertheless, phase transition can still be investigated without the need of a complete solution. Phase transition mainly affects the time-dependence of the temperature profiles. It has, however, an indirect impact on the steady state that stems from the difference between water and ice thermal conductivities. It creates a discontinuity in the temperature gradient. More precisely, the first law of thermodynamics imposes that the heat flux J(x, t) be continuous. The boundary conditions at the water/ice interface located in $x_0(t)$ with water and ice respectively on the left and right side of the interface.are then $u(x_0(t)) = 0$ and $k_{water}\frac{\partial u}{\partial x}(x_0^-(t))=k_{ice}\frac{\partial u}{\partial x}(x_0^+(t))$. As a result, at steady state, the temperature profile in the ice region is:

for $x\in[x_{0\infty}, L]: u_\infty^{tr}(x)=u_\infty(x)=\frac{J}{k_{ice}}(L-x)+T_c$. T_c is the temperature at x = L and symbolizes the cooler's temperature.

If a heat flux $J=1.2*10^4 W/cm^2$ calculated to impose T = 0°C in x = 100μm is used, the interface will be at $x=x_\infty=100\mu m$: the microchannel is still 100μm deep at steady state. Nonetheless, the temperature gradient in the liquid region is higher since the heat conductivity is lower. The steady state solution for $x\in[x_{0\infty}, L]$ is easily derived, and the complete steady state solution is:

$$u_\infty^{tr}(x)=\frac{J}{k_{water}}(x_{0\infty}-x)\ if\ x\in[0,\ x_{0\infty}[$$

$$u_\infty^{tr}(x)=\frac{J}{k_{ice}}(L-x)+T_c\ if\ x\in[x_{0\infty},\ L]$$

Essentially, the temperature gradient is higher in the liquid region. It leads to higher temperatures than in the case where only ice is considered. For instance it leads to a steady state temperature in x = 0 of value $T=u_\infty^{tr}(0)=2.8°C=276K$

Transient State. The influence of phase transition on the transient state is now addressed. The temperature and heat flux solution of the heat conduction problem are plotted (see Figure 1) to give more insight on the physics that take place. It can be seen that the temperature reaches the melting point after 0.8s. Up to that point a single phase solution to the heat conduction problem is accurate since no phase transition occurs. Figure 1 illustrates that the temperature and heat flux profile when melting begins are already close to steady state. In particular, heat flux is almost constant, which is detrimental to melting. An approximate calculation of the melting time can be performed based on those curves. Since the melting region is small, the problem can be linearized for a first-order approximation. The heat used for melting the 100μm region then results from the difference between the heat flux entering the region and the heat flux leaving the region. The heat flux J entering at x = 0 is constant and has been calculated previously. The heat flux exiting the region at x = 100 μm and t = 0.8s is inferred from the single phase heat conduction problem solution. Their difference equals 280W/m^2. With H the water enthalpy of fusion per unit volume, H = 3.32*10^{8}J.m^{-3}, and l the length of the melting zone, l = 100μm, the first-order approximation for the melting time is:

$$\tau'_{melting} = \frac{Hl}{J - J(x=100\mu m,\ t=0.8s)} = 119s \approx 2min$$

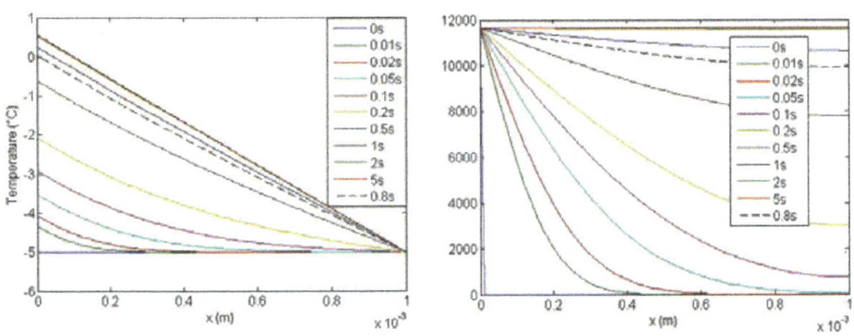

Figure 1. Evolution of temperature and heat flux profiles in the first 5s after heating starts.

Unfortunately, $\tau'_{melting}$ is much longer than the characteristic transient times calculated previously. Because meanwhile the temperature profile will evolve towards homongeneization of the heat flux, it entails that the real melting time will probably be longer. At this point, for better understanding and assessment of the problem, more qualitative insight is required. When melting starts, it creates an interface at the melting temperature that gradually progresses towards its steady state location. The speed of this progression is determined by the coupling of the latent heat of fusion and the derivative of the heat flux across the interface. Because the latent heat is much higher than the heat capacity, phase transition is a much slower process than the increase of temperature. It also imposes a fixed temperature. As a result, phase transition thwarts the heat flux homogenization process of heat conduction across the melting interface. The characteristic transient time for an ice region of 1mm and a water region of 100μm do not depend on the heat flux and are respectively $\tau_{ice}=0.4s$ and $\tau_{ice}=0.03s$. When comparing to $\tau'_{melting}$, it seems a reasonable assumption to consider that steady state is reached in a negligible time with regard to melting, both in the liquid region and the ice region. If the interface is represented by a small interval, the heat flux difference across such interval can be assessed. As a consequence, it turns out that the overall melting time can be calculated rather precisely using the following method. The overall melting region can be discretized in small intervals that each correspond to a value d(t). For each d(t), the steady state profiles in the water and in the ice regions are estimated. The temperature in the liquid region will be described as $u_{water}(x, t) = \frac{J}{k_{water}}(d(t) - x)$ while the temperature in the ice region will be $u_{ice}(x, t) = T_c \frac{L - x}{L - d(t)} - T_c$. Consequently the inward heat flux is J while the outward heat flux is $J_{out}(d(t)) = \frac{k_{ice} T_c}{L - d(t)}$. It entails that the melting time for each interval is $\delta\tau'' = \frac{H\varepsilon}{J - J_{out}(d(t))}$, where ε is the size on an interval. Since the interface progression slows down, the interval can be discretized more finely near the steady state position. Here, $\varepsilon = 10\mu m$ on the first 80μm, then $\varepsilon = 1\mu m$ from x = 80μm to x = 99μm and finally $\varepsilon = 100nm$from x = 99μm to x = 100μm. The overall time calculated through this method gives an estimate of the melting duration of a 100μm channel with a precision of 100nm. It gives

$$\tau''_{melting} = \sum \delta\tau'' = 181s \approx 3min$$

The movement of the solid liquid interface is plotted in Figure 2. When $\tau''_{melting}$ is added to the transient time, namely 2s, it becomes the time that a 100μm channel needs to reach stability with a precision of a hundred nanometers. The times to melt smaller regions with the same heat flux are also noteworthy, even thought the steady state melted region will still be a hundred micrometers long. For instance 50μm are melted in 18s, while 10μm are melted in 3s. Besides, 90μm and 99μm are reached respectively in 56s and 106s. Interestingly, the first approximation $\tau'_{melting}$ = 119s was not that far-fetched in comparison. Moreover, it means that two third of the time are spent melting the last 10μm. Varying the heat flux could therefore shorten melting times tremendously. For instance a high heat flux can first be used to reach the desired size in a short time; it can then be decreased to the value corresponding to the steady state related to that size.

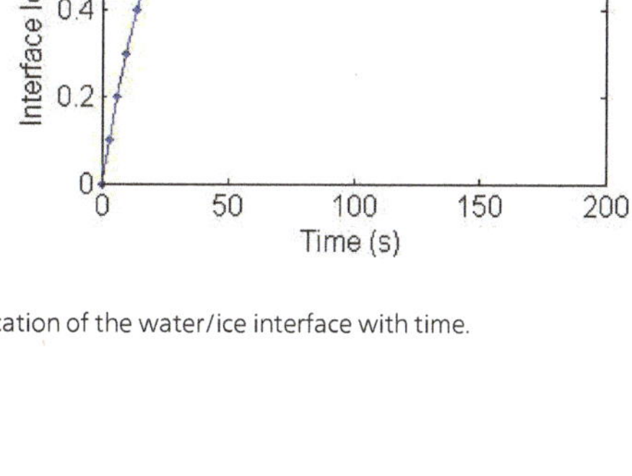

Figure 2. Evolution of the location of the water/ice interface with time.

Finally the case of freezing is also noteworthy. Microchannel freezing indeed happens at the very beginning of the cooling process: the temperature reaches 0°C within 0.1s. As a result, the temperature profile in the ice still imposes a high heat flux through Fourier's law while the heat flux at x = 0 is now zero. Since phase transition blocks the temperature evolution in the ice by fixing the temperature at the freezing point, freezing occurs at a heat flux value close to the highest one in the device: J. The freezing time can therefore be evaluated as $\tau_{ice}=0.4s$. Steady state will be reached two seconds later according to the last section, hence after about 2.5s. The times required to melt a microchannel and freeze a microchannel have very different values; this is due to their timing within the transient state. Melting happens at the end, when the derivative of the heat flux is already close to zero: little heat is available for phase transition. On the contrary freezing occurs at the very beginning of the transient phase and thus benefits from high heat flux derivatives: freezing times are short. This observation bolsters the strategy for fast melting described hereinbefore that consists in using a high heat flux at the beginning, and to decrease its value when the desired microchannel size is obtained. Indeed, melting will start earlier in the transient phase providing higher heat flux gradients for phase transition. Those higher heat flux gradients concentrate across the interface when heat flux homogenizes in the liquid and solid zones. Besides, by reducing heat flux abruptly, the slow convergence of the interface to its steady state position is skipped. Therefore, since varying the heat flux offers schemes for fast melting, this analysis tends to prove that fast reconfiguration can be achieved. Additionally, it clearly proves that melted microchannels will be stable. Finally, the question of flows has not yet been addressed. A fluid entering a channel will thermally interact with its environment. It will probably have a higher temperature and could cause the channel shapes to vary. Nevertheless, since the transient time for small volumes is very short this issue can probably be easily addressed. It should not be a major concern.

Theoretical models and simulations have proven the feasibility of microchannel reconfiguration by local melting. Preliminary experiments were also performed that corroborated the analysis and demonstrated liquid transport in such channels. They involved patterned metal electrodes that would locally thaw ice by resistive heating. Further work has characterized determined the output power of the phenomenon used in OERM, optoelectronic heating [10]. Building on those findings, we dealt with the complete platform for optoelectronic reconfigurable microchannels, the OERM platform. OERM consist in locally melting microchannels in a frozen working media with light patterns. Since it is a reversible process, microchannels can reconfigure when the actuating patterns change. Figure 3 illustrates this principle. The previous paragraphs have reported the power needed for such task: direct melting by illumination would require high power light sources. There are great advantages to low-power light for parallel manipulation, flexibility, and device integration. OERM thus rely on a transduction mechanism where low-power light is converted into high-power heat; it is achieved via optoelectronics and Joule effect. Therefore, such technology is built on the synergetic harnessing of several physical phenomena: light transmission, transduction of light into electrical currents, conversion of electrical currents into heat, melting, and micro flows: hence the complexity of the OERM platform. At its core lies a photoconductive material. Because of its high light absorption and very low thermal conductivity compared to other photoconductive materials such as crystalline silicon, hydrogenated amorphous silicon has proved the best candidate.

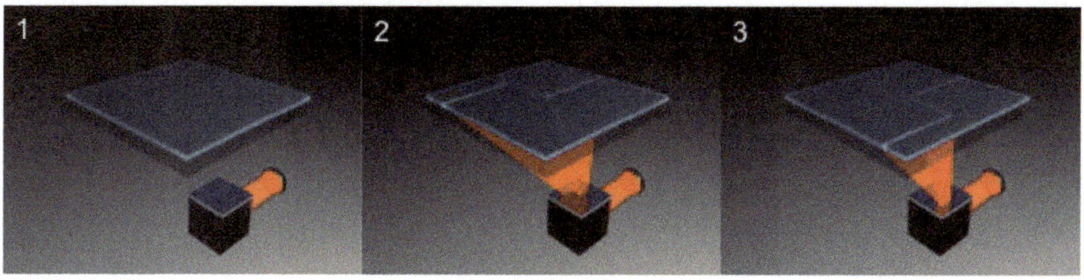

Figure 3. OERM principle: microchannel reconfiguration controlled by light patterns. 1) Working media is frozen. 2) A first light pattern is displayed, which creates a corresponding microchannel; through external pumping green particles (channel end, left-hand side of image) and purple particles (channel first branch, right-hand side of image) flow through. 3) A second light pattern is displayed; the microchannel reconfigures accordingly, and the purple particles (end of new channel) are directed towards another outlet.

OERM is a new microfluidic platform that can provide incomparable flexibility in the planar manipulation of bio-molecules in microchannels. Thanks to this new technology any microfluidic network can be manufactured and reconfigured within seconds. It will be used on the plasmonic sensor platform for cell, drug and sensing probe delivery (cf. Figure 5).

A schematic overview of the optoelectronic reconfigurable microchannel platform is shown in Figure 4. An optoelectronic chip is fabricated by successively depositing on a glass substrate a transparent conductive layer made of indium tin oxide (ITO), a photoconductive layer made of hydrogenated amorphous silicon (a-Si:H), and a highly reflective conductive layer made of titanium, platinum and silver or gold. A liquid layer is laid upon the optoelectronic chip and is confined between the chip and a cover slip. The system is placed on a transparent cooler to freeze the liquid. A voltage bias is then applied between the two conductive layers of the optoelectronic chip resulting in a constant electric field across the photoconductive layer. A light pattern is projected on the optoelectronic chip through the transparent cooler. It is absorbed by the photoconductive layer (a-Si:H). In the dark parts of the light pattern a-Si:H behaves as an insulator, hence no electrical current flows across it. On the contrary in the bright parts of the light pattern local currents flow across a-Si:H. Those currents thus create local Joule heating. This process is called optoelectronic heating and is schematically represented (cf. Figure 4). Setting the local Joule heating to the right power triggers local melting in the frozen liquid, thus forming microchannels corresponding to the light pattern. When light stops shining on a microchannel, local Joule heating stops and the fluid in the microchannel freezes thus closing the microchannel. If the light pattern has reconfigured, other microchannels will open in other locations on the chip. Flows can be produced in the microchannel network using an external pumping system so other phases or particles can be manipulated in the network. The melted liquid is kept as the main phase to allow the melted microchannels to freeze back when light patterns disappear or reconfigure.

The key novel feature of this technology is that the light pattern can be changed at will, hence making the microfluidic network reconfigure instantaneously. This potentially allows for any dynamic two-dimensional design of the microchannel network. The projected pattern is controlled by a computer, which entails that not only can the light pattern be pre-programmed, but also it can evolve on its own thanks to a feedback control loop. This is a real breakthrough for microfluidics and offers a whole new range of possible applications.

Last year a first version of the optoelectronic chip was manufactured that led to the first demonstrations of local melting controlled by low-power light images. The set-up consisting of the electrical controls, the cooling system and the optical projection system was assembled and optimized. It paved the way to this year's accomplishments.

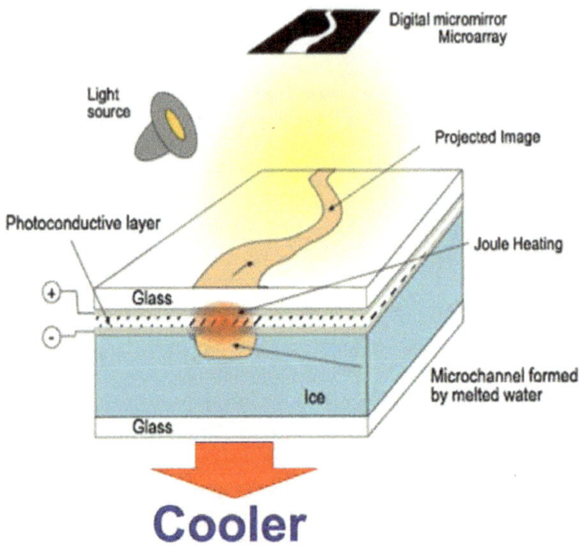

Figure 4. Working principle of OERM

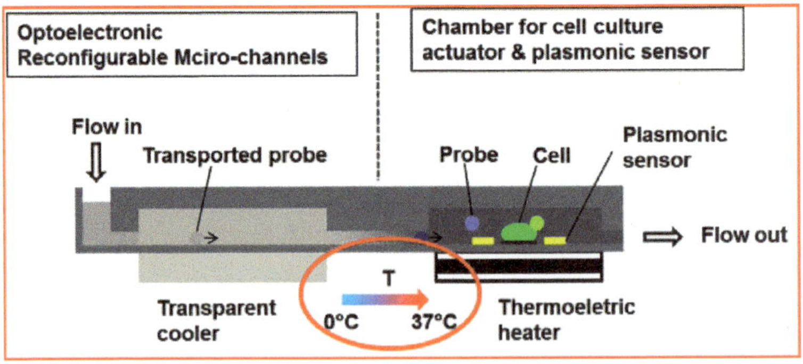

Figure 5. OERM integrated with plamsonic sensor platform

3. Cell interrogation platform

OERM is being integrated in a biosensing platform. The platform's objective is to achieve effective drug screening at the single cell level. The platform has mainly two parts as shown in Figure 5: one is reconfigurable delivery and the other is cell interrogation. In the reconfigurable delivery area, cells, drugs and molecules are prepared and supplied to the cell interrogation area by the reconfigurable microchannel. In the cell interrogation area, cells area attached on the substrate and plasmonic sensors are placed around cells to detect bio-markers from cells under drug stimulation.

To realize the platform, two different temperature conditions have to be realized in the platform. The reconfigurable delivery needs low temperature condition to keep working materials frozen. On the other hand, a high temperature condition (e.g. 37 ◦C) has to be kept in the cell interrogation part to keep cells alive. It is also important to supply drugs from reconfigurable microchannel. The flow from the reconfigurable microchannel has to be controlled to supply drugs to cells efficiently. This section presents mainly these two issues: temperature control and flow control for the cell interrogation platform.

3.1. Temperature control

In the cell interrogation platform, water (or culture media) is used as a working media of the reconfigurable microchannel for cell transportation. It is important to realize the temperature on the reconfigurable area below 0 ◦C to keep ice. On the other hand, temperature has to be kept at 37 ◦C to culture cells in the cell interrogation area. Thermoelectric devices are used to realize these two temperature conditions in the small platform.

At first, two thermoelectric devices area were used to check whether these devices can maintain desired temperatures. Thermoelectric devices are widely used to control temperature below 0 ◦C in a small area such as a few square millimeters [11, 12]. Thermoelectric devices have also been integrated in microfluidic devices to control temperature for microvalves operation [13], making ice in a microchannel [14], and DNA analysis [15]. Figure 6 shows the experimental setup of the thermoelectric devices. The thermoelectric cooler has a heat sink and a fan on the warm side to cool down the warm side and achieve low temperature on the cold side. A glass slide was placed on the thermoelectric devices to check the temperature on the substrate. The temperature on the glass substrate was measured by an infrared thermometer. 8 points as shown in Figure 6 were measured in this experiment. In the experiment, 12 V and 2.3 V were applied to the thermoelectric cooler and heater respectively. Table 1 shows the results. On the cooler, the temperature was below 0 ◦C and frosts grew on the cooler and the glass substrate. On the heater, the temperature showed around 37 °C. The results indicate that the thermoelectric devices can be used to control temperature on the cell interrogation platform.

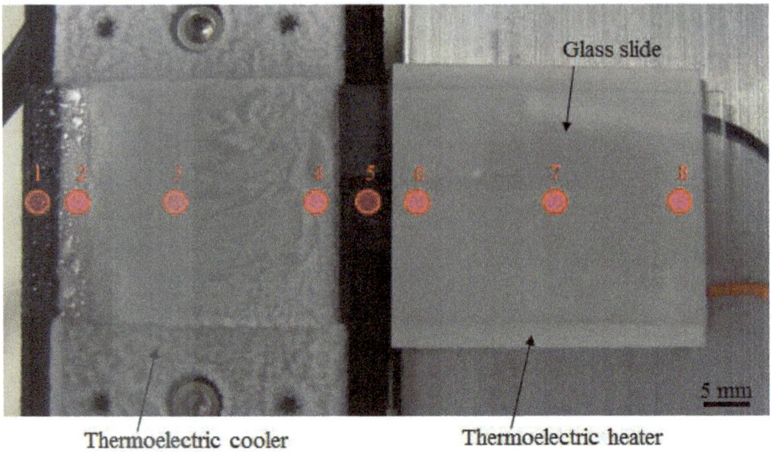

Figure 6. Experimental setup for temperature measurement on the thermoelectric devices

Measurement point	Temperature [◦C]
1	12
2	1
3	-6
4	-4
5	9
6	30
7	38
8	36

Table 1. Temperature measurement on the thermoelectric devices

The temperature variation has to be small to form uniform microchannels in the reconfigurable area. To check the temperature distributions on the platform, the simulation was conducted using a finite element analysis method. At first, the platform was directly put on the two thermoelectric coolers to check the temperature uniformity on the platform in this simple setting. Figure 7 (a) shows the constructed model. Two thermoelectric coolers are placed with some distance to project light patterns onto the platform in between the coolers. The platform is made of glass, and the thermoelectric coolers and room temperature are set to -5 ◦C and 20 ◦C respectively. The thermal conductivity of the glass substrate was set to 0.74 W/(m K) in the simulation. The simulation result is shown in Figure 7 (b). In this condition, the temperature variation on the glass substrate is large and the temperature cannot maintain below 0 ◦C in the reconfigurable area. In this experimental setup, ice cannot be formed on the reconfigurable area.

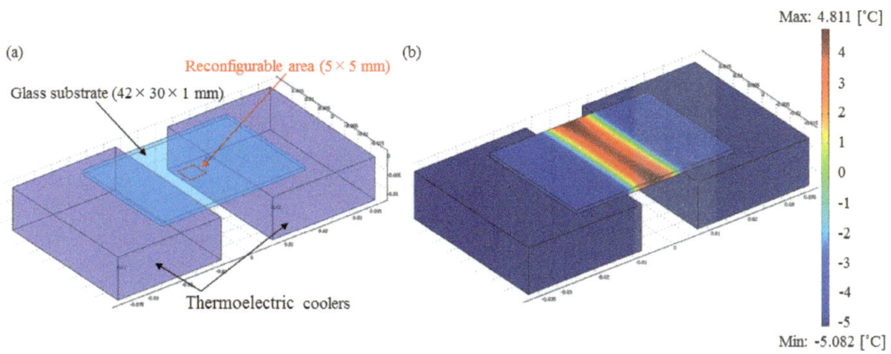

Figure 7. Simulation results when the platform is directly placed on the two thermoelectric cooler (a) A schematic of constructed model for the simulation (b) Simulation result.

To realize low temperature (below 0 ◦C) and small variation of the temperature on the platform, an aluminium plate was placed in between the thermoelectric coolers and the platform. Figure 8 (a) shows the constructed model. The aluminium has high thermal conductivity (225 W/(m K)). The aluminium plate can be cooled by the thermoelectric coolers precisely and fast. The aluminium plate has a 7 mm × 7 mm square hole to project light images from underneath onto the reconfigurable area. The platform can be cooled strongly because the bottom face of the platform can be attached to the aluminium plate outside of the reconfigurable area. As shown in Figure 8 (b) and (c), the temperature in the reconfigurable area was maintained below 0 ◦C and the temperature variation was less than 0.7 ◦C. This result indicates that the aluminium plate is effective for cooling.

For practical uses of this cooling system, condensation of the vapour is one of the significant problems. Light patterns from the DMD device come from underneath of the reconfigurable area. If there is condensation water on the platform, the light pattern is refracted by the condensation and the light pattern is changed. To prevent condensation on the platform, an acrylic bar is placed on the path of the light pattern from the DMD device to the reconfigurable area. Acrylic materials have low thermal conductivity (0.19 W/(m K)) and it is considered that the bottom face of the acrylic bar can be maintained close to the room temperature even if the top face is under 0 °C provided that the acrylic bar is long enough. In the simulation, the acrylic bar is placed in and under the hole of aluminium plate as shown in Figure 9 (a). The height of the acrylic bar was set on 35 mm.

Figure 9 (b) and (c) shows the simulation results. The temperature in the reconfigurable area is kept under 0 °C and the temperature variation decreased to less than 0.3 °C. On the other hand, the bottom face of the acrylic bar is more than 18 °C. The results show that the uniform low temperature in the reconfigure area can be realized and condensation under the reconfigurable area can be prevented by putting the acrylic bar under the reconfigurable area.

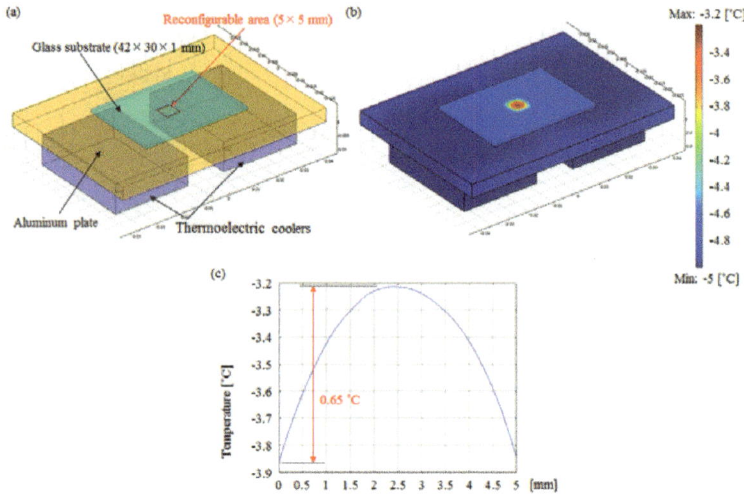

Figure 8. Simulation results when an aluminium plate was paced in between the cooler and platform (a) A schematic of constructed model (b) Simulation result (c) Temperature variation in the reconfigurable area (temperature on the red dash line in (b))

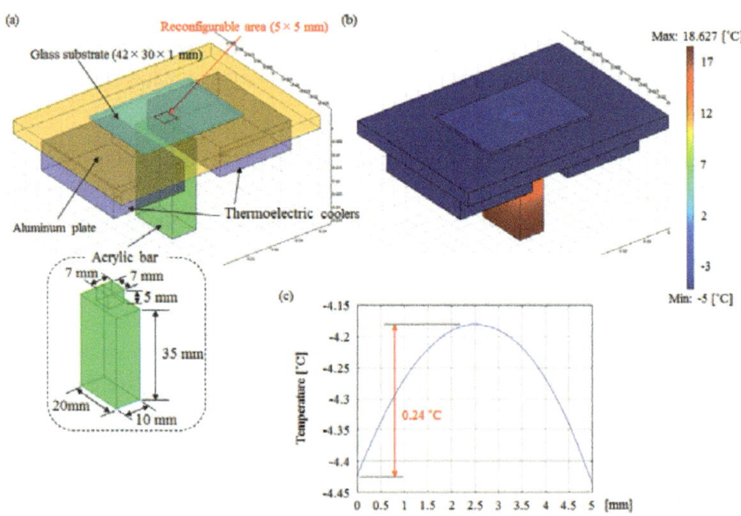

Figure 9. Simulation results when an acrylic bar was paced under the reconfigurable area (a) A schematic of constructed model (b) Simulation result (c) Temperature variation in the reconfigurable area (temperature on the red dash line in (b))

A thermoelectric heater is added in the model to realize cell interrogation area in the platform. Fig. 10 (a) shows the constructed model. the thermoelectric heater and coolers are set on 37 °C and -5 °C respectively. The distance between the coolers and heater were set on 10 mm.

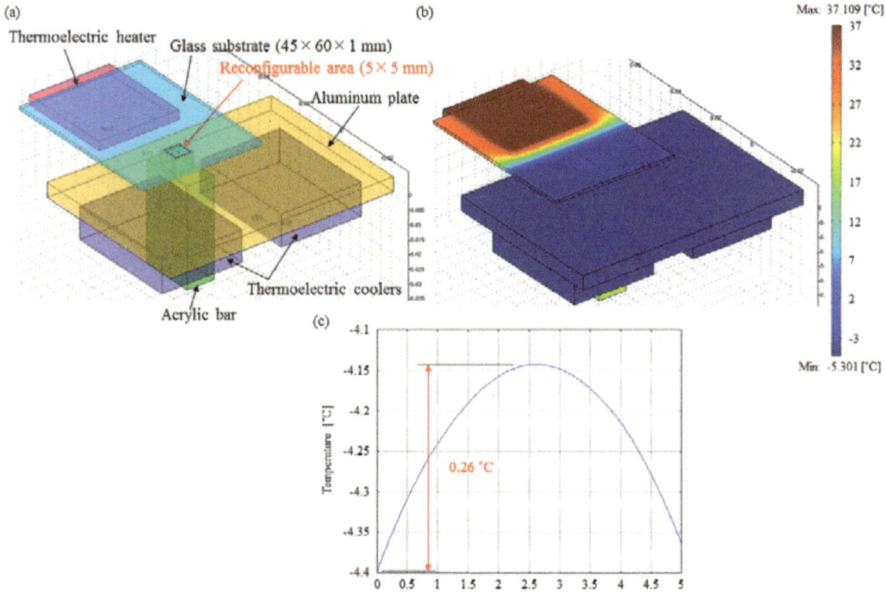

Figure 10. Simulation results when the thermoelectric coolers and heater are integrated for the platform. (a) A schematic of constructed model (b) Simulation result (c) Temperature variation in the reconfigurable area (temperature on the red line in (b))

The simulation results shows that the reconfigurable area is kept under 0 °C even if the heater is added.

3.2. Flow control

To realize the cell interrogation platform as shown in Figure 5, cells, drugs and molecules for cell interrogation have to be supplied from reconfigurable delivery area. To supply cells and drugs to the cell interrogation area efficiently, the spread angle of the flow from the reconfigurable area has to be small. To check the relationship in between the flow rate and the spread angle, a polydimetylsiloxane (PDMS) microchannel was used. The cross-sectional shape of the microchannel is about 200 μm×100 μm. In the experiment, flow rate was changed from 700 μl/min to 1000 μl/min by a syringe pump. To check the spread angle, blue dyed water flowed out from the microchannel and the outlet of the microchannel was observed under a microscope.

Figure 11 shows the experimental results. Blue-dyed water flowed out from the PDMS microchannel to the open space. The spread angle decreased when the flow rate increased, which reached 5 ° when the flow rate was set on 1000 μl/min. It is considered that the spread angle is small enough to supply cells and drugs to the target position in the platform.

To check the spread of cells by discharging from the microchannel, RAW 264.7 macrophage cells were used. Cell analysis has been achieved in microfluidic devices previously [16]. The glass substrate was coated by poly-L-lysine to attach cells on the substrate before experiment [18]. Polylysine is commonly used to improve cell adhesion on substrates and can be used to make arrays of cells and hydrogels for biological application [18, 19]. In the experiment, the cells were discharged from the microchannel in 5 minutes by the flow rate of 1000 μl/min. Figure 12 shows the experimental result. The cells were discharged from the microchannel and spread on the substrate. The spread angle of cells was large compared to the flow experiment as shown in Figure 3.7. When the syringe pump was stopped, the flow was not suddenly stopped but gradually slowed down. During the slowdown of the flow speed, the cells were spread throughout the substrate. It is expected that the cells not attached to the substrate can be flush out by a solution without cells when the patterning of the substrate is conducted to attach cells partially before experiment.

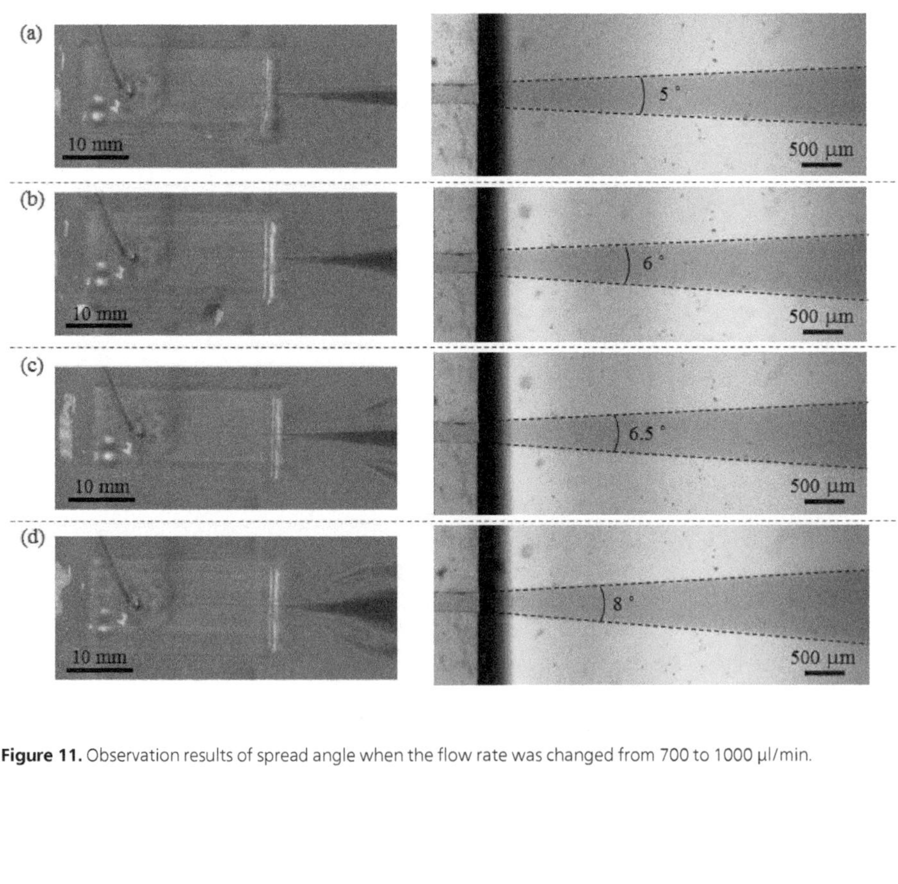

Figure 11. Observation results of spread angle when the flow rate was changed from 700 to 1000 μl/min.

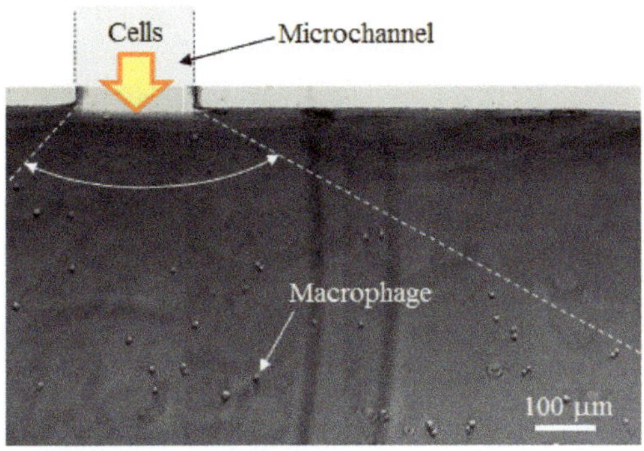

Figure 12. Discharge of macrophages from the microchannel

4. Conclusion

This chapter shows theoretical analysis about phase transition in the OERM device and an application for a biosensing platform. The OERM platform is currently being integrated within a cell interrogation platform. Temperature and flow control are conducted to realize the platform. The platform will realize sensing protocols automatically for lab-on-the-chip analysis applied to disease therapies.

Acknowledgements

This project is supported by NSF nano manufacture center SINAM (DMI-0327077)

Author details

Masaru Takeuchi[1], Gauvain Haulot[2] and Chih-Ming Ho[3]

1 Graduate School of Engineering, Nagoya University, Nagoya, Japan

2 Henry Samueli School of Engineering and Applied Science, University of California, Los Angeles, USA

3 Department of Mechanical & Aerospace Engineering, Department of Bioengineering, University of California, Los Angeles, USA

References

[1] Liu J, Tai Y. C, Pong K, Ho C. M. Micromachined Channel/Pressure Sensor Systems for Micro Flow Studies. June 1993, International Conference on Solid State Sensors and Actuators. 1993; 995-997.

[2] Ho C. M, Tai Y. C. Micro-Electro-Mechanical-Systems and Fluid Flows. Annual Review of Fluid Mechanics 1998;30 579-612.

[3] Lillehoj P. B, Wei F, Ho C. M. A self-pumping lab-on-a-chip for rapid detection of botulinum toxin. Lab on a chip 2010;10(17) 2265-70.

[4] Wei F, Lillehoj P. B, Ho C. M. DNA Diagnostics: Nanotechnology-Enhanced Electrochemical Detection of Nucleic Acids. Pediatric Research 2010;67(5) 458-468.

[5] Haulot G, Benahmed A. J, Ho C. M. Optoelectronic reconfigurable microchannels. In: Prodeedings of IEEE 24th International Conference on Micro Electro Mechanical Systems (MEMS). 23-27 January 2011.

[6] Bonnerot R, Jamet P. Numerical computation of the free boundary for the two-dimensional Stefan problem by space-time finite elements. Journal of Computational Physics 1977;25(2) 163-181.

[7] Crowley A. B. Numerical solution of stefan problems, International Journal of Heat and Mass Transfer 1978;21(2) 215-219.

[8] Ciavaldini J. F. Analyse Numerique D'un Probleme de Stefan A Deux Phases Par Une Methode D'elements Finis. SIAM Journal on Numerical Analysis 1975;12(3) 464-487.

[9] Chen S, Merriman B, Osher S, Smereka P. A Simple Level Set Method for Solving Stefan Problems, Journal of Computational Physics 1997;135(1) 8-29.

[10] Haulot G, Ho C. M. Optoelectronic Heating for Fabricating Microfluidic Circuitry. Advances in Opto Electronics 2011;2011(237026).

[11] Walle B. L, Gauthier M, Chaillet N. Principle of a Submerged Freeze Gripper for Microassembly. IEEE Transactions on Robotics 2008;24(4) 897-902.

[12] Liu J, Zhou Y. X, Yu T. H. Freeze tweezer to manipulate mini/micro objects. Journal of Micromechanics and Microengineering 2004;14 269-276.

[13] Zimmermann M, Bentley S, Schmid H, Hunziker P, Delamarche E. Continuous flow in open microfluidics using controlled evaporation. Lab on a Chip 2005;5 1355-1359.

[14] Li Y, Wang F, Wanga H. Cell death along single microfluidic channel after freeze-thaw treatments. Biomicrofluidics 2010;4 014111.

[15] Khandurina J, McKnight T. E, Jacobson S. C, Waters L. C, Foote R. S, Ramsey J. M. Integrated System for Rapid PCR-Based DNA Analysis in Microfluidic Devices. Analytical Chemistry 2000;72 2995-3000.

[16] Amatore C, Arbault S, Chen Y, Crozatier C, Tapsoba I. Electrochemical detection in a microfluidic device of oxidative stress generated by macrophage cells. Lab on a Chip 2007;7 233-238.

[17] Nilforoushan D, Gramoun A, Glogauer M, Manolson M. F. Nitric oxide enhances osteoclastogenesis possibly by mediating cell fusion. Nitric Oxide 2009;21 27-36.

[18] Chang W. C, Sretavan D. W. Novel High-Resolution Micropatterning for Neuron Culture Using Polylysine Adsorption on a Cell Repellant, Plasma-Polymerized Background. Langmuir 2008;24 13048-13057.

[19] Zhang H, Shepherdb J. N. H, Nuzzo R. G. Microfluidic contact printing: a versatile printing platform for patterning biomolecules on hydrogel substrates. Soft Matter 2010;6 2238-2245.

Chapter 17

Transferrin-Toxin Conjugates for Cancer

Daniel T. Kamei

Abstract

For several years, researchers have targeted receptors overexpressed on cancer cells to improve the selectivity of toxic anticancer agents. In addition to conjugating the toxin to an antibody for such a receptor, the natural ligand for the overexpressed receptor has also been used as a protein-based drug delivery vehicle. The therapeutic efficacy of such ligand-drug molecular conjugates, however, may be limited, since they naturally follow the intracellular trafficking pathways of the endogenous ligands, which have certainly not been optimized by nature for drug delivery efficacy. Accordingly, novel design criteria may be identified for these ligands through an understanding of their intracellular trafficking pathways. This short review briefly describes the transferrin ligand/transferrin receptor system, where intracellular trafficking considerations have led to improvements in the therapeutic efficacy of transferrin-toxin molecular conjugates.

1. Introduction

Cancer is the second leading cause of death in the United States [1, 2]. Unfortunately, current treatments involve invasive surgery followed by nonspecific radiation and chemotherapy that harm both healthy and cancer cells. Accordingly, much research has been dedicated to improving the tumor selectivity of chemotherapy treatments. Since the early 1980s, many receptors were found to have increased expression in cancer cells compared to their normal counterparts. These include transferrin receptor, interleukin-13 receptor, and growth factor receptors [3-7]. These receptors have been investigated for several years as cancer-selective targets for therapeutic purposes.

An understanding of the trafficking pathway of the natural ligand can lead to novel design criteria for engineering the ligand to be a more effective drug carrier [8-12]. This short review focuses on transferrin ligand-toxin molecular conjugates with particular emphasis given to the intracellular trafficking properties of the ligand. Moreover, although transferrin has been conjugated to liposomes [13] and nanoparticles [14] to enhance their targeting to cancer cells, they are not included in this review, since the intracellular trafficking properties of these drug delivery vehicles can vary from that of the ligand alone due to the large differences in size.

Human serum transferrin (Tf) has been investigated for several years as a targeting agent due to the overexpression of its receptor on cancer cells. Tf has a molecular weight of approximately 80 kDa and is responsible for transporting free iron from the circulation to cells. Each Tf molecule has the capability to bind to two ferric (Fe^{3+}) ions, one in the N-terminal lobe (N-lobe) and the other in the C-terminal lobe (C-lobe). Each lobe binds to a ferric ion with an equilibrium dissociation constant (K_D) of approximately 10^{-22} M [15, 16]. This iron-bound Tf, or holo-Tf, then binds to its cell-surface receptor (TfR). After binding, The Tf/TfR complex is internalized, and holo-Tf delivers its iron to the cell, promoting cellular growth and proliferation [17]. Since cancer cells require more iron to sustain their rapid proliferation, they have been found to overexpress TfR, and this high expression level of TfR has been exploited to achieve selective targeting of anticancer agents to cancer cells.

2. Intracellular trafficking

The intracellular trafficking pathway of Tf and its receptor has been studied for several years and has been reviewed in many journals [18-20]. After holo-Tf binds to TfR on the surface of cells with nanomolar affinity ($K_D \sim 10^{-9}$ M), the Tf/TfR complex is internalized as part of an endocytic vesicle. The endosome then matures and acidifies to a pH between 5 and 6 [21], causing iron to be released from Tf. Once iron is released, it is reduced to the ferrous (Fe^{2+}) form due to the presence of oxidoreductases in the endosome. A divalent metal transporter then shuttles Fe^{2+} into the cytosol. The iron-free Tf/TfR complex recycles back to the cell surface, and since iron-free Tf (apo-Tf) has a low binding affinity for TfR at the near neutral pH of the cell surface, apo-Tf quickly dissociates from the cell-surface receptor. This entire cycle of the Tf/TfR trafficking pathway lasts only about 5 minutes [22].

While the rapid recycling of Tf contributes to the efficient transport of iron, it also limits the ability of the ligand to deliver its payload. Accordingly, drug delivery efficacy may be improved through a better understanding of the kinetics involved in the intracellular trafficking pathway of Tf [23]. Through *in vitro* experiments and mathematical modeling, Murphy and coworkers previously demonstrated that monoclonal anti-transferrin receptor antibodies would be more effective drug delivery vehicles if they associated with cells for a greater period of time [24-26]. In other words, this increased cellular association would lead to an increase in the exposure of cancer cells to the conjugated drug. Instead of investigating antibodies for TfR, the Kamei laboratory studied the Tf ligand itself. Specifically, by deriving and analyzing a mathematical model for Tf/TfR trafficking, Kamei and coworkers identified a novel design criterion for engineering Tf to enhance its drug delivery efficacy [8-10], as discussed below.

3. Transferrin-diphtheria toxin conjugates

Diphtheria toxin (DT) is a protein toxin secreted by *Corynebacterium diphtheria*. DT acts by inhibiting protein synthesis through inactivation of elongation factor 2 (EF-2) [27]. Tf conjugated to a mutant of DT, known as CRM107, has been effective against malignant gliomas [28]. The mutation in CRM107 significantly inhibits the binding of the toxin to its native receptor, a heparin-binding epidermal growth factor-like growth factor precursor, thereby reducing the nonspecific toxicity of Tf-CRM107 [29]. Youle and Oldfield's group performed *in vivo* studies using Tf-CRM107 on solid human gliomas in the flanks of nude mice, and observed increased cytotoxicity exerted by the conjugate. The success of Tf-CRM107 eventually led to phase III clinical trials, which were unfortunately canceled in late 2006 following the results of a conditional power analysis suggesting that its efficacy would not significantly improve upon the current standard-of-care treatments.

As mentioned above, Tf recycles very rapidly, and this short duration inside the cell can limit the ability of Tf to deliver its cytotoxic payload. Therefore, to identify new approaches to improving the efficacy of Tf-CRM107, Kamei and coworkers used mass action kinetics to derive a mathematical model of the Tf/TfR trafficking pathway. Analysis of the model helped determine that, by reducing or inhibiting the iron release rate of Tf within the endosome, its drug delivery efficacy would be significantly improved [8]. In this scenario, iron would be retained by Tf upon recycling to the cell surface and the conjugate would be reinternalized to participate in another cycle of trafficking due to the preserved high affinity of holo-Tf for TfR, increasing its cellular association. The drug delivery efficacy of Tf can therefore be improved, since a single Tf-drug conjugate would undergo multiple trafficking cycles, thereby increasing the probability of delivering the drug. Kamei and coworkers engineered two Tf variants that satisfied the molecular level design criterion using site-directed mutagenesis [9]. These Tf mutants were conjugated to DT and were shown to be more effective than wild-type Tf in delivering DT *in vitro* to U87 and U251 human glioma cell lines [10]. Furthermore, Kamei and coworkers performed *in vivo* experiments, and demonstrated that both mutant Tf-DT conjugates were more effective than their wild-type counterpart in shrinking glioma tumors on the flanks of mice [10]. Studies are currently being performed with these mutant Tf molecules conjugated to CRM107.

4. Concluding remarks

Ligand-toxin molecular conjugates for cancer have been studied for several years. Though these conjugates have demonstrated some success, not many have obtained FDA approval for the treatment of cancer. The general lack of FDA-approved ligand-toxin conjugates may be attributed to the physiological behavior of the targeting ligand. Although ligands can effectively target cancer cells via their cell-surface receptors, they will follow the ligand's physiological pathway once bound to receptors. For instance, the rapid recycling rate of Tf which aids in iron delivery can restrict the ability to deliver a conjugated toxin [8]. Through quantitative experiments and modeling, a more thorough understanding of the intracellular trafficking properties of various ligands can be achieved. The Kamei laboratory has used this approach to find an innovative approach to manipulating the Tf trafficking pathway for the advancement of Tf-based cancer therapeutics [8, 9]. This example demonstrates that a systems analysis of intracellular trafficking properties can result in the engineering of effective targeting agents to cancers.

Author details

Daniel T. Kamei

Department of Bioengineering, University of California, Los Angeles, USA

References

[1] Minino A. M, Xu J, Kochanek K. D, Tejada-Vera B. Death in the United States, 2007, NCHS data brief; 2009.

[2] Jemal A, Siegel R, Xu J, Ward E. Cancer statistics, 2010, CA: a cancer journal for clinicians 2010;60 277-300.

[3] Mendelsohn J. Targeting the epidermal growth factor receptor for cancer therapy, Journal of Clinical Oncology 2002;20 1S-13S.

[4] Kobrin M. S, Yamanaka Y, Friess H, Lopez M. E, Korc M. Aberrant expression of type I fibroblast growth factor receptor in human pancreatic adenocarcinomas. Cancer research 1993;53 4741-4744.

[5] Low P. S, Kularatne S. A. Folate-targeted therapeutic and imaging agents for cancer. Current opinion in chemical biology 2009;13 256-262.

[6] Murata T, Obiri N. I, Debinski W, Puri R. K. Structure of IL-13 receptor: analysis of subunit composition in cancer and immune cells. Biochemical and biophysical research communications 1997;238 90-94.

[7] Recht L, Torres C. O, Smith T. W, Raso V, Griffin T. W. Transferrin receptor in normal and neoplastic brain tissue: implications for brain-tumor immunotherapy. Journal of neurosurgery 1990;72 941-945.

[8] Lao B. J, Tsai W. L, Mashayekhi F, Pham E. A, Mason A. B, Kamei D. T. Inhibition of transferrin iron release increases in vitro drug carrier efficacy. Journal of Controlled Release 2007;117 403-412.

[9] Yoon D. J, Chu D. S, Ng C. W, Pham E. A, Mason A. B, Hudson D. M, Smith V. C, MacGillivray R. T, Kamei D. T. Genetically engineering transferrin to improve its in vitro ability to deliver cytotoxins. Journal of Controlled Release 2009;133 78-184.

[10] Yoon D. J, Kwan B. H, Chao F. C, Nicolaides T. P, Phillips J. J, Lam G.Y, Mason A. B, Weiss W. A, Kamei D. T. Intratumoral therapy of glioblastoma multiforme using genetically engineered transferrin for drug delivery. Cancer research 2010;70 4520-4527.

[11] Lu Y, Low P. S. Folate targeting of haptens to cancer cell surfaces mediates immunotherapy of syngeneic murine tumors. Cancer Immunol Immunother 2002;51 153-162.

[12] Lu Y, Sega E, Low P. S. Folate receptor-targeted immunotherapy: induction of humoral and cellular immunity against hapten-decorated cancer cells. International journal of cancer 2005;116 710-719.

[13] Shmeeda H, Mak L, Tzemach D, Astrahan P, Tarshish M, Gabizon A. Intracellular uptake and intracavitary targeting of folate-conjugated liposomes in a mouse lymphoma model with up-regulated folate receptors. Molecular cancer therapeutics 2006;5 818-824.

[14] Davis M. E, Zuckerman J. E, Choi C. H, Seligson D, Tolcher A, Alabi C. A, Yen Y, Heidel J. D, Ribas A. Evidence of RNAi in humans from systemically administered siRNA via targeted nanoparticles. Nature 2010;464 1067-1070.

[15] Aisen P, Leibman A, Zweier J. Stoichiometric and site characteristics of the binding of iron to human transferrin. The Journal of biological chemistry 1978;253 1930-1937.

[16] Baker E. N, Baker H. M, Kidd R. D. Lactoferrin and transferrin: functional variations on a common structural framework. Biochemistry and cell biology = Biochimie et biologie cellulaire 2002;80 27-34.

[17] Cazzola M, Bergamaschi G, Dezza L, Arosio P. Manipulations of cellular iron metabolism for modulating normal and malignant cell proliferation: achievements and prospects. Blood 1990;75 1903-1919.

[18] Li H, Sun H, Qian Z. M. The role of the transferrin-transferrin-receptor system in drug delivery and targeting. Trends in pharmacological sciences 2002;23 206-209.

[19] Li H, Qian Z. M. Transferrin/transferrin receptor-mediated drug delivery. Medicinal research reviews 2002;22 225-250.

[20] Ponka P, Lok C. N. The transferrin receptor: role in health and disease. The international journal of biochemistry & cell biology 1999;31 1111-1137.

[21] Ciechanover A, Schwartz A. L, Dautry-Varsat A, Lodish H. F. Kinetics of internalization and recycling of transferrin and the transferrin receptor in a human hepatoma cell line. Effect of lysosomotropic agents. The Journal of biological chemistry 1983; 258 9681-9689.

[22] Klausner R. D, Van Renswoude J, Ashwell G, Kempf C, Schechter A. N, Dean A, Bridges K. R. Receptor-mediated endocytosis of transferrin in K562 cells. The Journal of biological chemistry 1983;258 4715-4724.

[23] Lao B. J, Kamei D. T. Improving therapeutic properties of protein drugs through alteration of intracellular trafficking pathways. Biotechnology progress 2008;24 2-7.

[24] Yazdi P. T, Murphy R. M. Quantitative analysis of protein synthesis inhibition by transferrin-toxin conjugates. Cancer research 1994;54 6387-6394.

[25] Wenning L. A, Yazdi P. T, Murphy R. M. Quantitative analysis of protein synthesis inhibition and recovery in CRM107 immunotoxin-treated HeLa cells. Biotechnology and bioengineering 1998;57 484-496.

[26] Yazdi P. T, Wenning L. A, Murphy R. M. Influence of cellular trafficking on protein synthesis inhibition of immunotoxins directed against the transferrin receptor. Cancer research 1995;55 3763-3771.

[27] Greenfield L, Johnson V. G, Youle R. J. Mutations in diphtheria toxin separate binding from entry and amplify immunotoxin selectivity. Science 1987;238 536-539.

[28] Weaver M, Laske D. W. Transferrin receptor ligand-targeted toxin conjugate (Tf-CRM107) for therapy of malignant gliomas. Journal of neuro-oncology 2003;65 3-13.

[29] Naglich J. G, Metherall J. E, Russell D. W, Eidels L. Expression cloning of a diphtheria toxin receptor: identity with a heparin-binding EGF-like growth factor precursor. Cell 1992;69 1051-1061.

Chapter 18

MEMS Sensors and Their Applications

Mitsuhiro Shikida

1. Introduction

Micro-electro-mechanical systems (MEMS) technology has opened the doorway to miniaturizing and integrating various mechanical and electrical components on the same chip. The principal fabrication technologies for producing MEMS devices, such as anisotropic wet etching for single crystal silicon, sacrificial etching for producing mechanical elements, and anodic bonding for the packaging, were invented in the 1960s. These technologies have been used for producing various miniaturized sensors from the 1970s onwards. Silicon diaphragm structures produced by anisotropic wet etching are used as sensing elements for pressure sensors. An ion sensitive field-effective-transistor was developed as a chemical sensor. MEMS sensors started to be integrated on silicon wafers in the middle of the 1970s. A pressure sensor and its electrical circuits are integrated on the same silicon wafer in order to reduce the parasitic capacitance and footprint. During this time, a miniaturized gas chromatograph for use in aerospace was fabricated on a 2-inch silicon wafer; it integrated a column, valve seat, and sensing heater. Moreover, a mass-flow controller driven by piezo-electric actuators for precisely regulating gas flow in a pipe was developed in the late 1980s. Since the 1990s, MEMS technologies have spread to many different fields, and these days there are miniaturized sensors, for example, for acceleration, gyroscopic, flow, infrared, and tactile sensors.

1.1. Attractive features of MEMS sensors

MEMS technologies can produce miniaturized sensors a few mm in size that offer the following very useful features for industry.

1. Lower costs: MEMS sensors are normally produced in batches like semiconductor chips; as a result, they offer lower fabrication costs.

2. High space and time resolution: MEMS sensors are typically measure less than a few mm. This means they can have very small footprints, which in turn raises the spatial resolution. The volume reduction afforded by them can be used to increase the time resolution of

thermal sensor applications. This is because the heat capacity decreases as the sensor volume decreases.

3. Small parasitic capacitance and dead space: MEMS sensors can be integrated with electrical circuits on the same silicon wafer, and thus, one can reduce the parasitic capacitance in the sensing circuit. MEMS technologies can integrate different components, for example, sensors and actuators, on the same wafer to build up the system, and thus, they make it possible to reduce the dead space in the system.

4. 2D analysis: MEMS sensors can be integrated on silicon wafer in the form of an array, and thus, one can analyze the distribution of the physical quantities, for example, pressure and force, acting on a plane with high spatial resolution.

1.2. Difficulties in developing MEMS sensors

Despite the above-mentioned advantages, MEMS sensors face a number of difficulties in regard to their continued development.

1. Large capital investment: The MEMS technologies, especially in those used in the fabrication process, are based on semiconductor technologies that have been developed for producing 3D microstructures from silicon wafers. A large amount of capital investment is generally required in the fabrication of MEMS sensors.

2. Variations in the fabrication process: MEMS sensors typically have cantilever, diaphragm, or comb structures. They are suspended and actuated to detect physical quantities. Many technologies are required for producing such structures.

    a. Sacrificial etching is used to make movable elements on a silicon wafer. Parts of the deposited layers are selectively etched away. This etching is normally done by isotropic wet etching, and it is especially used to produce comb-drive actuators. Sticking between the released movable element and the substrate underneath it is sometimes a problem, but it can be solved by using a specialized drying method and thin film deposition.

    b. Anisotropic wet etching is used to produce diaphragm and cantilever structures. The micro-structures are formed by orientation-dependent etching of single crystal silicon.

    c. Dry etching is normally used to make comb-drive actuators. To produce the designed structure precisely, one has to overcome loading and notch effects.

    d. Variation in packaging is an issue since each MEMS sensor requires an order-made packaging process. For instance, tactile force and flow velocity sensors have to be exposed to their environment in order to detect physical and chemical quantities of interest. On the other hand, acceleration and gyro sensors are completely encapsulated in a shell structure. The pressure inside the shell is kept at a vacuum level in order to increase the Q-factor at the resonant frequency.

   e. Compatibility of specialized fabrication processes, such those described in (a) to (c), with semiconductor electrical circuit fabrication processes has to be carefully considered when MEMS sensors are to be integrated with electrical circuits.

3. Multi-disciplinary know-how: MEMS sensors' development requires multi-physics knowledge. For example, developers of acceleration sensors need a thorough knowledge of the following areas:

   a. Mechanics for designing an adequate deformation vs. force relation and ensuring reliability,

   b. Vibration analysis for designing the resonant frequency of the device,

   c. Electrical circuit analysis for designing the sensing circuits and electrical properties (capacitance and resistance) of the sensing structures, and

   d. Fabrication technology for producing the sensors.

   The multi-physics approach is taken because the above points are related to each other.

4. Scaling effects: The sizes of structural elements such as cantilevers and diaphragms are on the order of 10 micrometers. Such a level of miniaturization means that forces that are proportional to the area of the structural element dominate the motion of the sensor. Other forces, such as magnetic and inertial forces which are proportional to the volume of elements, are greatly reduced as a result of miniaturization. The effects of the electrostatic force and surface tension have to be carefully considered when designing MEMS sensors.

1.3. Applications of MEMS sensors

MEMS technologies have a wide variety of applications, and there are many types (examples are shown in Table 1). In the following sections, tactile sensors for human interfaces and flow sensors for air-conditioning systems and medical applications will be focused.

2. Tactile sensors for human interfaces

The single crystal silicon is often used as a MEMS material because it is a high-quality semiconductor with excellent mechanical properties [1]. It is said that the yield strength of silicon is comparable to that of stainless steel. Silicon sensors for gyroscopes and for measuring pressure, acceleration, flow, and inclination have already been commercialized and are being used in the automotive and information industries [2-5]. Silicon has also been used to make miniaturized tactile sensors as interface devices in robotic automation systems since in the middle of the 1980s. Various tactile sensors based on silicon MEMS technologies have been developed [6-9]. In particular, Takao et al. integrated a tactile sensor with a CMOS device on the same chip [10]. Moreover, Shikida et al. proposed to integrate an actuator mechanism onto a tactile sensor as means of measuring the hardness of objects [11-13].

Name	Detection principle	Application
Pressure sensor	Capacitance, Piezo-resistance	Automotive (fuel, tire) Medical care (Blood pressure)
Acceleration sensor	Capacitance, Piezo-resistance	Automotive Information (Projector) Communication (Portable phone) Amusement
Gyro sensor	Capacitance	Automotive Information (Video camera)
Tactile sensor	Capacitance, Piezo-resistance	Communication (Robot systems)
Flow sensor	Heat transfer	Automotive (fuel) Production (Gas control in system) Medical equipment
Shear stress sensor	Heat transfer, Optical, Capacitance	Air-plane
IR sensor	Diode, Thermoelectric power	Automotive
Chemical sensor	FET, Resonance	Biochemical

Table 1. MEMS sensor applications

MEMS sensors are being used to study human activity by embedding them in clothes and everyday consumer products. Silicon tactile sensors have high space and time resolutions. However, it is difficult to make them wearable because silicon is so brittle. To overcome this problem, MEMS researchers have started using resin materials to fabricate flexible sensor structures [14-17]. Notably, Konishi et al. and N. Chen et al. devised knitted structures for electrical wiring [18, 19]. These structures have advantages whereby they can fit on arbitrary bendable surfaces. Shikida et al. proposed an artificial hollow fiber as a MEMS material to implement a tactile sensing function on a fabric and developed a tactile sensor by weaving together fibers [20-23]. They studied the unit's ability to detect force, cross-talk, etc., by applying concentrated loads to it.

The following sections focus on silicon-based and fabric tactile sensors.

2.1. Si-based tactile sensors

A schematic view of a magnetically driven tactile sensor is shown in Figure 1. It consists of a diaphragm with a mesa at the center, piezo-resistive strain sensors for detecting the displacement of the diaphragm at the periphery, a permanent magnet on the backside of the diaphragm, and a flat coil. The magnetic actuation system is formed by pairing the permanent magnet and the flat coil, and the sensing diaphragm is driven up and down by the magnetic force generated by this actuation system. The sensor is used in two different modes, as shown in Figure 2.

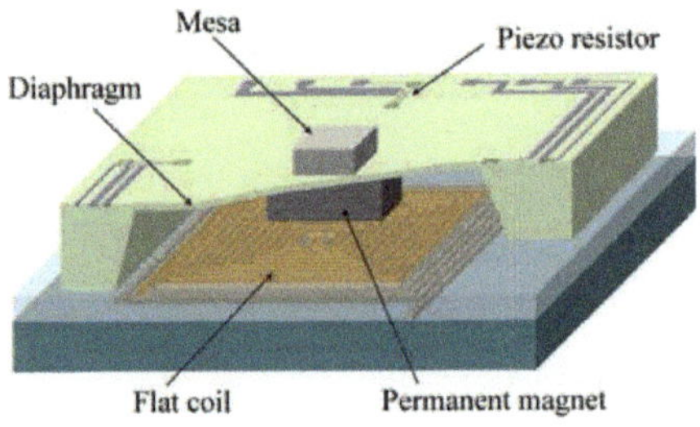

Figure 1. Magnetically driven Si-based tactile sensor. Republished with permission of IOP Published Ltd from Ref. [13].

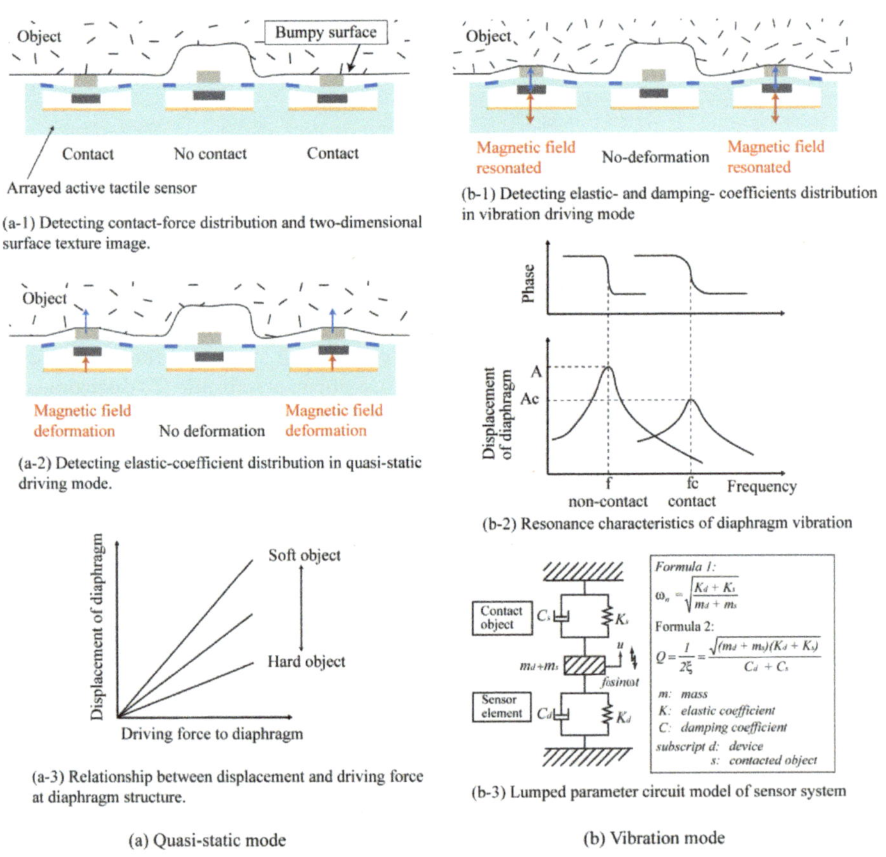

Figure 2. Detection principle of Si-based active tactile sensor. Republished with permission of IOP Published Ltd from Ref. [13].

Quasi-static mode

In this mode of operation, the tactile sensor detects two physical quantities, the contact force and the elasticity of the contacted object. The values are measured when the forces working on the object and the actuating magnetic force are in equilibrium. The detection principle is as follows. The mesa structure on the diaphragm moves only one direction in this mode. When the mesas on the diaphragm of the tactile sensor arrays come into contact with the object, some of the mesa structures are pushed inward by the bumpy surface of the object (Figure 2(a)). The contact force is detected by piezo-resistive strain sensors measuring the deflection of the diaphragm. The force distribution and a 2D surface image of the object can be generated by using an array of sensors. After the contact force detection, the contacting mesas are driven against the object by the magnetic actuation system to measure the hardness of the object. As a result, the contacting regions are deformed in a way that depends on the elasticity of the object. Thus, one can calculate the elasticity of the object by using the relationship between the displacement of the diaphragm and the applied force, which is measured quasi-statically.

Vibration mode

In this mode, the mesa structure is alternately driven up and down at the resonance frequency of the diaphragm. The sensor can estimate the elastic and damping coefficients of the object in contact. Each sensing diaphragm is vibrated at the resonant frequency by magnetic force, as shown in Figure 2(b). Sensor outputs, such as the amplitude of the diaphragm, resonant frequency, and phase shift, change in accordance with the mechanical properties of the contacted object. The lumped parameter circuit model of this sensor system and the formulas for the resonant frequency and Q-factor are shown in Figure 2(b-3). The resonant frequency and Q-factor depend on the elastic coefficient and equivalent mass of the object (see the formulas). The equivalent mass, m_s, depends on the vibration amplitude, which is controlled with the measurements. The detection procedure depends on the magnitude of difference between the device mass m_d, and object mass m_s, as follows:

1. $m_s << m_d$

In this case, the vibrating mass of the sample is much less than that of the mesa with the diaphragm. The mass of the mesa and diaphragm dominates the total mass in vibration; thus, one can estimate elastic- and damping- coefficients of the contacted object with the following procedure.

a. Elastic coefficient K_s

The resonant frequency depends only on the elastic coefficient of the object, and it increases as the elastic coefficient K_s increases, as shown in Formula 1. Thus, one can calculate K_s of the contacted object simply by measuring the change in the resonant frequency.

b. Damping coefficient C_s

The Q-factor of the vibrating diaphragm structure depends on the elastic and damping coefficients (K_s and C_s), as shown in Formula 2. Thus, one can estimate C_s of the contacted object from the Q-factor and K_s.

2. m_s is not negligible compared to m_d

In this case, the sample's equivalent mass is not negligible compared to that of the mesa with diaphragm. Thus, the resonance characteristics, such as the resonant frequency and the Q-factor, depend on the equivalent mass and the elastic coefficient of the contacted object. The following procedure is used to estimate physical quantities.

a. Elastic coefficient K_s

Both operation modes are required to estimate K_s and the equivalent mass m_s of the object. First, the sensor precisely detects the elastic coefficient of the object using quasi-static mode. Then, the equivalent mass of the object is estimated from the elastic coefficient of the object and resonant frequency obtained in vibration mode.

b. Damping coefficient C_s

C_s of the contacted object is estimated from K_s, m_s, and the Q-factor.

The fabrication process and the sensor characteristics driven by the two means of actuation are described in the references [11-13].

2.2. Fabric tactile sensors

Figure 3(a) shows the applications of fabric tactile sensors. These sensors can be used for monitoring the force distribution at human interfaces. A schematic view and operation principle are shown in Figures 3(b)-(d). An artificial hollow fiber is made from a single elastic hollow fiber covered with metal and insulation layers. The sensor is fabricated by weaving the artificial hollow fibers in a reticular pattern. The applied normal load is detected by measuring the change in capacitance between the warp and weft fibers at their intersection point. When a normal load is applied to an intersection point of the fabric tactile sensor, the resulting deformation increases the contact area between the warp and weft fibers. Thus, the capacitance between them increases as the applied load increases. The fabric tactile sensor is therefore able to detect the normal load by measuring this capacitance change. This fabric tactile sensor has the following advantages.

1. Tactile sensing at arbitrary surface

The fabric sensor can fit on any surface, and it can perform tactile sensing by weaving the fibers as if they were cloth.

2. Wearable sensors

Wearable tactile sensors can be made by directly weaving artificial hollow fibers into clothes, gloves, and shoes, or by patching them to these items.

3. Detecting 2D contact-force distributions

The sensor can detect the 2D normal-load distribution by sequentially measuring the capacitance changes at all intersection points. The resolution of the 2D normal-force distribution is

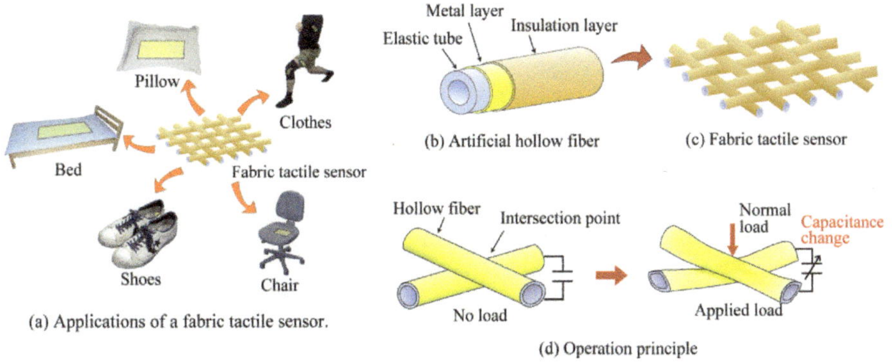

Figure 3. Fabric tactile sensor and its applications. Republished with permission of IOP Published Ltd from Ref. [21].

determined by the grid spacing between the warp and weft fibers, and it is easily adjusted for any purpose by combining the artificial hollow fibers with cotton yarn.

4. Tactile sensing over a large surface area

Silicon-based MEMS sensors have limited area coverage, because they use a small silicon wafer as a substrate. However, fabric tactile sensors consist of fibers; therefore, they can be made as big as needed for the application.

A large rectangular fabric tactile sensor was fabricated by weaving together hollow fibers, brass wires, and conventional cotton yarn (0.3 mm in diameter). The fabric sensor during the weaving process is shown in the right of Fig. 4(a). Figure 4(b) shows one of the produced fabric tactile sensors. The sensor measures 87.5 mm x 52.0 mm, and its sensing points are arranged in a 4 x 4 array. The pitch of detection points is 7.0 mm horizontally and vertically.

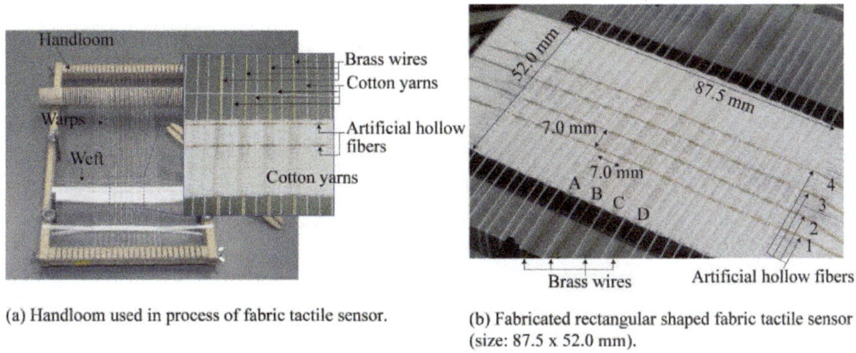

Figure 4. Fabrication of fabric tactile sensor. Republished with permission of The Institute of Engineering and Technology from Ref. [23].

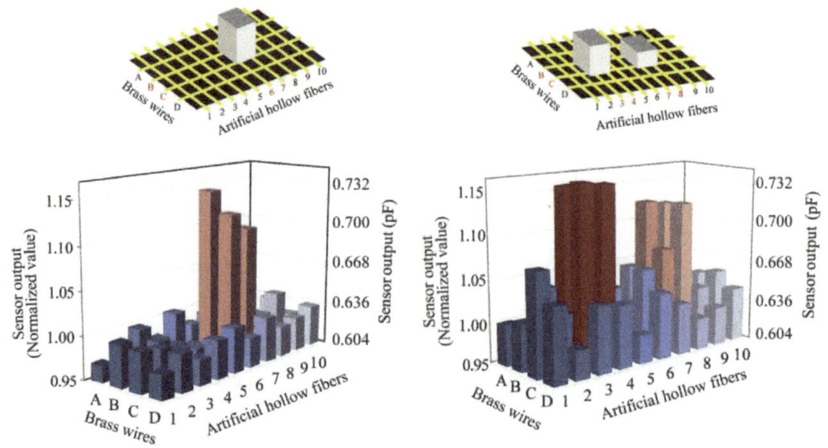

(a) Square plate (size: 20 x 20 mm, force: 2925 mN).

(b) Two different weight of square block (left: 269 g, right: 187 g, size: 24.2 x 24.2 mm).

Figure 5. Applied force and shape detection using a fabric tactile sensor. Republished with permission of The Institute of Engineering and Technology from Ref. [23].

A fabric tactile sensor measuring 180.0 mm x 56.0 mm was evaluated as to its shape detection capability. The unit had a 4 x 10 array of sensing points (total number: 40). The number of the warp wires and weft fibers varied from A to D and 1 to 10, respectively. Thus, the intersection points were each labeled with a letter and number combination (see Fig. 5). The pitch of the detection points was 18.0 mm horizontally and 14.0 mm vertically. Figure 5 shows how the applied pressure and the object's shape are detected. A square plate measuring 20 mm x 20 mm was placed on the sensor (the area from B3 to B4 and from C3 to C4), and a force of 2925 mN was applied to it (Figure 5(a)). Two different metal weights were placed on the sensor (Figure 5(b)). A heavy weight of 269 g and a light weight of 187 g were directly placed on the area from B3 to B4 and from C3 to C4 and the area from B7 to B8 and from C7 to C8. Both weights were the same size, i.e., 24.2 mm x 24.2 mm. The fabric sensor successfully detected the two-dimensional shape of the contacted objects.

3. Flow sensors and their applications

MEMS flow sensors fabricated on Si wafers have a long history, and various types have been developed. C. Liu et al. developed a thermal type of flow sensor to detect shear stress [24]. J. Zhe et al. used a micro-machined cantilever structure for measuring shear stress from the flow [25]. Si-MEMS flow sensors have excellent space and time resolutions, and they are used in semiconductor equipment and in automobiles for controlling gas flows precisely. Additionally, Unnikrishan et al. developed a MEMS-on-tube assembly to simplify the packaging process [26]. They integrated Si-MEMS devices directly on a glass tube, which is compatible with a Swagelok® connector, by fusion bonding. These and other Si-based flow sensors are summarized in the references [27, 28]. The sensors are assembled onto a rigid flat board in the packaging process, to compensate for the brittleness of the Si material. This means these sensors are difficult to put on curved surfaces.

Improved flexibility is the key to making MEMS sensors suitable for a larger range of applications, and this can be accomplished by using resin materials as a substrate. Zhu et al. fabricated a flow senor directly on a flexible printed circuit board with electrical circuits, and mounted it onto the curved surface of a wing to control an aircraft [29]. Ma et al., produced a flexible flow sensor made of polyimide film for detecting the dynamic wave flows in a water channel [30]. For medical applications, Li et al., used Kapton film as a substrate, and they integrated pressure, temperature, glucose, and oxygen sensors on it for blood analysis [31]. Naito et al. fabricated a flow sensor on a flexible 7.5-μm-thick polyimide film and produced a miniaturized on-wall in-tube flow sensor [32]. They used it to fabricate a catheter flow sensor for measuring aspirated- and inspired-air characteristics [33, 34]. Recently, flow sensors have been used to control air supplies precisely in large scale air-conditioning network systems for the purpose of reducing wasted energy.

3.1. Thermal flow sensors

Three principles, thermal anemometry, calorimetric flow sensing, and time-of-flight sensing, are used in thermal flow sensing. Thermal anemometry, i.e., hot-wire anemometry, detects the flow rate by its cooling effect on a heated element. A calorimetric flow sensor, on the other hand, uses a heater and two sensing elements; the heater is placed between the two sensing elements, and the flow is detected from the difference in heating effects at the two sensing elements. The time-of-flight method uses the travelling time of a thermal pulse from the heater to the sensing element for detecting flow. The methods are explained in detail in the references [27, 28]. Hot-wire anemometry flow sensors are widely used because they can measure high flow rates and have the simplest structure.

3.1.1. Hot-wire anemometry

Hot-wire anemometry uses the cooling effect of forced convection to detect changes in flow. When a heated wire is placed in a flow, the equilibrium temperature at the wire depends on the amount of convection in the fluid. On the other hand, the electrical resistance of the wire generally depends on its temperature, and it can be expressed as,

$$R_2 = R_1 \cdot \left[1 + \alpha \cdot \left(T_2 - T_1\right)\right] \tag{1}$$

where R_1 and R_2, are the electrical resistances of the wire at temperature T_1 and T_2, and α is the temperature coefficient of resistance (TCR). Flow changes are detected from changes in the electrical resistance of the wire. Therefore, in practice, this method requires a large and stable TCR value.

As stated above, the hot-wire method detects a flow by measuring the heat it dissipates to the fluid by forced convection. The relationship between the dissipated power Q and flow velocity U in a low Reynolds number flow is given by King's law [35]:

$$Q = \left(A + BU^n\right) \cdot \left(T_h - T_0\right) \tag{2}$$

where A, B, and n are constants depending on the geometry of the wire element. T_h and T_0 are the temperatures of the wire and in the fluid, respectively. If the wire were infinitely long, n would be 0.5. However, n differs form 0.5 in actual wires of finite length.

Two different operations are commonly used for heating the wire electrically. One is constant voltage, or current mode. The flow rate is calculated from the change in electrical resistance, and this method has the advantage of needing only relatively simple electrical circuits. However, it has a drawback in that it is difficult to shorten the response time. To improve the response of the sensing wire, another mode, i.e., constant temperature (CT), is frequently used in flow rate measurements. In this mode, the temperature of the wire is kept constant by using a feedback circuit to shorten the response time, and the flow rate is calculated from the feedback voltage. CT mode is normally used in practice. The electrical energy supplied to the hot wire in CT mode is equal to the power dissipated from the wire to the fluid. As a result, equation (2) is modified as follows:

$$\frac{V_h^2}{R_h} = \left(A + BU^n\right) \cdot \left(T_h - T_0\right) \tag{3}$$

where V_h and R_h are the voltage difference and electrical resistance at the wire, respectively. Here, the square of the voltage at the wire is proportional to the n-th power of the flow velocity.

3.1.2. On-wall in-tube flexible thermal flow sensors

The most common hot-wire anemometers are placed in the center of flow channel structures. Thus, the devices detect the flow rate through the center of a tube. This measurement method has the following drawbacks.

1. Requirement of large inlet length

The flow velocity at the center of tube depends on the flow conditions. For example, the flow distribution in the tube depends on the distance from the entrance. Thus, the sensor has to be placed in the hydraulically fully developed flow region to obtain a constant value. This means that it needs to be a certain distance from the tube entrance.

2. Requirement of position accuracy

The sensor output depends on the radial position in the tube, because of the flow rate distribution. This is true even in a fully developed flow.

3. Difficulty of measuring flow rates in curved tubes

The flow velocity distribution changes depending on the curvature radius and angle of the tube and on the distance from the end of the curved region. Thus, the sensor cannot output a constant value in a curved tube because the velocity distribution is asymmetric in the radial direction.

To overcome these problems, a wall-mounted in-tube thermal flexible sensor able to measure the flow rate under both hydraulically developing and fully developed conditions was developed [36, 37]. The sensor consists of a ring-shaped heater element on a flexible polyimide film (Figure 6), and it is fabricated on a polyimide film substrate and mounted on the inner wall surface of the tube to form a ring-shaped sensor structure.

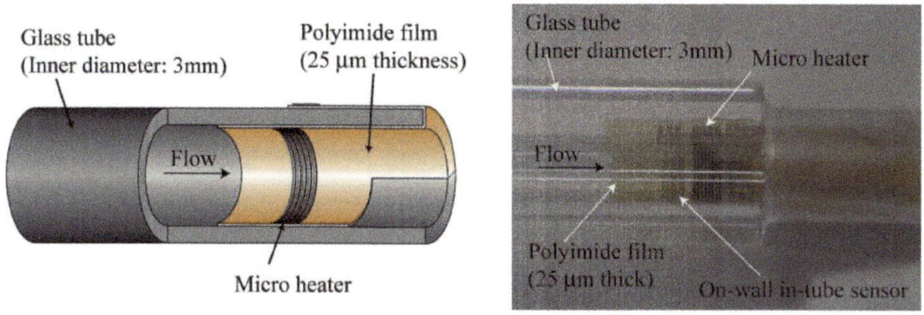

Figure 6. On-wall in-tube thermal flow sensor. Republished with permission of IOP Published Ltd from Ref. [36].

The advantages of the sensor are as follows.

1. It can measure flows in hydraulically developing regions

The flow develops in the direction from the inner wall surface to the center of the tube. The flow rate near the inner wall surface becomes steady first. Thus, this sensor can measure the flow even if the flow is not fully developed in all regions in the tube, because it measures the flow rate only near the inner wall. This means that the sensor can measure even in the hydraulically developing region of the tube, which shortens the distance from the entrance required by the sensor.

2. It can measure axially asymmetric flows.

The output signal from the sensor is insensitive to the velocity distribution in the tube, because ring-shaped sensing structure averages out the flow distribution. This means the sensor outputs a steady value even if the velocity distribution is not axial symmetric.

The on-wall in-tube flow sensor is described in the references [36, 37]. The following sections describe two of its applications.

3.2. Flow sensors for air-conditioning networks

Air conditioning systems are located outside of buildings and they deliver treated air to rooms through a network of ducts. These ducts usually have complicated configurations and a number of bends because they have to go inside a limited amount empty space between the beams of the building's ceiling. To reduce wasted energy, the air supplied to each room has to be controlled at an ideal sensing point near the outlet port. This means the flow rate must to be precisely measured downstream of the bent duct.

To measure the flow rate downstream of a bent duct, the following points have to be considered in the sensor's development.

1. The flow velocity distribution in the tube is expressed as a quadratic function. This means that the flow velocity reaches a maximum at the duct's center and a minimum at the inside surface of the duct. Thus, to reduce the air flow resistance of the sensor itself, the sensor has to be put on the inside surface of the duct, and its thickness has to be minimized as much as possible.

2. The flow sensor structure has to be flexible in order to fit on the rounded inside surface of the duct.

3. The air flow downstream of a bent duct is complicated because of the secondary flow caused by the bend. This means that it is difficult to measure the flow rate by using a single-point measurement. To overcome this problem, a number of sensors have to be attached to the inside surface, and their outputs should be averaged to reduce the effect of the secondary flow.

A patch-type flexible flow sensor was developed as a way to control the air supply precisely in large-scale air-conditioning network systems. The sensor is based on polyimide film, and it was fabricated by photolithography and thin-film deposition. First, the photoresist was patterned by photolithography, and Au/Cr film was deposited by sputtering. The metal was then patterned by selectively removing the photoresist (lift-off process). The Au and Cr thicknesses were 250 nm and 10 nm, respectively. Figure 7(a) shows flexible film sensors fabricated on a 3-inch wafer. Each sensor measures 10 mm x 10 mm. A flexible printed circuit was used for the electrical connection. The sensor was placed on the printed circuit, and it was bonded to the board by adhesive. The electrical connection was manually made with silver paste. The electrical contact area was covered with adhesive to increase its mechanical strength. To form a cavity for thermal isolation, the sensor was placed on a silicone rubber sheet with a 5.0 mm x 5.0 mm hole in it, and the sensor was fixed to the sheet with adhesive polyimide film. The rubber sheet was 0.5-mm thick. Figure 7(b) shows an assembled flexible flow sensor.

In an experimental evaluation, four flexible flow sensors were attached at an angle of 90° to the inside surface of an 8-inch duct, and the outputs from them were averaged. The relationship between the output and flow rate obeyed King's equation for flows of from 0 to 3000 m^3h^{-1}. The averaged sensor outputs depended on the distance of the sensor from the bend of the duct, and their values were slightly higher than those obtained in a straight duct. A conversion factor,

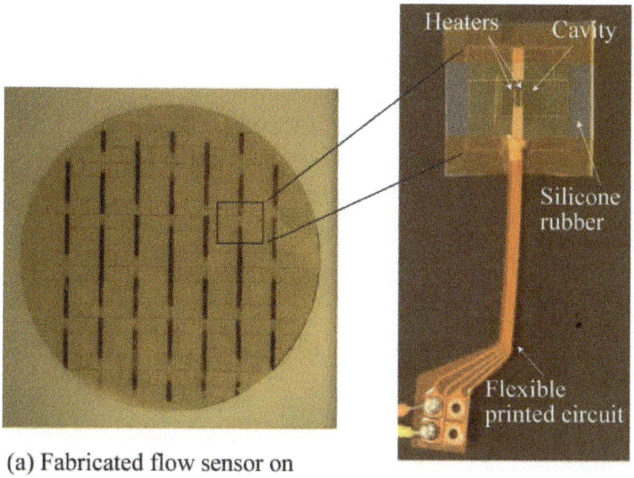

(a) Fabricated flow sensor on polyimide film

(b) Assembled patch type flow sensor

Figure 7. Patch-type flexible flow sensor for large scale air-conditioning network systems.

which enables us to calculate the flow rate values from the obtained sensor outputs in bent ducts, was derived [38].

3.3. Flow sensors for medical applications

The numbers of cardiac disease and cerebral stroke patients are gradually decreasing, thanks to developments in medical devices and improved health guidance. However, chronic obstructive pulmonary disease (COPD) is still on the increase. Spirometry is normally used to evaluate the progress of COPD. It measures the flow rate in the human mouth. The respiratory system consists of numerous levels of bronchi just like a tree, and the lung alveoli are located at the end of diverging bronchi. In the case of COPD, the alveoli structures gradually collapse as the patient ages and because of absorbed cigarette smoke or other air-polluting substances. Lesions develop at the ends of the diverging bronchi. The current method of measurement, which evaluates lesions in the mouth, cannot be used to evaluate such small lesions in the bronchi.

To overcome the above problem, Shikida et al. [34, 39] devised a catheter-type flow sensor that can measure aspirated- and inspired-air flow characteristics trans-bronchially. The flow sensor (Figure 8) can be inserted into a small bronchus for measuring aspirated and inspired air characteristics. An on-wall in-tube thermal flow sensor is mounted on the inside of the tube, and it is used in the bronchoscope. The external diameter of the tube is only a few mm, and therefore, it can reach into the small bronchus. Two heaters were formed on the film in order to detect the flow direction.

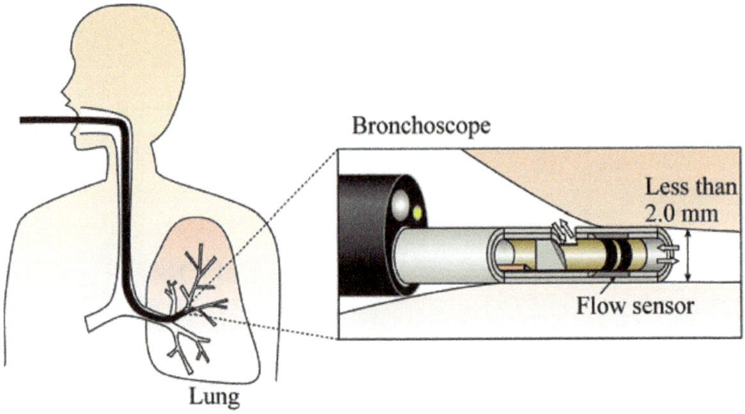

Figure 8. Concept of catheter flow sensor for trans-bronchial measurement. Republished with permission of IOP Published Ltd from Ref. [34].

Polymer-MEMS technologies and heat shrinkable tubes are used to produce the sensor. The fabrication process consists of two steps.

1. Flexible film sensor fabrication

Polyimide film is used as a substrate. The thickness of the film affects the thermal response performance (it has to be relatively thick; 7.5 μm). The metal film heater structures are formed on the film by using photolithography and sputtering, and they are patterned with a lift-off process. The typical size of the sensor is 2.8 mm x 5.5 mm. Two heaters are formed on the film in order to detect the flow direction. Each film sensor is mechanically cut before being mounted inside the tube.

2. Mounting film sensor inside a tube

The sensor is mounted on the inside surface of the tube as follows.

a. The sensor is inserted into a heat shrinkable tube made of Teflon.

b. The heat shrinkable tube is baked at 110°C. The Teflon tube shrinking to almost half its original size as a result of heating, and the film sensor is automatically mounted on the inner wall surface during the shrinking process and becomes fixed on the tube surface.

c. A cavity structure is formed under the heating element to improve thermal isolation. To produce the active structure, a slit is formed on the tube, and it is covered with a one-mode Teflon heat-shrinkable tube to seal it. The outer diameter of the inner tube is only a little bit larger than the inner diameter of the outer one, after the heat shrinking process. Thus, these two Teflon tubes are tightly fixed to one other.

Figures 9(a) and 9(b) show a schematic diagram and a photograph of the fabricated catheter-type flow sensor. The inner and outer diameters of the tube are 1.0mm and 1.8 mm. The temperature coefficient of resistance of the sensor is $0.0025K^{-1}$. The package method of the on-wall in-tube film mounting is suitable for miniaturization because the sensor structure itself does not disturb the flow stream.

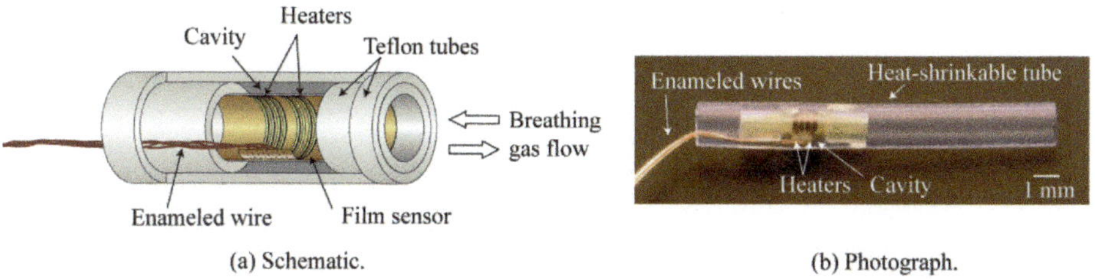

Figure 9. Fabricated catheter type flow sensor. Republished with permission of IOP Published Ltd from Ref. [39].

Experiments were done under the regulations set out in the Nagoya University Animal Experiments Guidelines, and they were approved by the animal ethics committee. The experiments were done under anesthesia. Thus, the following data are aspirated- and inspired-air characteristics under anesthesia. Infiltration of bodily fluids into the tube was prevented by using Teflon as the tube material. The catheter flow sensor is intended to be incorporated into a bronchoscope and to be inserted in the small airways from the mouth. Thus, the sensor was first tested on rats. An optical fiberscope with an outer diameter of 0.8 mm, instead of a bronchoscope, was used in the tests. Inserting the flow sensor into the airway from the mouth of the rat with the fiberscope involved three steps.

1. Only the fiberscope was inserted, from the mouth to the targeted location, by observing the inside of the airway. Then, the sensor was inserted to the targeted location with a Teflon tube guide.

2. The fiberscope was withdrawn when the flow sensor reached the target location.

3. The Teflon tube guide was carefully extracted and only the flow sensor remained at the location.

The breathing waveform of the rat is shown in Figure 10. A period of 820 ms for inspiration and aspiration was obtained. This means that the respiration frequency was 1.1 Hz. The ventilated air volume was calculated from this breathing waveform, and a value ranging from 1.01–1.09 cc was obtained. The known respiration frequency and ventilated air values of rats range from 1.1–1.9 for the former and from 0.60–1.25 for the latter. The measured values coincided with the physiological values in the literature.

The air was inspired for a short time period, suddenly becoming aspirated for a time period. In aspirated mode, a large amount of air was aspirated at the beginning, and the aspirated air gradually decreased afterwards. Inspiration and aspiration were done by moving the diaphragm. The air was inspired by expanding the thoracis. This was done by contracting the diaphragm. The air was simply aspirated by the restorative force of the thoracis. Thus, the air was inspired in a short time, and a large amount of air was aspirated at the beginning of the aspiration mode, and the amount gradually decreased after that. The sensor signal quantitatively corresponded to this natural respiratory mechanism. From these results, the catheter flow sensor will be useful for evaluating the flow characteristics in the small bronchus region in the future.

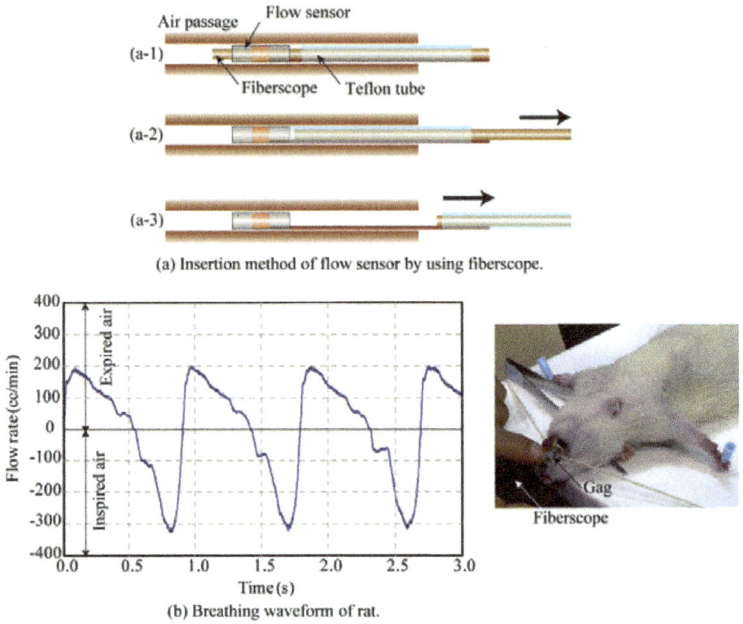

Figure 10. Breathing waveform of rat measured with intubated catheter flow sensor. Republished with permission of IOP Published Ltd from Ref. [39].

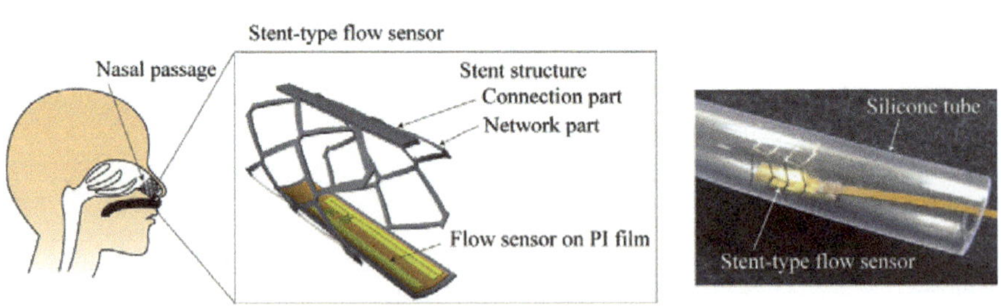

Figure 11. Stent flow sensor for evaluation of nasal respiration. Republished with permission of IOP Published Ltd from Ref. [40].

A human being breathes through the mouth and the nose. Bacteria and viruses are trapped in the nose cavity, and only clean inspired air is supplied to the hung. The inspired air is also humidified in the cavity in order for lungs to adsorb oxygen effectively. The inspired air is warmed inside the body is then expired outside; this is a method of heat exchange. The flow characteristics of the nose are deeply related to various health concerns. Thus, a stent-type of thermal flow sensor was also developed for measuring nasal respiration. The MEMS stent flow sensor is shown in Figure 11. It is a thermal flow sensor fabricated on polymer film and monolithically integrated on the stent structure [40].

4. Conclusion

This chapter described the fabrication and applications of micro-sensors produced by MEMS technologies. The attractive features and problems of MEMS sensors were discussed. In particular, MEMS sensors have low costs, high space and time resolutions, and reduced dead space. The applications of MEMS sensors include tactile Si-solid and fabric sensors for human interfaces and flow sensors for air-conditioning systems and medicine.

Acknowledgements

The research was supported by the Centre of Excellence (COE) for Education and Research on Micro-Nano Mechatronics and a Grant-in-Aid for Scientific Research (B) No. 23310091 from the Ministry of Education, Culture, Sports, Science and Technology (MEXT), Japan.

Author details

Mitsuhiro Shikida

Department of Micro-Nano Systems Engineering, Nagoya University, Nagoya, Japan

References

[1] Petersen K. E. Silicon as a mechanical material. Proceedings of the IEEE 1982;70(5) 420-457.

[2] Muller R. S, Howe R. T, Senturia S. D, Smith R. L, White R. M. Microsensor. USA: IEEE PRESS; 1990.

[3] Yamasaki H., editor. Handbook of sensors and actuators – Intelligent sensors (vol. 3). The Netherlands: Elsevier Science B. V.; 1996. 3.

[4] Ristic L. Sensor technology and devices. USA: Artech House, Inc.; 1994.

[5] Maluf N. An introduction to microelectromechanical systems engineering. USA: Artech House, Inc.; 2000.

[6] Chu Z, Sarro P. M, Middlehoek S. Silicon three-axial tactile sensor, In: Technical Digest of The 8th International Conference on Solid-State Sensors and Actuators, June 1995, Stockholm, Sweden; 1995.

[7] Kane B. J, Cutkosky M. R, Kovacs G. T. A. A traction stress sensor array for use in high-resolution robotic tactile imaging. Journal of microelectromechanical systems 2000;9(4) 425-434.

[8] Kasten K, Amelung J, Mokwa W. CMOS-compatible capacitive high temperature pressure sensors. Sensors and Actuators A: Physical 2000;85(1-3) 147-152.

[9] Kim K, Lee K. R, Kim Y. K, Lee D. S, Cho N. K, Kim W. H, Park K. B, Park H. D, Park Y. K, Kim J. H, Pak J. J. 3-axes flexible tactile sensor fabricated by Si micromachining and packaging technology. In: Proceedings of IEEE MEMS'06. January 2006, Istanbul, Turkey; 2006.

[10] Takao H, Yawata M, Sawada K, Ishida M. A robust and sensitive silicon-MEMS tactile-imager with scratch resistant surface and over-range protection. Technical Digest of The 14th International Conference on Solid-State Sensors, Actuators and Microsystems. June 2007, Lyon, France; 2007.

[11] Shikida M, Shimizu T, Sato K, Itoigawa K. Active tactile sensor for detecting contact force and hardness of an object. Sensors and Actuators A: Physical 2003;103(1-2) 213-218.

[12] Hasegawa Y, Shimizu T, Miyaji T, Shikida M, Sasaki H, Sato K, Itoigawa K. A micromachined active tactile sensor for hardness detection. Sensors and Actuators A : Physical 2004;114(2-3) 141-146.

[13] Hasegawa Y, Shikida M, Sasaki H, Sato K, Itoigawa K. An active tactile sensor for detecting mechanical characteristics of contacted objects. Journal of micromechanics and microengineering 2006;16(8) 1625-1632.

[14] Engel J, Chen J, Liu C. Development of polyimide flexible tactile sensor skin. Journal of micromechanics and microengineering 2003;13(3) 359-366.

[15] Kim S-H, Engel J, Liu C, Jones D. L. Texture classification using a polymer-based MEMS tactile sensor. Journal of micromechanics and microengineering 2005;15(5) 912-920.

[16] Lee H-K, Chang S-I, Yoon E. A flexible polymer tactile sensor: Fabrication and modular expandability for large area deployment. Journal of microelectromechanical systems 2006;15(6) 1681–1686.

[17] Hwang E-S, Seo J-H, Kim Y-J. A polymer-based flexible tactile sensor for both normal and shear load detections and its application for robotics. Journal of microelectromechanical systems 2007;16(3) 556–563.

[18] Konishi S, Maeda H, Ezaki T, Kawato M, Asajima S, Makikawa M. A selective stimulation of nerve bundles using a flexible multielectrode. In: Technical Digest of The 10th International Conference on Solid-State Sensors and Actuators. June 1999, Sendai, Japan; 1999.

[19] Chen N, Engel J, Pandya S, Liu C. Flexible skin with two-axis bending capability made using weaving by-lithography fabrication method. In: Proceedings of IEEE MEMS'06, January, 2006, Istanbul, Turkey; 2006.

[20] Hasegawa Y, Shikida M, Ogura D, Sato K. Glove type of wearable tactile sensor produced by artificial hollow fiber. In: Technical Digest of The 14th International Conference on Solid-State Sensors, Actuators and Microsystems. June 2007, Lyon, France; 2007.

[21] Hasegawa Y, Shikida M, Ogura D, Suzuki Y, Sato K. Fabrication of a wearable fabric tactile sensor produced by artificial hollow fiber. Journal of micromechanics and microengineering 2008;18(8) 085014 .

[22] Kita G, Suzuki Y, Shikida M, Sato K. Fabric tactile sensor composed of ball-shaped umbonal fiber for detecting normal and lateral force. Micro & Nano Letters 2010;5(4) 211-214.

[23] Kita G, Shikida M, Suzuki Y, Tsuji Y, Sato K. Large-sized fabric tactile sensors for detecting contacted objects. Micro & Nano Letters 2010;5(6) 389-392.

[24] Liu C, Huang J-B, Zhu Z, Jiand F, Tung S, Tai Y-C, Ho C-M. A micromachined flow shear-stress sensor based on thermal transfer principle. Journal of Microelectromechanical systems 1999;8(1) pp. 90-99.

[25] Zhe J, Midi V, Farmer Jr K. R. A microfabricated wall shear-stress sensor with capacitative sensing. Journal of Microelectromechanical systems 2005;14(1) 167-175.

[26] Unnikrishnan S, Jansen H. V, Berenschot J. W, Mogulkoc B, Elwenspoek M. C. MEMS within a Swagelok®: a new platform for microfluidic devices. Lab on a Chip 2009;9(13) 1966-1969.

[27] Gianchandani Y, Tabata O, Zappe H. Comprehensive MEMS, Elsevier Science B. V. 2008.

[28] Elwenspoek M, Wiegerink R. Mechanical microsensors. Germany, Springer; 2001.

[29] Zhu R, Liu P, Liu X. D, Zhang F. X, Zhou Z. Y. A low-cost flexible hot-film sensor system for flow sensing and its application to aircraft. Technical Digest IEEE Micro Electro Mechanical Systems Conference. January 2009, Sorrento, Italy; 2009.

[30] Ma B, Ren J, Deng J, Yuan W. Flexible thermal sensor array on PI film substrate for underwater applications. In: Technical Digest IEEE Micro Electro Mechanical Systems Conference. J anuary 2010, Hong Kong; 2010.

[31] Li C, Wu P-M, Jung W, Ahn C. H, Shutter L. A, Narayan R. K. A novel lab-on-a-tube for multimodal monitoring of patients with traumatic brain injury. In: Technical digest of The 15th International Conference on Solid-State Sensors and Actuators. June 2009, Denver, USA; 2009.

[32] Naito J, Shikida M, Hirota M, Tan Z, Sato K. Miniaturization of on-wall in-tube flexible thermal flow sensor using heat shrinkable tube. In: Technical Digest IEEE Micro Electro Mechanical Systems Conference. January 2008, Tucson, USA; 2008.

[33] Yokota T, Naito J, Shikida M, Kawabe T, Hayashi Y, Sato K. Catheter type of flow sensor for trans-bronchial measurement for lung functions. Technical digest of The 15th International Conference on Solid-State Sensors and Actuators. June, 2009, Denver, USA; 2009.

[34] Shikida M, Naito J, Yokota T, Kawabe T, Hayashi Y, Sato K. A Catheter-type Flow Sensor for Measurement of Aspirated- and Inspired-air Characteristics in Bronchial Region. Journal of Micromechnics and Microengineering 2009;19(10) 105027

[35] King L. V. On the Convection of Heat from Small Cylinders in a Stream of Fluid: Determination of the Convection Constants of Small Platinum Wires with Applications to Hot-Wire Anemometry. Philosophical Transactions of the Royal Society of London. Series A, Containing Papers of a Mathematical or Physical Character 1914;90(622) 563-570.

[36] Tan Z, Shikida M, Hirota M, Sato K, Iwasaki T, Iriye Y. Experimental and theoretical study of an on-wall in-tube flexible thermal sensor. Journal of Micromechnics and Microengineering 2007;17(4) 679-686.

[37] Tan Z, Shikida M, Hirota M, Xing Y, Sato K, Iwasaki T, Iriye Y. Characteristics of on-wall in-tube thermal flow sensor under radially asymmetric flow condition. Sensors and Actuators A 2007;138(1) 87-96.

[38] Yoshikawa K, Iwai S, Shikida M, Sato K. Attached-type flexible flow sensor for air conditioning network. Technical digest of The 16th International Conference on Solid-State Sensors and Actuators. June 2011, Beijing, China; 2011.

[39] Shikida M, Yokota T, Kawabe T, Funaki T, Matsushima M, Iwai S, Matsunaga N, Sato K. Characteristics of an optomized catheter-type thermal flow sensor for measuring reciprocating airflows in bronchial pathways. Journal of Micromechnics and Microengineering 2010;20(12) 125030.

[40] Shikida M, Yokota T, Naito J, Sato K. Fabrication of a stent-type thermal flow sensor for measuring nasal respiration. Journal of Micromechnics and Microengineering 2010;20(5) 055029.

Chapter 19

Single Cell Nanosurgery System

Toshio Fukuda, Masahiro Nakajima,
Yajing Shen and Masaru Kojima

1. Introduction

A model organism is one of the species that is extensively studied to understand particular biological phenomena, with the expectation that discoveries made in the organism model will provide insight into the workings of other organisms [1]. In particular, model organisms are widely used to explore potential causes and treatments for human disease when human experimentation would be unfeasible or unethical [2]. It shows some of the model organisms that have been used in a biomedical research.

The first and foremost consideration in the selection of any model organisms before conducting any bio-related research is how relevance the selected model organisms to human. If the first consideration is justified, then the second consideration which needs to be addressed is how practical and easy the selected model organisms for experimental endeavors.

Yeast is one of the simplest eukaryotic organisms (organism whose cells contain a clear defining membrane-bound structure of nucleus) but many essential cellular processes are conserved between yeast and humans. There are genes in yeast and mammals that encode very similar proteins [3]. Comparison of the yeast and human genomes, reported in 1997, revealed that 30% of known genes involved in human disease have yeast orthologs (i.e. functional homologs) [4]. Furthermore, hundreds of yeast genes exhibit a link to human disease genes as reported by [5].

Yeast is a good experimental tool for molecular and cellular biology studies. Yeast growth and division can be controlled efficiently and effectively by adjusting environmental conditions. Furthermore, yeast cells divide in a similar manner to human cells.

Because of these advantageous features, yeast has become the model organism of choice for medicine-related research. For example, studies with yeast have contributed greatly to our knowledge of the regulation of eukaryotic cell division, including the cancer-related disturbances [6]. Up to now, yeast has maintained its role as a useful model system in fundamental studies of disease processes.

2. Single cell nanosurgery system based on nanorobotic manipulation system

2.1. Single cell nanosurgery system

We have been proposed a "Nanolaboratory" based on nanorobotic manipulation system from around 2000 [7]. It is one of the systems to realize various nanoscale fabrication and assembly to develop novel nanodevices to integrate borderless technologies based on nanorobotic manipulation system. It is readily applied to the scientific exploration of macroscopic phenomena and the construction of prototype nanodevices. It would be one of the most significant enabling technologies to realize the manipulation and fabrication technology with individual atoms and molecules for the assembly of devices. Recently, the investigation of Nanoelectromechanical Systems (NEMS) has attracted much attention [8-11]. It is expected to realize high integrated, miniaturized, and multi-functional devices for various applications. One of the effective ways is the direct usage of the bottom-up fabricated nanostructures.

Nanolaboratory can be applied for the single cell analysis and manipulations. As show in Figure 1, the integration is important for the single cell nanosurgery system between micro and nanorobotic manipulators under various microscopes. The applications under dry, semi wet, and wet conditions can be done under TEM/SEM, E-SEM and optical microscope (OM). The nanomanipulation system inside TEM/SEM is a fundamental technology for property characterization of nano materials, structures and mechanisms, with the fabrication of nano building blocks, and for the assembly of nano devices. The nanomanipulation system inside E-SEM provides single cell manipulation and analysis under nano-scale high resolution images for the application of nanodevices or nanotools assembled under dry condition. OM micromanipulation system is used under water, hence the biological cells can be cultured with medium.

2.2. Nanorobotic manipulations

Nanorobotic manipulation; nanomanipulation, has been received much more attention, because it is an effective strategy for the property characterizations of individual nano-scale materials and the construction of nano-scale devices [12]. They might finally be the core-most part of nanotechnology. One of the attractive future applications of nanomanipulation is to realize the ultimate goal of nanotechnology, or nanomanipulation, is considered.

To manipulate nano-scale objects, it is needed to observe them with a resolution higher than nano-scale. Hence, the manipulators and observation systems, microscopes in general, are necessary for nanomanipulations. Figure 1 shows the strategies of nanomanipulations with various kinds of microscopes. The nanomanipulation under various microscopes for 2D/3D nanomanipulations. Optical microscope (OM) is one of the most historical and basic micro-

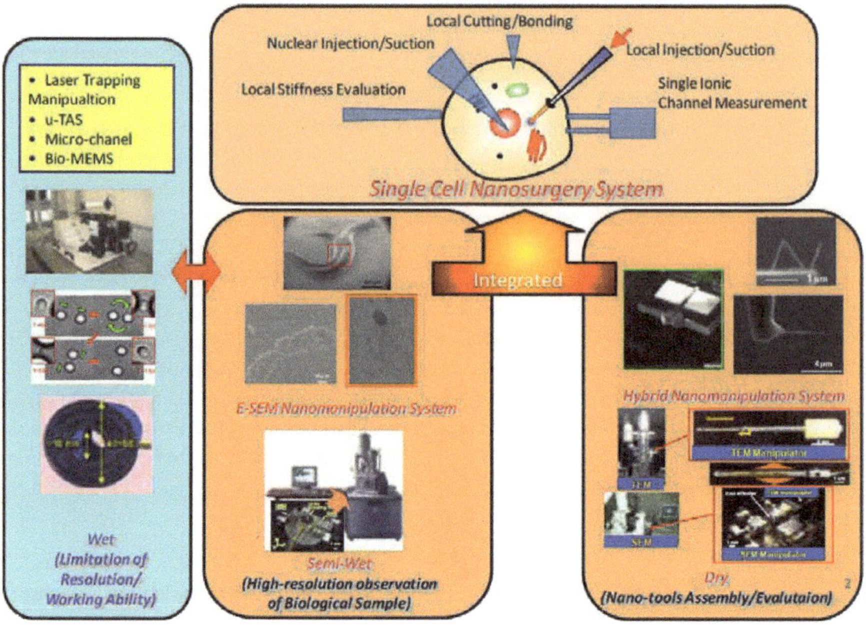

Figure 1. Single cell nanosurgery system based on micro/nanomanipulators under various microscopes (wet/semi-wet/dry conditions).

scope. However, its resolution is limited to ~100 nm because of the diffraction limit of optical wavelength (~400 - ~800 nm) explained by the well-known Abbe's law [13].

The scanning tunnelling microscopes (STMs) or atomic force microscopes (AFMs), have functions of both observation and manipulation in nano-scale. Their high resolution makes them capable of atomic manipulation. In 1990, Eigler and Schweize demonstrated that the first atom practice nanomanipulation with scanning tunnelling microscope (STM) [14]. Avouris et al. applied an AFM to bend and translate carbon nanotubes (CNTs) on a substrate [15]. They combined the techniques with an inverse process, namely straightening, by pushing along a bent tube, and realized the translation of a tube to another place. Ning Xi et. al at Michigan State University, developed AFM based nanomanipulation system with interactive operation system [16, 17]. The system realized a real-time visual feedback during AFM based nanomanipulation.

Normally, SPM systems have limit for the observation in 2D plane with quite smooth surface. Moreover, the observation area is limited and long time is needed to get one image (more than mins). This limitation rises up as 3D nanomanipulation of nanostructures. On the other hand, the electron microscopes (EMs) provide atomic scale resolution with the electron beam which wave length is less than ~0.1 Å. EMs are divided mainly two types as scanning electron microscopes (SEMs) and transmission electron microscopes (TEMs). For example, M. F. Yu et. al presented the tensile strength of individual CNTs inside a SEM [18]. However the resolution of SEM, generally ~1 nm resolution, is approximately one order in magnitude lower than that of a TEM. High resolution and transmission image of TEMs are useful for measurement and evaluation of nano-scale objects. Kizuka et.al proposed the manipulation holder inside high-resolution transmission electron microscope (HR-TEM). The manipulator was specially designed with atomic level positioning resolution [19].

However, the specimen chamber and observation area of TEM are too narrow to contain manipulators with complex functions. Hence, special sample preparation techniques are also needed. We proposed a hybrid nanorobotic manipulation system which is integrated TEM and SEM nanorobotic manipulators as core system for the Nanolaboratory [20, 21]. The strategy is named as hybrid nanomanipulation so as to differentiate it from those with only an exchange-able specimen holder. The most important feature of the manipulator is that it contains several passive DOFs, which makes it possible to perform relatively complex manipulations whereas to keep compact volume to be installed inside the narrow vacuum chamber of a TEM [22].

Recently single cells analysis has been much more attentions because of the progress of the micro/nano scale techniques on the local environmental measurements and controls [23]. Under conventional SEMs and TEMs, the sample chambers of these electron microscopes are set under the high vacuum (HV) to reduce the disturbance of electron beam for observation. To observe water-containing samples, for example bio-cells, the appropriate drying and dying treatments are needed before observations. Hence, direct observations of water-containing samples are normally quite difficult through these electron microscopes.

On the other hand, the environmental-SEM (E-SEM) can be realized the direct observation of water-containing samples with nanometer high resolution by specially built secondly electron detector [24]. The evaporation of water is controlled by the sample temperature (~0 - ~40 °C) and sample chamber pressure (10 - 2600 Pa). The unique characteristic of the E-SEM is the direct observation of the hydroscopic samples with non-drying treatment. Hence, the nano-manipulation inside the E-SEM is considered to be an effective tool for a water-containing sample with nanometer resolution [25-28].

3. Single cell analysis based on an E-SEM nanorobotic manipulation system

Single cells analysis needs to be investigated through the micro/nano scale techniques based on the local environmental measurements and controls. We developed the Environmental-SEM (E-SEM) nanorobotic manipulation system to manipulate and control the local environ-ments for biological samples in nano scale (Figure 2). It realized that direct observation and manipulation of water-containing biological samples under nanometer high resolution imaging.

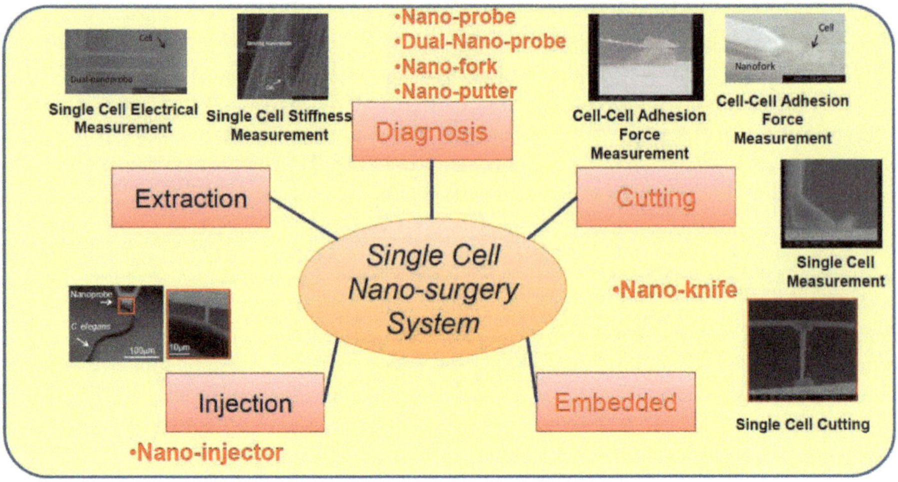

Figure 2. Single Cell Nano-surgery System with various nano-tools

In this chapter, the novel local stiffness evaluation, local cutting, and local extraction of biological organism are presented by micro-nanoprobes based on the E-SEM nanorobotic manipulation system for future cell diagnosis and surgery system.

3.1. Adhesion force measurement of single cell using nano-putter

Cell activities, such as embryogenesis, mitosis, morphogenesis, cell orientation, cell motility, and survival depend on attachment to neighboring cells and the extracellular matrix. Cellular attachment to extracellular matrices influences cell morphology, cell function, and signaling mechanisms that direct cellular proliferation and differentiation [29]. Cell-surface interaction is important in the development of any material or device for biomedical applications, since the performance of a medical device in the body must be compatible with the surrounding tissue [30, 31]. Understanding of the cell adhesion process would benefit the development of suitable biomaterials or device for both tissue engineering and medical fields.

Cell adhesion processes are influenced by numerous parameters, such as the nature of the biomaterial and its surface characteristics (roughness, topography, chemical composition, surface wettability, surface charge and surface treatments) have been investigated. However, the effect of ambient humidity on cell adhesion has had less attention, especially at the single cell level. Understanding the adhesion force at various humidity conditions could help us to better understand the processes of cell-directed integration during water evaporation. The understanding is also useful in controlling yeast infections at wet environment. Moreover, cell adhesion is influenced by the surface energy of substrate strongly, but the mechanism is still not clear [32]. The study of cell adhesion on substrates with different surface energy could help us to understand the adhesion mechanism better.

We presented a yeast cell adhesion force measurement performed using the nanorobotic manipulation system inside the ESEM [33, 34, 35]. Figure 3 (A) shows a typical force-displacement curve during the single cell adhesion force measurement. Figure 3 (B) shows the initial position of the micro putter and the single cell. The micro putter was driven by the nanorobotic manipulation system. First, it was moved towards the cell until it contacted the cell (Figure 3 (C)). Then, a continuous movement was applied to the micro putter by the nanomanipulator. The micro putter beam deflected owing to the increasing pushing force. Figure 3 (D) shows the deflection of the micro putter during the adhesion force measurement. Finally, the cell was detached from its initial position under a certain force (Figure 3 (E)). The maximum force during this manipulation procedure was defined as the adhesion force.

Single yeast cell adhesion force measurement was performed at three humidity conditions, i.e. 100%, 70% and 40%. The mean adhesion force and the deviation are with 95% confidence at each humidity conditions. It demonstrates that the yeast cell adhesion forces range from 10 to 25 μN at various humidity conditions. The adhesion forces were 11.0 ± 5.1 μN, 17.4 ± 4.7 μN and 23.5 ± 6.1 μN at 100%, 70% and 40% relative humidity conditions respectively. It showed clearly that the cell adhesion was affected by the ambient humidity. The cell adhesion force is larger at low humidity than at high humidity. For example, the cell adhesion force was 23.5 μN at a humidity of 40%, which was 1.14 times larger than the force 11.0 μN at humidity 100%.

3.2. Single cell cutting using nano-knife

Cell cutting is an important step in cell analysis processes. For instance, it was widely used to prepare cell specimen slices for the observation of an inner structure [36]. Different to group cells analysis, research on individual cells could give accurate data rather than average results. Single cell analysis can help us to understand the biological processes more accurately. In-situ single cell cutting technique could potentially benefit cell analysis, such as single cell operation and disease treatment.

Recently, a nano knife fabricated from a carbon nanotube (CNT) has been developed for the purpose of cell cutting [37]. The nano knife was designed by welding a CNT across two tungsten needles inside a scanning electron microscopy (SEM). This device can reduce the angle by which the sample is bent during cutting, due to the small diameter of the CNT. It can be seen clearly that the nano knife can leave a mark on the epon resin surface, which means it can cut very thin slices of cells. However, the bonding force between the CNT and tungsten probes by using electron beam induced deposition (EBID) method is quite small. There are still certain types of hard specimens such as bone, plants, and thick-walled spores. The CNT based nano knife may not be able to deal with such samples, since a larger cutting force is required, especially when the sample size is large.

We presented a nano knife with a buffering beam was designed for an in-situ single cell cutting purpose [38]. A schematic drawing of the single cell cutting using a nano knife is shown in Figure 4. The nano knife was immobilized to the nanomanipulator by using the electrical conductive tape inside the ESEM chamber. Under the driving of the nanorobotic manipulation system, the nano knife can move towards and cut the single cell finally. The cutting force can be calculated based on the deformation of the nanokinfe's beam.

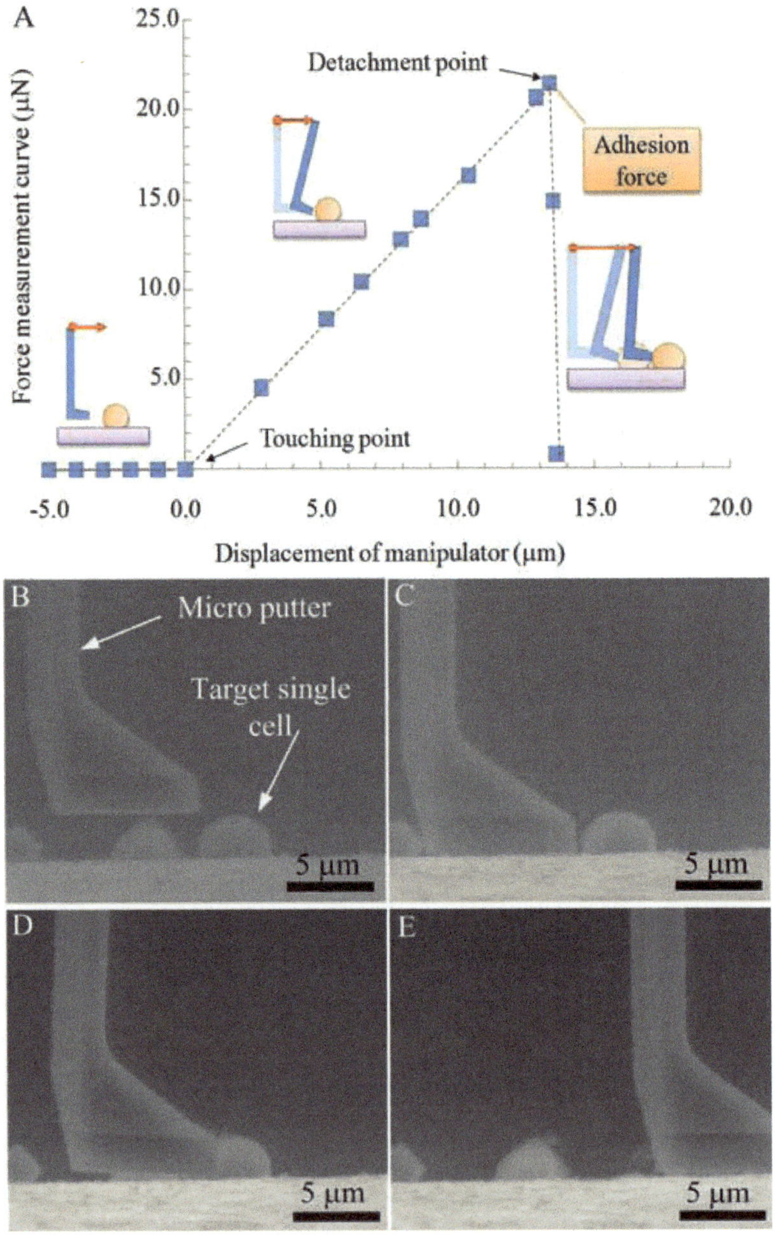

Figure 3. Single cell adhesion force measurement using micro putter inside ESEM.

The in situ single cell cutting experiment was performed using these three nano knives. Figure 5(A) shows the initial position of the nano knife and a single cell. Figure 5(B) shows the touching between the nano knife tip and the single cell. The deformation of the nano knife beam and the single cell during the cutting is shown in Figure 5(C). The deformation of the beam can be measured from the ESEM image directly using image analysis software. Therefore, the cutting force can be calculated based on Hooke's law. The separated single cell after cutting is shown in Figure 5(D). The sample slice angle can be measured from the ESEM image directly as well. Figure 5(E) shows the single cell cutting image using the 25° knife. Figure 5(F) shows the sample slice angle after cutting. The images of single cell cutting and sample slice angle after cutting using 45° knife are shown in Figure 5 (G) and Figure 5 (H).

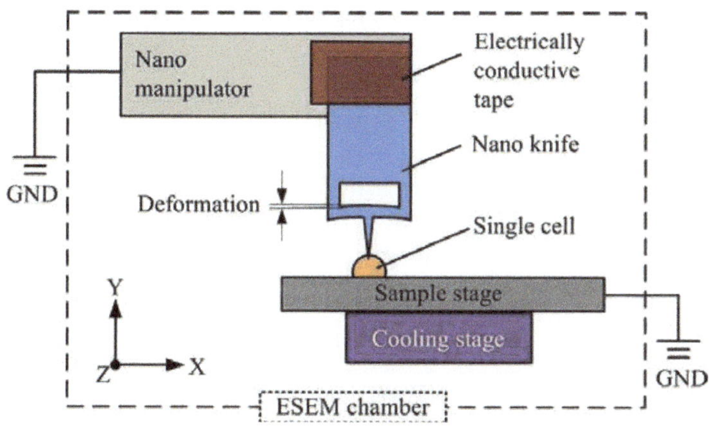

Figure 4. Schematic drawing of the single cell cutting experiment using nano knife inside ESEM.

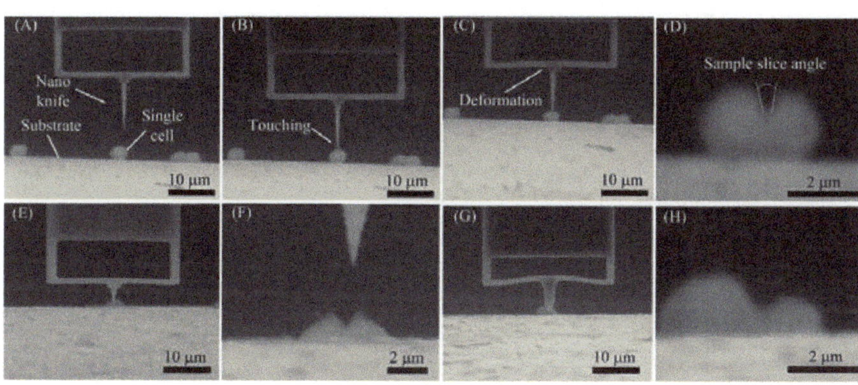

Figure 5. Single cell cutting using nano knife inside ESEM.

4. Conclusion

This chapter presents the single cell nanosurgery system based on nanomanipulation techniques. The micro-nano tools have been proposed to investigate single cell analysis to manipulate and control the local environment in micro-nano scale. The E-SEM nanomanipulation system was constructed to realize the local stiffness evaluation, local cutting, and local extraction of biological organism in nano-meter scale. The adhesion force measurement was presented by micro-putter for single cells. The single cell cutting was also described using nano-knife. As future direction, the multiple micro-nanotools are used continuously depending on the purposes by exchanging machinery system (NTExS: Nanotool Exchanger System) [39]. We are investigating on the nanoinjection applications for the Caenorhabditis elegans (C. elegans) as one of the model organisms [40, 41]

Acknowledgements

The authors are grateful to Prof. T. Inada N. Uozumi for providing with W303 yeast cells. This work was partially supported by a Grant-in-Aid for Scientific Research from the Ministry of Education, Culture, Sports, Science and Technology of Japan.

Author details

Toshio Fukuda, Masahiro Nakajima, Yajing Shen and Masaru Kojima

Department of Micro-Nano Systems Engineering, Nagoya University, Nagoya, Japan

References

[1] Fields S, Johnston M. Cell biology: Whither model organism research? Science 2005;307 1885 - 1886.

[2] Fox M. A. The case for animal experimentation: An evolutionary and ethical perspective: University of California Press; 1986.

[3] Botstein D, Chervitz S. A, Cherry J. M. Yeast as a model organism. Science 1997;277 1259 - 1260.

[4] Foury F. Human genetic diseases ? a cross-talk between man and yeast. Gene 1997;195 1-10.

[5] Mager W. H, Winderickx J. Yeast as a model for medical and medicinal research. TRENDS in Pharmacological Sciences 2005;26 265-273.

[6] Hartwell L. H. Yeast and cancer," Bioscience Reports 2002;22 373-394.

[7] Dong L. X, Arai F, Fukuda T. Destructive Constructions of Nanostructures with Carbon Nanotubes Through Nanorobotic Manipulation. IEEE Transaction on Mechatronics 2004;9 350-357.

[8] Siegel R. W, Hu E, Roco M. C. Nanostructure Science and Technology. Kluwer Academic Publishers; 1999.

[9] Craighead H. G. Nanoelectromechanical Systems. Science 2000; 290 1532-1535.

[10] Staples M, Daniel K, Sima M. J, Langer R. Applications of Micro- and Nano-Electromechanical Devices to Drug Delivery. Pharnaceutical Research 2006;23 847-863.

[11] Leary S. P, Liu C. Y, Apuzzo M. L. J. Toward the Emergence of Nanoneurosurgery: Part III - Nanomedicine: Targeted Nanotherapy, Nanosurgery, and Progress Toward the Realization of Nanoneurosurgery. Neurosurgery 2006;58 1009-1026.

[12] Du E, Cui H, Zhu Z. Review of Nanomanipulators for Nanomanufacturing. International Journal of Nanomanufacturing 2006;1 83-104.

[13] Hell S. W. Far-Field Optical Nanoscopy. Science 2007;316 1153-1158.

[14] Eigler D. M, Schweizer E. K. Positioning Single Atoms with a Scanning Electron Microscope. Nature 1990;344 524-526.

[15] Hertel T, Martel R, Avouris P. Manipulation of Individual Carbon Nanotubes and their Interaction with Surfaces. The Journal of Physical Chemistry B 1998;102 910-915.

[16] Li G, Xi N, Chen H, Poneroy C, Prokos M. Videolized Atomic Force Microscopy for Interactive Nanomanipulation and Nanoassembly. IEEE Transaction on Nanotech 2005;4 605-615.

[17] Li G, Xi N, Yu M, Fung W-K. Development of Augmented Reality System for AFM based Nanomanipualiton. IEEE Transaction on Mechatronics 2004;9 358-365.

[18] Yu M. F, Lourie O, Dyer M. J, Moloni K, Kelley T. F, Ruoff R. S. Strength and Breaking Mechanism of Multiwalled Carbon Nanotubes under Tensile Load. Science 2000;287 637-640.

[19] Kizuka T, Yamada K, Deguchi S, Naruse M, Tanaka N. Cross-sectional Time Resolved High-Resolution Transmission Electron Microscopy of Atomic-Scale Contact and Noncontact-Type Scannings on Gold Surfaces. Physical Review B 1997;55 7398-7401.

[20] Nakajima M, Arai F, Fukuda T. In situ Fabrication and Electric Actuation of Telescoping Nanotube inside TEM through Hybrid Nanorobotic Manipulation System. In: Proceedings of the 2006 IEEE/RSJ International Conference on Intelligent Robotics and Systems (IROS 2006); 2006., 1915-1920.

[21] Nakajima M, Arai F, Fukuda T. In situ Measurement of Young's Modulus of Carbon Nanotube inside TEM through Hybrid Nanorobotic Manipulation System. IEEE Transsaction on Nanotechnology 2006;5 243-248.

[22] Leary S. P, Liu C. Y, Apuzzo M. L. J. Toward the Emergence of Nanoneurosurgery. Neurosurgery 2006;58 1009-1026.

[23] Donald A. M. The use of environmental scanning electron microscopy for imaging wet and insulating materials. Nature Materials 2003;2 511-516.

[24] Nakajima M, Arai F, Fukuda T. Nanofixation with Low Melting Metal Based on Nanorobotic Manipulation. Proceedings of the 6th IEEE International Conference on Nanotechnology (IEEE-Nano 2006)., 925-928.

[25] Ahmad M. R, Nakajima M, Kojima S, Homma M, Fukuda T. The effects of cell sizes, environmental conditions and growth phases on the strength of individual w303 yeast cells inside esem. IEEE Transactions on Nanobioscience 2008;7 185-193.

[26] Ahmad M. R, Nakajima M, Kojima S, Homma M, Fukuda T. In-situ Single Cell Mechanics Characterization of Yeast Cells using Nanoneedles inside Environmental-SEM. IEEE Transactions on Nanotechnology 2008;7(5) 607-616.

[27] Ahmad M. R, Nakajima M, Kojima S, Homma M, Fukuda T. Nanoindentation Methods to Measure Viscoelastic Properties of Single Cells using Sharp, Flat and Buckling Tips inside ESEM. IEEE Transactions on Nanobioscience 2010;9(1) 12-23.

[28] Ahmad M. R, Nakajima M, Kojima S, Homma M, Fukuda T. Buckling Nanoneedle for Characterizing Single Cells Mechanics inside Environmental SEM. IEEE Transactions on Nanotechnology 2011;10(2) 226-236.

[29] Gumbiner B. M. Cell adhesion: the molecular basis of tissue architecture and morphogenesis. Cell 1996., 84 345.

[30] Lelievre S, Weaver V. M, Bissell M. J. Extracellular matrix signaling from the cellular membrane skeleton to the nuclear skeleton: a model of gene regulation. Recent Progress in Hormone Research (1996)., 51 417.

[31] Hong-Geller E. A role for cell adhesion in beryllium-mediated lung disease. Journal of Occupational and Environmental Hygiene 2009;6 727-731.

[32] Marcotte L, Tabrizian M. Sensing surfaces: Challenges in studying the cell adhesion process and the cell adhesion forces on biomaterials. Irbm 2008;29 77-88.

[33] Shen, Y, Nakajima, M, Kojima, S, Homma, M, Kojima, M, & Fukuda, T., Single cell adhesion force measurement for cell viability identification by using AFM cantilever based micro putter., Measurement science and technology, (2011), 22(11) 115802

[34] Shen, Y, Ridzuan, M, Nakajima, M, Kojima, S, Homma, M, & Fukuda, T., Evaluation of the single yeast cell's adhesion to ITO substrates with various surface energies via

ESEM nanorobotic manipulation system. IEEE Transactions on Nanobioscience, (2011)., 10(4), 217-224.

[35] Shen, Y, Nakajima, M, Ridzuan, M, Kojima, S, Homma, M, & Fukuda, T. Effect of ambient humidity on the strength of the adhesion force of single yeast cell inside environmental-SEM. Ultramicroscopy (2011). , 111(8), 1176-1183.

[36] Mobius, W. Cryopreparation of biological specimens for immunoelectron microscopy. Annals of Anatomy-Anatomischer Anzeiger (2009)., 191-231.

[37] Singh, G, Rice, P, Mahajan, R, & Mcintosh, J. Fabrication and characterization of a carbon nanotube-based nanoknife. Nanotechnology (2009). 20 (9), 095701.

[38] Shen, Y, Nakajima, M, Yang, Z, Kojima, S, Homma, M, & Fukuda, T. Design and characterization of nanoknife with buffering beam for in situ single cell cutting. Nanotechnology (2011), 22 (30), 305701.

[39] Nakajima, M, Kawamoto, T, Hirano T, Kojima, M. & Fukuda, T. Nanotool Exchanger System based on E-SEM Nanorobotic Manipulation System. Proceedings of the 2012 IEEE International Conference on Robotics and Automation (ICRA 2012), (2012), 2773-2778.

[40] Nakajima, M, Hirano, T, Kojima, M, Hisamoto, N, Homma, M, & Fukuda, T. Direct Nano-injection Method by Nanoprobe Insertion based on E-SEM Nanorobotic Manipulation under Hybrid Microscope. Proceedings of the 2011 IEEE International Conference on Robotics and Automation (ICRA 2011), (2011), 4139-4144.

[41] Nakajima, M, Hirano, T, Kojima, M, Hisamoto, N, Nakanishi, N, Tajima, H, Homma, M, & Fukuda T. Local Nano-injection of Fluorescent Nano-beads inside C. elegans based on Nanomanipulation. Proceedings of the 2012 IEEE/RSJ International Conference on Intelligent Robotics and Systems (IROS 2012) (2012), 3241-3246.

Permissions

The contributors of this book come from diverse backgrounds, making this book a truly international effort. This book will bring forth new frontiers with its revolutionizing research information and detailed analysis of the nascent developments around the world.

We would like to thank Toshio Fukuda, for lending his expertise to make the book truly unique. He has played a crucial role in the development of this book. Without his invaluable contribution this book wouldn't have been possible. He has made vital efforts to compile up to date information on the varied aspects of this subject to make this book a valuable addition to the collection of many professionals and students.

This book was conceptualized with the vision of imparting up-to-date information and advanced data in this field. To ensure the same, a matchless editorial board was set up. Every individual on the board went through rigorous rounds of assessment to prove their worth. After which they invested a large part of their time researching and compiling the most relevant data for our readers. Conferences and sessions were held from time to time between the editorial board and the contributing authors to present the data in the most comprehensible form. The editorial team has worked tirelessly to provide valuable and valid information to help people across the globe.

Every chapter published in this book has been scrutinized by our experts. Their significance has been extensively debated. The topics covered herein carry significant findings which will fuel the growth of the discipline. They may even be implemented as practical applications or may be referred to as a beginning point for another development. Chapters in this book were first published by InTech; hereby published with permission under the Creative Commons Attribution License or equivalent.

The editorial board has been involved in producing this book since its inception. They have spent rigorous hours researching and exploring the diverse topics which have resulted in the successful publishing of this book. They have passed on their knowledge of decades through this book. To expedite this challenging task, the publisher supported the team at every step. A small team of assistant editors was also appointed to further simplify the editing procedure and attain best results for the readers.

Our editorial team has been hand-picked from every corner of the world. Their multi-ethnicity adds dynamic inputs to the discussions which result in innovative outcomes. These outcomes are then further discussed with the researchers and contributors who give their valuable feedback and opinion regarding the same. The feedback is then collaborated with the researches and they are edited in a comprehensive manner to aid the understanding of the subject.

Apart from the editorial board, the designing team has also invested a significant amount of their time in understanding the subject and creating the most relevant covers. They scrutinized every image to scout for the most suitable representation of the subject and create an appropriate cover for the book.

The publishing team has been involved in this book since its early stages. They were actively engaged in every process, be it collecting the data, connecting with the contributors or procuring relevant information. The team has been an ardent support to the editorial, designing and production team. Their endless efforts to recruit the best for this project, has resulted in the accomplishment of this book. They are a veteran in the field of academics and their pool of knowledge is as vast as their experience in printing. Their expertise and guidance has proved useful at every step. Their uncompromising quality standards have made this book an exceptional effort. Their encouragement from time to time has been an inspiration for everyone.

The publisher and the editorial board hope that this book will prove to be a valuable piece of knowledge for researchers, students, practitioners and scholars across the globe.

List of Contributors

Toshio Fukuda and Masaru Kojima
Department of Micro-Nano Systems Engineering, Nagoya University, Nagoya, Japan

Masahiro Nakajima
Graduate School of Engineering Center For Micro-nano Mechatronics, Nagoya University, Nagoya, Japan

Tomohide Niimi
Department of Micro-Nano Systems Engineering, Nagoya University, Nagoya, Japan

Eiji Shamoto, Norikazu Suzuki, Takashi Kato and Burak Sencer
Department of Mechanical Science and Engineering, Nagoya University, Nagoya, Japan

Tomohiro Kawahara and Fumihito Arai
Department of Micro-Nano Systems Engineering, Nagoya University, Nagoya, Japan

Goro Obinata
EcoTopia Science Institute, Nagoya University, Nagoya, Japan

Hitoshi Hirata, Shigeru Kurimoto, Shuichi and Tomonori Nakano
Graduate School of Medicine, Nagoya University, Nagoya, Japan

Chikara Nagai
Graduate School of Engineering, Nagoya University, Nagoya, Japan

Osamu Takai, Maria Antoaneta Bratescu and Nagahiro Saito
EcoTopia Science Institute, Nagoya University, Nagoya, Japan

Tomonaga Ueno
Green Mobility Collaborative Research Center, Nagoya University, Nagoya, Japan

Nobutada Ohno, Dai Okumura and Yusuke Kinoshita
Department of Computational Science and Engineering, Graduate School of Engineering, Nagoya University, Nagoya, Japan

Minoru Ueda
Department of Oral and Maxillofacial Surgery, Graduate School of Medicine, Nagoya University, Nagoya, Japan

Kenji Fukuzawa
Department of Micro-Nano Systems Engineering, Nagoya University, Nagoya, Japan

Noritsugu Umehara, Takayuki Tokoroyama and Hiroyuki Kousaka
Department of Mechanical Science and Engineering, Nagoya University, Japan

Ken-ichi Isobe, Naomi Nishio, Thanasegaran Suganya, Zhao Cheng and Sachiko Ito
Department of Immunology, Graduate School of Medicine, Nagoya University, Nagoya, Japan

Yang Ju
Department of Mechanical Science and Engineering, Nagoya University, Japan

Akihiro Sasoh
Department of Aerospace Engineering, Nagoya University, Nagoya, Japan

Kensuke Kuroda and Masazumi Okido
Department of Materials Science and Engineering, Nagoya University, Nagoya, Japan

Hideharu Hibi and Minoru Ueda
Department of Oral & Maxillofacial Surgery, Graduate School of Medicine, Nagoya University, Nagoya, Japan

Masaru Takeuchi
Graduate School of Engineering, Nagoya University, Nagoya, Japan

Gauvain Haulot
Henry Samueli School of Engineering and Applied Science, University of California, Los Angeles, USA

Chih-Ming Ho
Department of Mechanical & Aerospace Engineering, Department of Bioengineering, University of California, Los Angeles, USA

Daniel T. Kamei
Department of Bioengineering, University of California, Los Angeles, USA

Mitsuhiro Shikida
Department of Micro-Nano Systems Engineering, Nagoya University, Nagoya, Japan

Toshio Fukuda, Masahiro Nakajima, Yajing Shen and Masaru Kojima
Department of Micro-Nano Systems Engineering, Nagoya University, Nagoya, Japan